An Introduction to the Theory of Numbers

Books by Ivan Niven

Calculus: An Introductory Approach (Van Nostrand)

Diophantine Approximations (Wiley)

Irrational Numbers (Wiley)

Mathematics of Choice (Random House)

Numbers: Rational and Irrational (Random House)

An Introduction to the Theory of Numbers

Third Edition

IVAN NIVEN

University of Oregon

HERBERT S. ZUCKERMAN

University of Washington

JOHN WILEY & SONS INC.

New York London Sydney Toronto

Library of Congress Catalog Card Number: 73-178149

ISBN 0-471-64154-5

Printed in the United States of America.

10 9 8 7 6 5 4 3 2

Preface

In this third edition Sections 4.4, 6.3, 6.4 and 9.10 have been added, and Sections 5.7 and 10.3 have been modified considerably. Following the last chapter a set of miscellaneous problems has been added, and also a collection of eleven special topics in outline form but with sufficient explanation for easy filling in of details.

Sincere thanks are expressed to Paul T. Bateman, Bruce C. Berndt, Charles W. Curtis, Edwin Hewitt, Ralph D. James, Emma Lehmer, Roy W. Ryden, Sigmund Selberg and John Steinig for helpful suggestions since the first version. Herbert S. Zuckerman shared in the planning of this edition; following his untimely death in 1970 I have missed very much his able collaboration in seeing it through the press.

February 1972 IVAN NIVEN

Preface to the First Edition

Our purpose is to present a reasonably complete introduction to the theory of numbers within the compass of a single volume. The basic concepts are presented in the first part of the book, followed by more specialized material in the final three chapters. Paralleling this progress from general topics to more particular discussions, we have attempted to begin the book at a more leisurely pace than we have followed later. Thus the later parts of the book are set forth in a more compact and sophisticated presentation than are the earlier parts.

The book is intended for seniors and beginning graduate students in American and Canadian universities. It contains at least enough material for a full year course; a short course can be built by the use of Sections 1.1 to 1.3, 2.1 to 2.4, 3.1, 3.2, 4.1, 5.1 to 5.3, 5.5, 6.1, and 6.2. Various other arrangements are possible because the chapters beyond the fourth are, apart from a very few exceptions, independent of one another. The final three chapters are entirely independent of each other.

To enable the student to deepen his understanding of the subject, we have provided a considerable number of problems. The variety of these exercises is extensive, ranging from simple numerical problems to additional developments of the theory. The beginner at number theory should take warning that the subject is noted for the difficulty of its problems. Many an innocent-looking problem gives, by the very simplicity of its statement, very little notion of the considerable ingenuity or depth of insight required for its solution. As might be expected, the more difficult problems are placed toward the ends of the sets. In many instances three or four consecutive problems constitute a related series in which the last ones can be solved more readily by use of information from the first ones. As a matter of principle we have made the text itself entirely independent of the problems. In no place does the proof of a theorem depend on the results of any problem.

In choosing methods of proof, we have tried to include as many methods as possible. We have tried to state the proofs accurately, avoiding statements that could be misleading and also avoiding unduly long discussions of unimportant details. As the reader progresses he will become familiar with more and more methods, and he should be able to construct accurate proofs by patterning them after our proofs.

The reader interested in further exploration of the subject will find the bibliography at the end of the book of considerable use. In particular, anyone interested in the history of the subject should consult O. Ore, *Number Theory and Its History*, and, for more specific information, L. E. Dickson, *History of the Theory of Numbers*. Our approach is analytical, not historical, and we make no attempt to attribute various theorems and proofs to their original discoverers. However, we do wish to point out that we followed the suggestion of Peter Scherk that we use F. J. Dyson's formulation of the proof of Mann's $\alpha\beta$ Theorem. Our proof is based on notes graciously placed at our disposal by Peter Scherk. For permission to use several problems from the *American Mathematical Monthly*, we are indebted to the editors. We also appreciate the careful reading of the manuscript by Margaret Maxfield, whose efforts resulted in numerous improvements. Finally we would like to record our deep appreciation of and our great debt to the mathematicians whose lectures were vital to our introduction to the theory of numbers: L. E. Dickson, R. D. James, D. N. Lehmer, and Hans Rademacher.

June 1960

IVAN NIVEN
HERBERT S. ZUCKERMAN

Contents

An Introduction to the Theory of Numbers

1

Divisibility

1.1 Introduction

The theory of numbers is primarily concerned with the properties of the natural numbers, $1, 2, 3, 4, \cdots$, also called the positive integers. However, the theory is not strictly confined to just the natural numbers or even to the set of all integers: $0, \pm 1, \pm 2, \pm 3, \cdots$. In fact, some theorems of number theory are most easily proved by making use of the properties of real or complex numbers even though the statement of the theorems may involve only natural numbers. Also, there are theorems concerning real numbers that depend so heavily on the properties of integers that they are properly included in the theory of numbers.

An integer n greater than 1 is called a prime if it has no divisor d such that $1 < d < n$. The fact that for every given positive integer m there is a prime greater than m is stated in terms of integers, and it can be proved from the properties of the natural numbers alone. The fact that every natural number can be expressed as a sum of, at most, fifty-four fifth powers of integers is also stated in terms of natural numbers, but any known proof depends on properties of complex numbers. Finally, the question as to how many primes there are that do not exceed x clearly belongs to the theory of numbers but its answer involves the function $\log x$ and is well outside of the realm of the natural numbers. The last two examples are beyond the scope of this book. However, we do not restrict ourselves to the integers but will use real and complex numbers when it is convenient. The questions discussed in this book are not numerical computations or numerical curiosities, except insofar as these are relevant to general propositions. Nor do we discuss the foundations of the number system; it is assumed that the reader is familiar not only with the integers, but also with the rational and real numbers. However, a rigorous logical analysis of the real-number system is not prerequisite to the study of number theory.

The theory of numbers relies for proofs on a great many ideas and methods. Of these, there are two basic principles to which we draw especial attention. The first is that any set of positive integers has a smallest element if it contains any members at all. In other words, if a set S of positive integers is not empty, then it contains an integer s such that for any member a of S, the relation $s \leqq a$ holds. The second principle, mathematical induction, is a logical consequence of the first.* It can be stated as follows: if a set S of positive integers contains the integer 1, and contains $n + 1$ whenever it contains n, then S consists of all the positive integers.

It may be well to point out that a negative assertion such as, for example, "Not every positive integer can be expressed as a sum of the squares of three integers," requires only that we produce a single example—the number 7 cannot be so expressed. On the other hand, a positive assertion such as "Every positive integer can be expressed as a sum of the squares of four integers," cannot be proved by producing examples, however numerous. This result is Theorem 5.6 in Chapter 5, where a proof is supplied.

Finally, it is presumed that the reader is familiar with the usual formulation of mathematical propositions. In particular, if A denotes some assertion or collection of assertions, and B likewise, the following statements are logically equivalent—they are just different ways of saying the same thing.

A implies B.
If A is true, then B is true.
In order that A be true it is necessary that B be true.
B is a necessary condition for A.
A is a sufficient condition for B.

If A implies B and B implies A, then one can say that B is a necessary and sufficient condition for A to hold.

In general, we shall use letters of the roman alphabet, $a, b, c, \cdots, m, n, \cdots, x, y, z$, to designate integers unless otherwise specified.

1.2 Divisibility

Definition 1.1 *An integer b is divisible by an integer a, not zero, if there is an integer x such that $b = ax$, and we write $a \mid b$. In case b is not divisible by a we write $a \nmid b$.*

* Compare G. Birkhoff and S. MacLane, *A Survey of Modern Algebra*, Macmillan, third edition, 1965, pp. 10–13.

Other language for the divisibility property $a \mid b$ is that a divides b, that a is a divisor of b, and that b is a multiple of a. If $a \mid b$ and $0 < a < b$ then a is called a proper divisor of b. It is understood that we never use 0 as the left member of the pair of integers in $a \mid b$. On the other hand, not only may 0 occur as the right member of the pair, but also in such instances we always have divisibility. Thus $a \mid 0$ for every integer a not zero. The notation $a^K \parallel b$ is sometimes used to indicate that $a^K \mid b$ but $a^{K+1} \nmid b$.

Theorem 1.1

(1) *$a \mid b$ implies $a \mid bc$ for any integer c;*

(2) *$a \mid b$ and $b \mid c$ imply $a \mid c$;*

(3) *$a \mid b$ and $a \mid c$ imply $a \mid (bx + cy)$ for any integers x and y;*

(4) *$a \mid b$ and $b \mid a$ imply $a = \pm b$;*

(5) *$a \mid b$, $a > 0$, $b > 0$, imply $a \leqq b$.*

(6) *if $m \neq 0$, $a \mid b$ implies and is implied by $ma \mid mb$.*

Proof. The proofs of these results follow at once from the definition of divisibility. Property 3 admits an obvious extension to any finite set, thus:

$$a \mid b_1,\ a \mid b_2,\ \cdots,\ a \mid b_n \text{ imply } a \,\Big|\, \sum_{j=1}^{n} b_j x_j \text{ for any integers } x_j.$$

Property 2 can be extended similarly.

Theorem 1.2 *The division algorithm. Given any integers a and b, with $a > 0$, there exist unique integers q and r such that $b = qa + r$, $0 \leqq r < a$. If $a \nmid b$, then r satisfies the stronger inequalities $0 < r < a$.*

Proof. Consider the arithmetic progression

$$\cdots,\ b - 3a,\ b - 2a,\ b - a,\ b,\ b + a,\ b + 2a,\ b + 3a,\ \cdots$$

extending indefinitely in both directions. In this sequence, select the smallest non-negative member and denote it by r. Thus by definition r satisfies the inequalities of the theorem. But also r, being in the sequence, is of the form $b - qa$, and thus q is defined in terms of r.

To prove the uniqueness of q and r, suppose there is another pair q_1 and r_1 satisfying the same conditions. First we prove that $r_1 = r$. For if not, we may presume that $r < r_1$ so that $0 < r_1 - r < a$, and then we see that $r_1 - r = a(q - q_1)$ and so $a \mid (r_1 - r)$, a contradiction to Theorem 1.1, part (5). Hence $r = r_1$, and also $q = q_1$.

We have stated the theorem with the assumption $a > 0$. However, this hypothesis is not necessary, and we may formulate the theorem without it: given any integers a and b, with $a \neq 0$, there exist integers q and r such that $b = qa + r$, $0 \leqq r < |a|$.

Theorem 1.2 is called the division algorithm. An algorithm is a mathematical procedure or method to obtain a result. We have stated Theorem 1.2 in the form "there exist integers q and r," and this wording suggests that we have a so-called existence theorem rather than an algorithm. However, it may be observed that the proof does give a method for obtaining the integers q and r, because the infinite arithmetic progression $\cdots, b - a, b, b + a, \cdots$ need be examined only in part to yield the smallest positive member r.

In actual practice the quotient q and the remainder r are obtained by the arithmetic division of a into b.

Definition 1.2 *The integer a is a* common divisor *of b and c in case $a \mid b$ and $a \mid c$. Since there is only a finite number of divisors of any non-zero integer, there is only a finite number of common divisors of b and c, except in the case $b = c = 0$. If at least one of b and c is not 0, the greatest among their common divisors is called the* greatest common divisor *of b and c and is denoted by (b, c). Similarly we denote the greatest common divisor g of the integers $b_1, b_2, \cdots, b_n$, not all zero, by $(b_1, b_2, \cdots, b_n)$.*

Thus the greatest common divisor (b, c) is defined for every pair of integers b, c except $b = 0, c = 0$, and we note that $(b, c) \geqq 1$.

Theorem 1.3 *If g is the greatest common divisor of b and c, then there exist integers x_0 and y_0 such that $g = (b, c) = bx_0 + cy_0$.*

Proof. Consider the linear combinations $bx + cy$, where x and y range over all integers. This set of integers $\{bx + cy\}$ includes positive and negative values, and also 0 by the choice $x = y = 0$. Choose x_0 and y_0 so that $bx_0 + cy_0$ is the least positive integer l in the set; thus $l = bx_0 + cy_0$.

Next we prove that $l \mid b$ and $l \mid c$. We establish the first of these, and the second follows by analogy. We give an indirect proof that $l \mid b$, that is, we assume $l \nmid b$ and obtain a contradiction. From $l \nmid b$ it follows that there exist integers q and r, by Theorem 1.2, such that $b = lq + r$ with $0 < r < l$. Hence we have $r = b - lq = b - q(bx_0 + cy_0) = b(1 - qx_0) + c(-qy_0)$, and thus r is in the set $\{bx + cy\}$. This contradicts the fact that l is the least positive integer in the set $\{bx + cy\}$.

Now since g is the greatest common divisor of b and c, we may write $b = gB$, $c = gC$, and $l = bx_0 + cy_0 = g(Bx_0 + Cy_0)$. Thus $g \mid l$, and so by part 5 of Theorem 1.1, we conclude that $g \leqq l$. Now $g < l$ is impossible, since g is the *greatest* common divisor, and so $g = l = bx_0 + cy_0$.

Theorem 1.4 *The greatest common divisor g of b and c can be characterized in the following two ways: (1) it is the least positive value of $bx + cy$ where x and y range over all integers; (2) it is the positive common divisor of b and c which is divisible by every common divisor.*

Proof. Part 1 follows from the proof of Theorem 1.3. To prove part 2, we observe that if d is any common divisor of b and c, then $d \mid g$ by part 3 of Theorem 1.1. Moreover, there cannot be two distinct integers with property 2, because of Theorem 1.1, part 4.

Theorem 1.5 *Given any integers $b_1, b_2, \cdots, b_n$ not all zero, with greatest common divisor g, there exist integers $x_1, x_2, \cdots, x_n$ such that*

$$g = (b_1, b_2, \cdots, b_n) = \sum_{j=1}^{n} b_j x_j.$$

Furthermore, g is the least positive value of the linear form $\sum_{j=1}^{n} b_j y_j$ where the y_j range over all integers; also g is the positive common divisor of $b_1, b_2, \cdots, b_n$ which is divisible by every common divisor.

Proof. This result is a straightforward generalization of the preceding two theorems, and the proof is analogous without any complications arising in the passage from two integers to n integers.

Theorem 1.6 *For any positive integer m,*

$$(ma, mb) = m(a, b).$$

Proof. By Theorem 1.4 we have

$$\begin{aligned}(ma, mb) &= \text{least positive value of } max + mby \\ &= m \cdot \{\text{least positive value of } ax + by\} \\ &= m(a, b).\end{aligned}$$

Theorem 1.7 *If $d \mid a$ and $d \mid b$ and $d > 0$ then*

$$\left(\frac{a}{d}, \frac{b}{d}\right) = \frac{1}{d}(a, b).$$

If $(a, b) = g$, then

$$\left(\frac{a}{g}, \frac{b}{g}\right) = 1.$$

Proof. The second assertion is the special case of the first obtained by using the greatest common divisor g of a and b in the role of d. The first assertion in turn is a direct consequence of Theorem 1.6 obtained by replacing m, a, b in that theorem by d, (a/d), (b/d) respectively.

Theorem 1.8 *If $(a, m) = (b, m) = 1$, then $(ab, m) = 1$.*

Proof. By Theorem 1.3 there exist integers x_0, y_0, x_1, y_1 such that $1 = ax_0 + my_0 = bx_1 + my_1$. Thus we may write $(ax_0)(bx_1) = (1 - my_0)(1 - my_1) = 1 - my_2$ where y_2 is defined by the equation $y_2 = y_0 + y_1 - my_0y_1$.

From the equation $abx_0x_1 + my_2 = 1$ we note, by part 3 of Theorem 1.1, that any common divisor of ab and m is a divisor of 1, and hence $(ab, m) = 1$.

Definition 1.3 *We say that a and b are* relatively prime *in case* $(a, b) = 1$, *and that* $a_1, a_2, \cdots, a_n$ *are* relatively prime *in case* $(a_1, a_2, \cdots, a_n) = 1$. *We say that* $a_1, a_2, \cdots, a_n$ *are* relatively prime in pairs *in case* $(a_i, a_j) = 1$ *for all* $i = 1, 2, \cdots, n$ *and* $j = 1, 2, \cdots, n$ *with* $i \neq j$.

The fact that $(a, b) = 1$ is sometimes expressed by saying that a and b are coprime, or by saying that a is prime to b.

Theorem 1.9 *For any* x, $(a, b) = (b, a) = (a, -b) = (a, b + ax)$.

Proof. Denote (a, b) by d and $(a, b + ax)$ by g. It is clear that $(b, a) = (a, -b) = d$.

By application of Theorem 1.1, parts 3 and 4, we obtain $d \mid g$, $g \mid d$, and hence $d = g$.

Theorem 1.10 *If* $c \mid ab$ *and* $(b, c) = 1$, *then* $c \mid a$.

Proof. By Theorem 1.6, $(ab, ac) = a(b, c) = a$. But $c \mid ab$ and $c \mid ac$, and so $c \mid a$ by Theorem 1.4.

Given two integers b and c, how can the greatest common divisor g be found? Definition 1.2 gives no answer to this question, and neither does Theorem 1.3 which merely asserts the existence of a pair of integers x_0 and y_0 such that $g = bx_0 + cy_0$. If b and c are small, values of g, x_0, and y_0 can be found by inspection. For example, if $b = 10$ and $c = 6$, it is obvious that $g = 2$, and one pair of values for x_0, y_0 is $2, -3$. But if b and c are large, inspection is not adequate except in rather obvious cases like $(963, 963) = 963$ and $(1000, 600) = 200$.

Consider the case $b = 963$, $c = 657$. If we divide c into b we get a quotient $q = 1$, and remainder $r = 306$. Thus $b = cq + r$, or $r = b - cq$, in particular $306 = 963 - 1 \cdot 657$. Now $(b, c) = (b - cq, c)$ by replacing a and x by c and $-q$ in Theorem 1.9, so we see that

$$(963, 657) = (963 - 1 \cdot 657, 657) = (306, 657).$$

The integer 963 has been replaced by the smaller integer 306, and this suggests that the procedure be repeated. So we divide 306 into 657 to get a quotient 2 and a remainder 45, and

$$(306, 657) = (306, 657 - 2 \cdot 306) = (306, 45).$$

Next 45 is divided into 306 with quotient 6 and remainder 36, then 36 is divided into 45 with quotient 1 and remainder 9. We conclude that

$$(963, 657) = (306, 657) = (306, 45) = (36, 45) = (36, 9).$$

Thus $(963, 657) = 9$, and we can express 9 as a linear combination of 963 and 657 by eliminating the remainders 36, 45, and 306 as follows:

$$\begin{aligned} 9 &= 45 - 36 \\ &= 45 - (306 - 45 \cdot 6) \\ &= -306 + 7 \cdot 45 \\ &= -306 + 7(657 - 306 \cdot 2) \\ &= 7 \cdot 657 - 15 \cdot 306 \\ &= 7 \cdot 657 - 15(963 - 657) \\ &= 22 \cdot 657 - 15 \cdot 963. \end{aligned}$$

In terms of Theorem 1.3 where $g = (b, c) = bx_0 + cy_0$, beginning with $b = 963$ and $c = 657$ we have used a procedure, called the "Euclidean algorithm," to find $g = 9$, $x_0 = -15$, $y_0 = 22$. Of course these values for x_0 and y_0 are not unique: $-15 + 657k$ and $22 - 963k$ will do where k is any integer.

In general, to find the greatest common divisor (b, c) of b and c, and also integers x_0 and y_0 satisfying $g = (b, c) = bx_0 + cy_0$, we generalize what is done in the special case above. By Theorem 1.9, $(b, c) = (b, -c)$, and hence we may presume c positive, because the case $c = 0$ is very special: $(b, 0) = |b|$.

Theorem 1.11 *The Euclidean algorithm. Given integers b and $c > 0$, we make a repeated application of the division algorithm, Theorem 1.2, to obtain a series of equations*

$$\begin{aligned} b &= cq_1 + r_1, & 0 &< r_1 < c, \\ c &= r_1q_2 + r_2, & 0 &< r_2 < r_1, \\ r_1 &= r_2q_3 + r_3, & 0 &< r_3 < r_2, \\ &\cdots & &\cdots \\ r_{j-2} &= r_{j-1}q_j + r_j, & 0 &< r_j < r_{j-1}, \\ r_{j-1} &= r_jq_{j+1}. \end{aligned}$$

The greatest common divisor (b, c) of b and c is r_j, the last non-zero remainder in the division process. Values of x_0 and y_0 in $(b, c) = bx_0 + cy_0$ can be obtained by eliminating $r_{j-1}, \cdots, r_2, r_1$ from the set of equations.

Proof. The chain of equations is obtained by dividing c into b, r_1 into c, r_2 into $r_1, \cdots, r_j$ into r_{j-1}. The process stops when the division is exact, that is when the remainder is zero. Thus in our application of Theorem 1.2 we have written the inequalities for the remainder without an equality sign. Thus, for example, $0 < r_1 < c$ in place of $0 \leqq r_1 < c$, because if r_1 were equal to zero, the chain would stop at the first equation $b = cq_1$, in which case the greatest common divisor of b and c would be c.

We now prove that r_j is the greatest common divisor g of b and c. By Theorem 1.9 we observe that

$$(b, c) = (b - cq_1, c) = (r_1, c) = (r_1, c - r_1q_2)$$
$$= (r_1, r_2) = (r_1 - r_2q_3, r_2) = (r_3, r_2).$$

Continuing by mathematical induction we get $(b, c) = (r_{j-1}, r_j) = r_j$ because r_j is a divisor of r_{j-1}, and r_j is positive.

To see that r_j is expressible as a linear combination of b and c, we need merely eliminate r_{j-1} from the second and third last equations of the chain, then eliminate r_{j-2} from the third and fourth last, and so on until all of $r_{j-1}, r_{j-2}, \cdots, r_2, r_1$ are eliminated. Thus we get r_j in the form $bx_0 + cy_0$.

Definition 1.4 *The integers $a_1, a_2, \cdots, a_n$, all different from zero, have a* common multiple *b if $a_i \mid b$ for $i = 1, 2, \cdots, n$. (Note that common multiples do exist; for example the product $a_1a_2 \cdots a_n$ is one.) The least of the positive common multiples is called the least common multiple, and it is denoted by $[a_1, a_2, \cdots, a_n]$.*

Theorem 1.12 *If b is any common multiple of $a_1, a_2, \cdots, a_n$, then $[a_1, a_2, \cdots, a_n] \mid b$. This is the same as saying that if h denotes $[a_1, a_2, \cdots, a_n]$, then $0, \pm h, \pm 2h, \pm 3h, \cdots$ comprise all the common multiples of $a_1, a_2, \cdots, a_n$.*

Proof. Let m be any common multiple and divide m by h. By Theorem 1.2 there is a quotient q and a remainder r such that $m = qh + r$, $0 \leqq r < h$. We must prove that $r = 0$. If $r \neq 0$ we argue as follows. For each $i = 1, 2, \cdots, n$ we know that $a_i \mid h$ and $a_i \mid m$, so that $a_i \mid r$. Thus r is a positive common multiple of $a_1, a_2, \cdots, a_n$ contrary to the fact that h is the least positive of all the common multiples.

Theorem 1.13 *If $m > 0$, $[ma, mb] = m[a, b]$. Also $[a, b] \cdot (a, b) = |ab|$.*

Proof. Since $[ma, mb]$ is a multiple of ma, it is a *fortiori* a multiple of m, and so can be written in the form mh_1. Denoting $[a, b]$ by h_2, we note that $a \mid h_2$, $b \mid h_2$, $am \mid mh_2$, $bm \mid mh_2$, and so $mh_1 \mid mh_2$ by Theorem 1.12. Thus $h_1 \mid h_2$. On the other hand, $am \mid mh_1$, $bm \mid mh_1$, $a \mid h_1$, $b \mid h_1$ and so $h_2 \mid h_1$. We conclude that $h_1 = h_2$ and thus the first part of the theorem is established.

It will suffice to prove the second part for positive integers a and b, since $[a, -b] = [a, b]$. We begin with the special case where $(a, b) = 1$. Now $[a, b]$ is a multiple of a, say ma. Then $b \mid ma$ and $(a, b) = 1$, so by Theorem 1.10 we conclude that $b \mid m$. Hence $b \leqq m$, $ba \leqq ma$. But ba, being a positive common multiple of b and a, cannot be less than the least common multiple, and so $ba = ma = [a, b]$.

Turning to the general case where $(a, b) = g > 1$, we have $((a/g), (b/g)) = 1$ by Theorem 1.7. Applying the result of the preceding paragraph, we obtain

$$\left[\frac{a}{g}, \frac{b}{g}\right]\left(\frac{a}{g}, \frac{b}{g}\right) = \frac{a}{g}\frac{b}{g}.$$

Multiplying by g^2 and using Theorem 1.6 as well as the first part of the present theorem, we get $[a, b](a, b) = ab$.

PROBLEMS

1. By using the Euclidean algorithm find the greatest common divisor (g.c.d.) of
(*a*) 7469 and 2464; (*b*) 2689 and 4001;
(*c*) 2947 and 3997; (*d*) 1109 and 4999.

2. Find the greatest common divisor g of the numbers 1819 and 3587, and then find integers x and y to satisfy
$1819x + 3587y = g$.

3. Find values of x and y to satisfy
(*a*) $243x + 198y = 9$;
(*b*) $71x - 50y = 1$;
(*c*) $43x + 64y = 1$;
(*d*) $93x - 81y = 3$;
(*e*) $6x + 10y + 15z = 1$.

4. Find the least common multiple (l.c.m.) of (*a*) 482 and 1687, (*b*) 60 and 61.

5. How many integers between 100 and 1000 are divisible by 7?

6. Prove that the product of three consecutive integers is divisible by 6; of four consecutive integers by 24.

7. Exhibit three integers that are relatively prime but not relatively prime in pairs.

8. Two integers are said to be of the same parity if they are both even or both odd: if one is even and the other odd, they are said to be of opposite parity, or of different parity. Given any two integers, prove that their sum and their difference are of the same parity.

9. Show that if $ac \mid bc$ then $a \mid b$.

10. Given $a \mid b$ and $c \mid d$, prove that $ac \mid bd$.

11. Prove that $4 \nmid (n^2 + 2)$ for any integer n.

12. Given that $(a, 4) = 2$ and $(b, 4) = 2$ prove that $(a + b, 4) = 4$.

13. Prove that $n^2 - n$ is divisible by 2 for every integer n; that $n^3 - n$ is divisible by 6; that $n^5 - n$ is divisible by 30.

14. Prove that if n is odd, $n^2 - 1$ is divisible by 8.

15. Prove that if x and y are odd, then $x^2 + y^2$ is even but not divisible by 4.

16. Prove that if a and b are positive integers satisfying $(a, b) = [a, b]$ then $a = b$.

17. Evaluate $(n, n+1)$ and $[n, n+1]$ where n is a positive integer.
18. Find the values of (a, b) and $[a, b]$ if a and b are positive integers such that $a \mid b$.
19. Prove that any set of integers that are relatively prime in pairs are relatively prime.
20. Given integers a and b, a number n is said to be of the form $ak + b$ if there is an integer k such that $ak + b = n$. Thus the numbers of the form $3k + 1$ are $\cdots -8, -5, -2, 1, 4, 7, 10, \cdots$. Prove that every integer is of the form $3k$ or of the form $3k + 1$ or of the form $3k + 2$.
21. Prove that if an integer is of the form $6k + 5$ then it is necessarily of the form $3k - 1$, but not conversely.
22. Prove that the square of any integer of the form $5k + 1$ is of the same form.
23. Prove that the square of any integer is of the form $3k$ or $3k + 1$ but not of the form $3k + 2$.
24. Prove that no integers x, y exist satisfying $x + y = 100$ and $(x, y) = 3$.
25. Prove that there are infinitely many pairs of integers x, y satisfying $x + y = 100$ and $(x, y) = 5$.
26. Let s and $g > 0$ be given integers. Prove that integers x and y exist satisfying $x + y = s$ and $(x, y) = g$ if and only if $g \mid s$.
27. Find positive integers a and b satisfying the equations $(a, b) = 10$ and $[a, b] = 100$ simultaneously. Find all solutions.
28. Find all triples of positive integers a, b, c satisfying $(a, b, c) = 10$ and $[a, b, c] = 100$ simultaneously.
29. Let g and l be given positive integers. Prove that integers x and y exist satisfying $(x, y) = g$ and $[x, y] = l$ if and only if $g \mid l$.
30. Let b and $g > 0$ be given integers. Prove that the equations $(x, y) = g$ and $xy = b$ can be solved simultaneously if and only if $g^2 \mid b$.
31. Let $n \geqq 2$ and k be any positive integers. Prove that $(n - 1) \mid (n^k - 1)$.
32. Let $n \geqq 2$ and k be any positive integers. Prove that $(n - 1)^2 \mid (n^k - 1)$ if and only if $(n - 1) \mid k$. *Suggestion:* $n^k = \{(n - 1) + 1\}^k$.
33. Prove that $(a, b, c) = ((a, b), c)$.
34. Extend Theorems 1.6, 1.7, and 1.8 to sets of more than two integers.
35. Prove that if $(b, c) = 1$ and $r \mid b$, then $(r, c) = 1$.
36. Prove that if $m > n$ then $a^{2^n} + 1$ is a divisor of $a^{2^m} - 1$. Show that if a, m, n are positive integers with $m \neq n$, then

$$(a^{2^m} + 1, a^{2^n} + 1) = \begin{cases} 1 & \text{if } a \text{ is even,} \\ 2 & \text{if } a \text{ is odd.} \end{cases}$$

1.3 Primes

Definition 1.5 *An integer $p > 1$ is called a* prime number, *or a* prime, *in case there is no divisor d of p satisfying $1 < d < p$. If an integer $a > 1$ is not a prime, it is called a* composite *number.*

Thus, for example, 2, 3, 5, and 7 are primes, whereas 4, 6, 8, and 9 are composite.

Theorem 1.14 *Every integer n greater than* 1 *can be expressed as a product of primes* (*with perhaps only one factor*).

Proof. If the integer n is a prime, then the integer itself stands as a "product" with a single factor. Otherwise n can be factored into, say, n_1n_2, where $1 < n_1 < n$ and $1 < n_2 < n$. If n_1 is a prime, let it stand; otherwise it will factor into, say, n_3n_4 where $1 < n_3 < n_1$ and $1 < n_4 < n_1$; similarly for n_2. This process of writing each composite number that arises as a product of factors must terminate because the factors are smaller than the composite number itself, and yet each factor is an integer greater than 1. Thus we can write n as a product of primes, and since the prime factors are not necessarily distinct, the result can be written in the form

$$n = p_1^{\alpha_1}p_2^{\alpha_2}\cdots p_r^{\alpha_r}$$

where $p_1, p_2, \cdots, p_r$ are distinct primes and $\alpha_1, \alpha_2, \cdots, \alpha_r$ are positive.

This expression is called a representation of n as a product of primes, and it turns out that the representation is unique in the sense that, for fixed n, any other representation is merely a permutation of the factors. It may appear obvious to the reader that the representation of an integer as a product of primes is unique, but it is a fact requiring proof. Indeed there are mathematical situations where it might appear equally "obvious" that factorization will be unique, but where in fact it is not. We digress from our main theme to discuss two of these situations where factorization is not unique.

First consider the class E of positive even integers, so that the elements of E are 2, 4, 6, 8, 10, $\cdots$. Note that E is a multiplicative system, the product of any two elements in E being again in E. Now let us confine our attention to E in the sense that the only "numbers" we know are members of E. Then $8 = 2 \cdot 4$ is "composite," whereas 10 is a "prime" since 10 is not the product of two or more "numbers." The "primes" are 2, 6, 10, 14, $\cdots$, the "composite numbers" are 4, 8, 12, $\cdots$. Now the "number" 60 has two factorings into "primes," namely $60 = 2 \cdot 30 = 6 \cdot 10$, and so factorization is not unique.

A somewhat less artificial, but also rather more complicated, example is obtained by considering the class C of numbers $a + b\sqrt{-6}$ where a and b range over all integers. We say that this system C is *closed* under addition and multiplication, meaning that the sum and product of two elements in C are elements of C. By taking $b = 0$ we note that the integers form a subset of the class C.

First we establish that there are primes in C, and that every number in C can be factored into primes. For any number $a + b\sqrt{-6}$ in C it will be convenient to have a norm, $N(a + b\sqrt{-6})$, defined as

$$N(a + b\sqrt{-6}) = (a + b\sqrt{-6})(a - b\sqrt{-6}) = a^2 + 6b^2.$$

Thus the norm of a number in C is the product of the complex number $a + b\sqrt{-6}$ and its conjugate $a - b\sqrt{-6}$. Another way of saying this, perhaps in more familiar language, is that the norm is the square of the absolute value. Now the norm of every number in C is a positive integer greater than 1, except for the numbers 0, 1, -1 for which we have $N(0) = 0$, $N(1) = 1$, $N(-1) = 1$. We say that we have a factoring of $a + b\sqrt{-6}$ if we can write

$$a + b\sqrt{-6} = (x_1 + y_1\sqrt{-6})(x_2 + y_2\sqrt{-6}), \tag{1.1}$$

where $N(x_1 + y_1\sqrt{-6}) > 1$ and $N(x_2 + y_2\sqrt{-6}) > 1$. This restriction on the norms of the factors is needed to rule out such trivial factorings as $a + b\sqrt{-6} = (1)(a + b\sqrt{-6}) = (-1)(-a - b\sqrt{-6})$. The norm of a product can be readily calculated to be the product of the norms of the factors, so that in the factoring (1.1) we have $N(a + b\sqrt{-6}) = N(x_1 + y_1\sqrt{-6})N(x_2 + y_2\sqrt{-6})$. It follows that

$$1 < N(x_1 + y_1\sqrt{-6}) < N(a + b\sqrt{-6}),$$
$$1 < N(x_2 + y_2\sqrt{-6}) < N(a + b\sqrt{-6}),$$

and so any number $a + b\sqrt{-6}$ will break up into only a finite number of factors since the norm of each factor is an integer.

We remarked above that the norm of any number in C, apart from 0 and ± 1, is greater than 1. More can be said. Since $N(a + b\sqrt{-6})$ has the value $a^2 + 6b^2$, we observe that

$$N(a + b\sqrt{-6}) \geqq 6 \qquad \text{if } b \neq 0, \tag{1.2}$$

that is, the norm of any complex number in C is not less than 6.

A number of C having norm > 1, but which cannot be factored in the sense of (1.1), is called a prime in C. For example 5 is a prime in C. For in the first place 5 cannot be factored into real numbers in C. In the second place, if we had a factoring $5 = (x_1 + y_1\sqrt{-6})(x_2 + y_2\sqrt{-6})$ into complex numbers, we could take norms to get

$$25 = N(x_1 + y_1\sqrt{-6})N(x_2 + y_2\sqrt{-6}),$$

which contradicts (1.2). Thus 5 is a prime in C, and a similar argument establishes that 2 is a prime.

We are now in a position to show that not all numbers of C factor uniquely into primes. Consider the number 10 and its two factorings:

$$10 = 2 \cdot 5 = (2 + \sqrt{-6})(2 - \sqrt{-6}).$$

The first product $2 \cdot 5$ has factors that are prime in C, as we have seen above. Thus we can conclude that there is not unique factorization of the number 10 in C. Note that this conclusion does not depend on our knowing that $2 + \sqrt{-6}$ and $2 - \sqrt{-6}$ are primes; they actually are, but it is unimportant in our discussion.

We now return to the discussion of unique factorization in the ordinary integers $0, \pm 1, \pm 2, \cdots$. It will be convenient to have the following result.

Theorem 1.15 *If $p \mid ab$, p being a prime, then $p \mid a$ or $p \mid b$. More generally, if $p \mid a_1 a_2 \cdots a_n$, then p divides at least one factor a_i of the product.*

Proof. If $p \nmid a$, then $(a, p) = 1$ and so by Theorem 1.10, $p \mid b$. We may regard this as the first step of a proof of the general statement by mathematical induction. So we assume that the proposition holds whenever p divides a product with fewer than n factors. Now if $p \mid a_1 a_2 \cdots a_n$, that is, $p \mid a_1 c$ where $c = a_2 a_3 \cdots a_n$, then $p \mid a_1$ or $p \mid c$. If $p \mid c$ we apply the induction hypothesis to conclude that $p \mid a_i$ for some subscript i.

Theorem 1.16 *The fundamental theorem of arithmetic, or the unique factorization theorem. The factoring of any integer $n > 1$ into primes is unique apart from the order of the prime factors.*

First proof. Suppose that there is an integer n with two different factorings. Dividing out any primes common to the two representations, we would have an equality of the form

$$p_1 p_2 \cdots p_r = q_1 q_2 \cdots q_s \tag{1.3}$$

where the factors p_i and q_j are primes, not necessarily all distinct, but where no prime on the left side occurs on the right side. But this is impossible because $p_1 \mid q_1 q_2 \cdots q_s$, and so by Theorem 1.15, p_1 is a divisor of at least one of the q_j. That is, p_1 must be identical with at least one of the q_j.

Second proof. Suppose that the theorem is false and let n be the smallest positive integer having more than one representation as the product of primes, say

$$n = p_1 p_2 \cdots p_r = q_1 q_2 \cdots q_s. \tag{1.4}$$

It is clear that r and s are greater than 1. Now the primes $p_1, p_2, \cdots, p_r$ have no members in common with $q_1, q_2, \cdots, q_s$ because if, for example, p_1 were a common prime, then we could divide it out of both sides of (1.4) to get two distinct factorings of n/p_1. But this would contradict our assumption that all integers smaller than n are uniquely factorable.

Next, there is no loss of generality in presuming that $p_1 < q_1$, and we define the positive integer N as

$$N = (q_1 - p_1)q_2q_3 \cdots q_s = p_1(p_2p_3 \cdots p_r - q_2q_3 \cdots q_s). \tag{1.5}$$

It is clear that $N < n$, so that N is uniquely factorable into primes. But $p_1 \nmid (q_1 - p_1)$, so (1.5) gives us two factorings of N, one involving p_1 and the other not, and thus we have a contradiction.

In the application of the fundamental theorem we frequently write any integer $a > 1$ in the form, sometimes called the "canonical form,"

$$a = p_1^{\alpha_1}p_2^{\alpha_2} \cdots p_r^{\alpha_r}$$

where the primes p_i are distinct and the exponents α_i are positive. However, it is sometimes convenient to use a slight variation of the canonical form and to permit some exponents to be zero. For example, if we want to describe the greatest common divisor g of a and b in terms of the prime factors of a and b, we would write

$$a = p_1^{\alpha_1}p_2^{\alpha_2} \cdots p_r^{\alpha_r}, \qquad b = p_1^{\beta_1}p_2^{\beta_2} \cdots p_r^{\beta_r} \tag{1.6}$$

where $\alpha_i \geqq 0$ and $\beta_i \geqq 0$. Then the greatest common divisor is seen to be

$$g = (a, b) = p_1^{\min(\alpha_1,\beta_1)}p_2^{\min(\alpha_2,\beta_2)} \cdots p_r^{\min(\alpha_r,\beta_r)}$$

where min (α, β) denotes the minimum of α and β. In case $a = 108$ and $b = 225$, we would have

$$a = 2^2 3^3 5^0, \qquad b = 2^0 3^2 5^2, \qquad g = 2^0 3^2 5^0 = 9.$$

Similarly the least common multiple of a and b is seen to be

$$[a, b] = p_1^{\max(\alpha_1,\beta_1)}p_2^{\max(\alpha_2,\beta_2)} \cdots p_r^{\max(\alpha_r,\beta_r)}$$

where max (α, β) denotes the maximum of α and β. If no α_i is greater than 1, we say that a is *square-free*.

Theorem 1.17 *Euclid. The number of primes is infinite. That is, there is no end to the sequence of primes*

$$2, 3, 5, 7, 11, 13, \cdots.$$

Proof. Suppose there were only a finite number of primes $p_1, p_2, \cdots, p_r$. Then form the number

$$n = 1 + p_1p_2 \cdots p_r.$$

Note that n is not divisible by p_1 or p_2 or $\cdots$ or p_r. Hence any prime divisor p of n is a prime distinct from $p_1, p_2, \cdots, p_r$. Since n is either a prime or has a prime factor p, this implies that there is a prime distinct from $p_1, p_2, \cdots, p_r$.

Theorem 1.18 *There are arbitrarily large gaps in the series of primes. Stated otherwise, given any positive integer k, there exist k consecutive composite integers.*

Proof. Consider the integers

$$(k+1)! + 2, (k+1)! + 3, \cdots, (k+1)! + k, (k+1)! + k + 1.$$

Every one of these is composite because j divides $(k+1)! + j$ if $2 \leqq j \leqq k+1$.

The primes are spaced rather irregularly, as the last theorem suggests. If we denote the number of primes that do not exceed x by $\pi(x)$, we may ask about the nature of this function. Because of the irregular occurrence of the primes, we cannot expect a simple formula for $\pi(x)$. However, one of the most striking results in advanced number theory, the prime number theorem, gives an asymptotic approximation for $\pi(x)$. It states that

$$\lim_{x \to \infty} \pi(x) \frac{\log x}{x} = 1,$$

that is, that the ratio of $\pi(x)$ to $x/\log x$ approaches 1 as x becomes indefinitely large.

PROBLEMS

1. With a and b as in (1.6) what conditions on the exponents must be satisfied if $a \mid b$? What conditions if $(a, b) = 1$?

2. Observe that the definition that a is square-free amounts to stating that a is divisible by the square of no integer greater than 1. What is the largest number of consecutive square-free positive integers? What is the largest number of consecutive cube-free positive integers, where a is cube-free if it is divisible by the cube of no integer greater than 1.

3. In any positive integer, such as 8347, the last digit is called the units digit, the next the tens digit, the next the hundreds digit, etc. In the example 8347, the units digit is 7, the tens digit is 4, the hundreds digit is 3, and the thousands digit is 8. Prove that a number is divisible by 2 if and only if its units digit is divisible by 2; that a number is divisible by 4 if and only if the integer formed by its tens digit and its units digit is divisible by 4; that a number is divisible by 8 if and only if the integer formed by its last three digits is divisible by 8.

4. Prove that an integer is divisible by 3 if and only if the sum of its digits is divisible by 3. Prove that an integer is divisible by 9 if and only if the sum of its digits is divisible by 9.

5. Prove that an integer is divisible by 11 if and only if the difference between the sum of the digits in the odd places and the sum of the digits in the even places is divisible by 11.
6. Show that every positive integer n has a unique expression of the form $n = 2^r m$, $r \geqq 0$, m a positive odd integer.
7. Let a/b and c/d be fractions in lowest terms, so that $(a, b) = (c, d) = 1$. Prove that if their sum is an integer, then $|b| = |d|$.
8. Prove that any prime of the form $3k + 1$ is of the form $6k + 1$.
9. Prove that any positive integer of the form $3k + 2$ has a prime factor of the same form; similarly for each of the forms $4k + 3$ and $6k + 5$.
10. If x and y are odd, prove that $x^2 + y^2$ cannot be a perfect square.
11. If x and y are prime to 3, prove that $x^2 + y^2$ cannot be a perfect square.
12. Prove that $(a, b) = (a, b, a + b)$, and more generally that $(a, b) = (a, b, ax + by)$ for all integers x, y.
13. Prove that $(a, a + k) \mid k$ for all integers a, k not both zero.
14. Prove that $(a_1, a_2, \cdots, a_n) = ((a_1, a_2, \cdots, a_{n-1}), a_n)$.
15. Prove that $(a, a + 2) = 1$ or 2 for every integer a.
16. If $(a, b) = p$, a prime, what are the possible values of (a^2, b)? Of (a^3, b)? Of (a^2, b^3)?
17. Evaluate (ab, p^4) and $(a + b, p^4)$ given that $(a, p^2) = p$ and $(b, p^3) = p^2$ where p is a prime.
18. If a and b are represented by (1.6), what conditions must be satisfied by the exponents if a is to be a perfect square? A perfect cube? For $a \mid b$? For $a^2 \mid b^2$?
19. Prove the second part of Theorem 1.13 by use of the g.c.d. and l.c.m. formulas following (1.6).
20. Prove that $(a^2, b^2) = c^2$ if $(a, b) = c$.
21. Let a and b be positive integers such that $(a, b) = 1$ and ab is a perfect square. Prove that a and b are perfect squares. Prove that the result generalizes to kth powers.
22. Given $(a, b, c)[a, b, c] = abc$, prove that $(a, b) = (b, c) = (a, c) = 1$.
23. Prove that $[a, b, c](ab, bc, ca) = |abc|$.
24. Determine whether the following assertions are true or false. If true, prove the result, and if false, give a counterexample.

(1) If $(a, b) = (a, c)$ then $[a, b] = [a, c]$.
(2) If $(a, b) = (a, c)$ then $(a^2, b^2) = (a^2, c^2)$.
(3) If $(a, b) = (a, c)$ then $(a, b) = (a, b, c)$.
(4) If p is a prime and $p \mid a$ and $p \mid (a^2 + b^2)$ then $p \mid b$.
(5) If p is a prime and $p \mid a^7$ then $p \mid a$.
(6) If $a^3 \mid c^3$ then $a \mid c$.
(7) If $a^3 \mid c^2$ then $a \mid c$.
(8) If $a^2 \mid c^3$ then $a \mid c$.
(9) If p is a prime and $p \mid (a^2 + b^2)$ and $p \mid (b^2 + c^2)$ then $p \mid (a^2 - c^2)$.
(10) If p is a prime and $p \mid (a^2 + b^2)$ and $p \mid (b^2 + c^2)$ then $p \mid (a^2 + c^2)$.
(11) If $(a, b) = 1$ then $(a^2, ab, b^2) = 1$.
(12) $[a^2, ab, b^2] = [a^2, b^2]$.

(13) If $b \mid (a^2 + 1)$ then $b \mid (a^4 + 1)$.
(14) If $b \mid (a^2 - 1)$ then $b \mid (a^4 - 1)$.
(15) $(a, b, c) = ((a, b), (a, c))$.

25. For which positive integers n is it true that

$$\sum_{j=1}^{n} j \mid \prod_{j=1}^{n} j?$$

26. Given positive integers a and b such that $a \mid b^2, b^2 \mid a^3, a^3 \mid b^4, b^4 \mid a^5, \cdots$, prove that $a = b$.

27. Given integers a, b, c, d, m, n, u, v satisfying $ad - bc = \pm 1, u = am + bn$, $v = cm + dn$, prove that $(m, n) = (u, v)$.

28. Prove that if n is composite it has a prime divisor p satisfying $p \leqq \sqrt{n}$.

29. Obtain a complete list of the primes between 1 and n, with $n = 200$ for convenience, by the following method, known as the "sieve of Eratosthenes." By the "proper" multiples of k we mean all positive multiples of k except k itself. Write all numbers from 2 to 200. Cross out all proper multiples of 2, then of 3, then of 5. At each stage the next larger remaining number is a prime. Thus 7 is now the next remaining larger than 5. Cross out the proper multiples of 7. The next remaining number larger than 7 is 11. Continuing, we cross out the proper multiples of 11 and then of 13. Now we observe that the next remaining number greater than 13 exceeds $\sqrt{200}$, and hence by the previous problem all the numbers remaining in our list are prime.

30. Consider the set S of integers $1, 2, 3, \cdots, n$. Let 2^k be the integer in S which is the highest power of 2. Prove that 2^k is not a divisor of any other integer in S.

31. Prove that $\sum_{j=1}^{n} 1/j$ is not an integer if $n > 1$.

32. Consider the set T of integers $1, 3, 5, \cdots, 2n - 1$. Let 3^r be the integer in T which is the highest power of 3. Prove that 3^r is not a divisor of any other integer of T.

33. Prove that $\sum_{j=1}^{n} 1/(2j - 1)$ is not an integer if $n > 1$.

34. Say that a positive integer n is a sum of consecutive integers if there exist positive integers m and k so that $n = m + (m + 1) + \cdots + (m + k)$. Prove that n is so expressible if and only if it is not a power of 2.

35. If $2^n + 1$ is an odd prime, prove that n is a power of 2.

36. If $2^n - 1$ is a prime, prove that n is a prime. (Numbers of the form $M_p = 2^p - 1$, where p is a prime, are called *Mersenne numbers*, because Mersenne stated that the only primes p for which M_p is a prime are $p = 2$, 3, 5, 7, 13, 17, 19, 31, 67, 127, 257. However, M_{67} and M_{257} are not primes, for example, and M_{61} is a prime. The question of which numbers M_p are primes is not settled.)

37. If a and $b > 2$ are any positive integers, prove that $2^a + 1$ is not divisible by $2^b - 1$.

38. Let positive integers g and l be given with $g \mid l$. Prove that the number of pairs of positive integers x, y satisfying $(x, y) = g$ and $[x, y] = l$ is 2^k, where

k is the number of distinct prime factors of l/g. (Count x_1, y_1 and x_2, y_2 as different pairs if $x_1 \neq x_2$ or $y_1 \neq y_2$.)

39. Let $k \geqq 3$ be a fixed integer. Find all sets $a_1, a_2, \cdots, a_k$ of positive integers such that the sum of any triplet is divisible by each member of the triplet.

40. Prove that $2 + \sqrt{-6}$ and $2 - \sqrt{-6}$ are primes in the class C of numbers $a + b\sqrt{-6}$.

41. Prove Theorem 1.13 by use of the fundamental theorem.

42. Prove that every positive integer is uniquely expressible in the form

$$2^{j_0} + 2^{j_1} + 2^{j_2} + \cdots + 2^{j_m}$$

where $m \geqq 0$ and $0 \leqq j_0 < j_1 < j_2 < \cdots < j_m$.

43. Prove that any positive integer a can be uniquely expressed in the form

$$a = 3^m + b_{m-1}3^{m-1} + b_{m-2}3^{m-2} + \cdots + b_0,$$

where each $b_j = 0$, 1, or -1.

44. Prove that there are no positive integers $a, b, n > 1$ such that $(a^n - b^n) \mid (a^n + b^n)$.

45. Prove that no polynomial $f(x)$ of degree >1 with integral coefficients can represent a prime for every positive integer x. *Suggestion:* if $f(j) = p$ then $f(j + kp) - f(j)$ is a multiple of p for every k, and so $f(j + kp)$ has the same property.

46. Prove that there are infinitely many primes by considering the sequence $2^{2^1} + 1, 2^{2^2} + 1, 2^{2^3} + 1, 2^{2^4} + 1, \cdots$. *Suggestion:* use Problem 36 of Section 1.2.

47. The numbers in the sequence in the preceding problem are called the *Fermat numbers*, after Pierre Fermat who thought they might all be primes. Verify that the fifth number in the sequence is not a prime by multiplying out

$$(2^9 + 2^7 + 1)(2^{23} - 2^{21} + 2^{19} - 2^{17} + 2^{14} - 2^9 - 2^7 + 1)$$

to get $2^{32} + 1$.

48. Prove that there are infinitely many primes of the form $4n + 3$; of the form $6n + 5$.

Remark. The last problem can be stated thus: each of the arithmetic progressions 3, 7, 11, 15, 19, $\cdots$, and 5, 11, 17, 23, 29, $\cdots$ contains an infinitude of primes. One of the famous theorems of number theory (the proof of which lies deeper than the methods of this book), due to Dirichlet, is that the arithmetic progression $a, a + b, a + 2b, a + 3b, \cdots$ contains infinitely many primes if the integers a and $b > 0$ are relatively prime, that is if $(a, b) = 1$.

NOTES ON CHAPTER 1

The symbol Z is widely used to denote the set of all integers, positive, negative, and zero. Likewise the set of all rational numbers is denoted by Q, presumably because a rational number is a quotient a/b of integers.

It can be noted that the second proof of Theorem 1.16 does not depend on Theorem 1.15 or indeed on any previous theorem. Thus the logical arrangement of this chapter could be altered considerably by putting Theorems 1.14 and 1.16 in an early position, and then using the formulas for (b, c) and $[b, c]$ following (1.6) to prove such results as Theorems 1.6, 1.7, 1.8, 1.10, 1.13, and 1.15.

The prime number theorem, stated at the end of Section **1.3**, is not proved in this book. A weaker version due to Tchebychef is given in Theorem 8.1. For a proof of the prime number theorem itself, the reader is referred to the excellent accounts by Hardy and Wright (listed in the General References on page 269) and by Norman Levinson, *Amer. Math. Monthly*, **76**, 225–244 (1969).

2

Congruences

2.1 Congruences

It is apparent from Chapter 1 that divisibility is a fundamental concept of number theory, one that sets it apart from many other branches of mathematics. In this chapter we continue the study of divisibility, but from a slightly different point of view. A congruence is nothing more than a statement about divisibility. However, it is more than just a convenient notation. It often makes it easier to discover proofs, and we will see that congruences can suggest new problems that will lead us to new and interesting topics.

Definition 2.1 *If an integer m, not zero, divides the difference $a - b$, we say that a is congruent to b modulo m and write $a \equiv b \pmod{m}$. If $a - b$ is not divisible by m, we say that a is not congruent to b modulo m, and in this case we write $a \not\equiv b \pmod{m}$.*

Since $a - b$ is divisible by m if and only if $a - b$ is divisible by $-m$, we can generally confine our attention to a positive modulus. Indeed, we shall assume throughout the present chapter that the modulus m is a positive integer.

Congruences have many properties in common with equalities. Some properties that follow easily from the definition are listed in the following theorem.

Theorem 2.1 *Let a, b, c, d, x, y denote integers. Then:*

(*a*) $a \equiv b \pmod{m}$, $b \equiv a \pmod{m}$, *and* $a - b \equiv 0 \pmod{m}$ *are equivalent statements.*

(*b*) *If* $a \equiv b \pmod{m}$ *and* $b \equiv c \pmod{m}$, *then* $a \equiv c \pmod{m}$.
(*c*) *If* $a \equiv b \pmod{m}$ *and* $c \equiv d \pmod{m}$, *then* $ax + cy \equiv bx + dy \pmod{m}$.
(*d*) *If* $a \equiv b \pmod{m}$ *and* $c \equiv d \pmod{m}$, *then* $ac \equiv bd \pmod{m}$.
(*e*) *If* $a \equiv b \pmod{m}$ *and* $d \mid m$, $d > 0$, *then* $a \equiv b \pmod{d}$.

Theorem 2.2 *Let f denote a polynomial with integral coefficients. If* $a \equiv b \pmod{m}$ *then* $f(a) \equiv f(b) \pmod{m}$.

Proof. We can suppose $f(x) = c_0x^n + c_1x^{n-1} + \cdots + c_n$ where the c_i are integers. Since $a \equiv b \pmod{m}$ we can apply Theorem 2.1*d* repeatedly to find $a^2 \equiv b^2, a^3 \equiv b^3, \cdots, a^n \equiv b^n \pmod{m}$, and then $c_ja^{n-j} \equiv c_jb^{n-j} \pmod{m}$, and finally $c_0a^n + c_1a^{n-1} + \cdots + c_n \equiv c_0b^n + c_1b^{n-1} + \cdots + c_n \pmod{m}$, by Theorem 2.1*c*.

The reader is, of course, well aware of the property of real numbers that if $ax = ay$ and $a \neq 0$ then $x = y$. More care must be used in dividing a congruence through by a.

Theorem 2.3

(*a*) $ax \equiv ay \pmod{m}$ *if and only if* $x \equiv y \left(\bmod \dfrac{m}{(a, m)}\right)$.
(*b*) *If* $ax \equiv ay \pmod{m}$ *and* $(a, m) = 1$, *then* $x \equiv y \pmod{m}$.
(*c*) $x \equiv y \pmod{m_i}$ *for* $i = 1, 2, \cdots, r$ *if and only if* $x \equiv y \pmod{[m_1, m_2, \cdots, m_r]}$.

Proof. (*a*) If $ax \equiv ay \pmod{m}$ then $ay - ax = mz$ for some integer z. Hence we have

$$\frac{a}{(a, m)}(y - x) = \frac{m}{(a, m)} z$$

and thus

$$\frac{m}{(a, m)} \,\Big|\, \frac{a}{(a, m)}(y - x).$$

But $(a/(a, m), m/(a, m)) = 1$ by Theorem 1.7 and therefore $\{m/(a, m)\} \mid (y - x)$ by Theorem 1.10. This implies

$$x \equiv y \pmod{m/(a, m)}.$$

If $x \equiv y \pmod{m/(a, m)}$ then $\{m/(a, m)\} \mid (y - x)$ and hence $m \mid (a, m)(y - x)$. Using Theorem 1.1 we get $m \mid a(y - x)$, and then $ax \equiv ay \pmod{m}$ follows.

(*b*) This is a special case of part (*a*). It is listed separately because we will use it very often.

(*c*) If $x \equiv y \pmod{m_i}$ for $i = 1, 2, \cdots, r$, then $m_i \mid (y - x)$ for $i = 1, 2, \cdots, r$. That is, $y - x$ is a common multiple of $m_1, m_2, \cdots, m_r$, and therefore (see Theorem 1.12) $[m_1, m_2, \cdots, m_r] \mid (y - x)$. This implies $x \equiv y \pmod{[m_1, m_2, \cdots, m_r]}$.

If $x \equiv y \pmod{[m_1, m_2, \cdots, m_r]}$ then $x \equiv y \pmod{m_i}$ by Theorem 2.1*e*, since $m_i \mid [m_1, m_2, \cdots, m_r]$.

In dealing with integers modulo m we are essentially performing the ordinary operations of arithmetic but are disregarding multiples of m. In a sense we are not distinguishing between a and $a + mx$, where x is any integer. Given any integer a, let q and r be the quotient and remainder upon division by m; thus $a = qm + r$ by Theorem 1.2. Now $a \equiv r \pmod{m}$ and, since r satisfies the inequalities $0 \leqq r < m$, we see that every integer is congruent modulo m to one of the values $0, 1, 2, \cdots, m - 1$. Also it is clear that no two of these m integers are congruent modulo m. These m values constitute a complete residue system modulo m, and we now give a general definition of this term.

Definition 2.2 *If $x \equiv y \pmod{m}$ then y is called a residue of x modulo m. A set $x_1, x_2, \cdots, x_m$ is called a complete residue system modulo m if for every integer y there is one and only one x_j such that $y \equiv x_j \pmod{m}$.*

It is obvious that there are infinitely many complete residue systems modulo m, the set $1, 2, \cdots, m - 1, m$ being another example.

A set of m integers forms a complete residue system modulo m if and only if no two integers in the set are congruent modulo m.

Theorem 2.4 *If $x \equiv y \pmod{m}$ then $(x, m) = (y, m)$.*

Proof. We have $y - x = mz$ for some integer z. Since $(x, m) \mid x$ and $(x, m) \mid m$, we have $(x, m) \mid y$ and hence $(x, m) \mid (y, m)$. In the same way we find $(y, m) \mid (x, m)$ and then have $(x, m) = (y, m)$ by Theorem 1.1 since (x, m) and (y, m) are positive.

Definition 2.3 *A reduced residue system modulo m is a set of integers r_i such that $(r_i, m) = 1$, $r_i \not\equiv r_j \pmod{m}$ if $i \neq j$, and such that every x prime to m, is congruent modulo m to some member r_i of the set.*

In view of Theorem 2.4 it is clear that a reduced residue system modulo m can be obtained by deleting from a complete residue system modulo m those members that are not relatively prime to m. Furthermore, all reduced residue systems modulo m will contain the same number of members, a number that is denoted by $\phi(m)$. This function is called *Euler's ϕ-function*, sometimes the *totient*. By applying this definition of $\phi(m)$ to the complete residue system $1, 2, \cdots, m$ mentioned in the paragraph following Definition 2.2, we can get what amounts to an alternative definition of $\phi(m)$, as given in the following theorem.

Theorem 2.5 *The number $\phi(m)$ is the number of positive integers less than or equal to m that are relatively prime to m.*

Theorem 2.6 *Let $(a, m) = 1$. Let $r_1, r_2, \cdots, r_n$ be a complete, or a reduced, residue system modulo m. Then $ar_1, ar_2, \cdots, ar_n$ is a complete, or a reduced, residue system, respectively, modulo m.*

Proof. If $(r_i, m) = 1$ then $(ar_i, m) = 1$ by Theorem 1.8.

There are the same number of $ar_1, ar_2, \cdots, ar_n$ as of $r_1, r_2, \cdots, r_n$. Therefore we need only show that $ar_i \not\equiv ar_j \pmod m$ if $i \neq j$. But Theorem 2.3*b* shows that $ar_i \equiv ar_j \pmod m$ implies $r_i \equiv r_j \pmod m$ and hence $i = j$.

Theorem 2.7 *Fermat's theorem. Let p denote a prime. If $p \nmid a$ then $a^{p-1} \equiv 1 \pmod p$. For every integer a, $a^p \equiv a \pmod p$.*

We will postpone the proof of this theorem and will obtain it as a corollary to Theorem 2.8.

Theorem 2.8 *Euler's generalization of Fermat's theorem. If $(a, m) = 1$ then*

$$a^{\phi(m)} \equiv 1 \pmod m.$$

Proof. Let $r_1, r_2, \cdots, r_{\phi(m)}$ be a reduced residue system modulo m. Then by Theorem 2.6, $ar_1, ar_2, \cdots, ar_{\phi(m)}$ is also a reduced residue system modulo m. Hence corresponding to each r_i there is one and only one ar_j such that $r_i \equiv ar_j \pmod m$. Furthermore, different r_i will have different corresponding ar_j. This means that the numbers $ar_1, ar_2, \cdots, ar_{\phi(m)}$ are just the residues modulo m of $r_1, r_2, \cdots, r_{\phi(m)}$, but not necessarily in the same order. Multiplying and using Theorem 2.1*d* we obtain

$$\prod_{j=1}^{\phi(m)} (ar_j) \equiv \prod_{i=1}^{\phi(m)} r_i \pmod m$$

and hence

$$a^{\phi(m)} \prod_{j=1}^{\phi(m)} r_j \equiv \prod_{j=1}^{\phi(m)} r_j \pmod m.$$

Now $(r_j, m) = 1$ so we can use Theorem 2.3*b* to cancel the r_j and we obtain $a^{\phi(m)} \equiv 1 \pmod m$.

Corollary. *Proof of Theorem* 2.7. If $p \nmid a$ then $(a, p) = 1$ and $a^{\phi(p)} \equiv 1 \pmod p$. To find $\phi(p)$ we refer to Theorem 2.5. All the integers $1, 2, \cdots, p - 1, p$ with the exception of p are relatively prime to p. Thus we have $\phi(p) = p - 1$, and the first part of Fermat's theorem follows. The second part is now obvious.

Corollary 2.9 *If* $(a, m) = 1$ *then* $ax \equiv b \pmod{m}$ *has a solution* $x = x_1$. *All solutions are given by* $x = x_1 + jm$ *where* $j = 0, \pm 1, \pm 2, \cdots$.

Proof. Since $(1, m) = 1$ and $1 \leqq m$, we see that $\phi(m) \geqq 1$. Then we need merely set $x_1 = a^{\phi(m)-1}b$.

If x is any solution then $ax - ax_1 \equiv b - b \equiv 0 \pmod{m}$ and hence $a(x - x_1) \equiv 0 \pmod{m}$. Using Theorem 2.3*b* we get $x - x_1 \equiv 0 \pmod{m}$ which implies $x = x_1 + jm$. The fact that all these actually are solutions follows from Theorem 2.2.

Euler's function $\phi(m)$ is of considerable interest. We will consider it further in section 2.4 and section 4.2.

Theorem 2.10 *Wilson's theorem. If* p *is a prime then* $(p - 1)! \equiv -1 \pmod{p}$.

Proof. If $p = 2$ or $p = 3$ the congruence is easily verified.

Now we can suppose $p \geqq 5$. The idea behind the proof is quite simple, but we must use a little care. We consider the integers whose product is $(p - 1)!$ and try to pair them off in such a way that the product of the two members of each pair is congruent to 1 modulo p.

Given an integer j satisfying $1 \leqq j \leqq p - 1$, then $(j, p) = 1$ and we see by Corollary 2.9 that there is exactly one integer i such that $ji \equiv 1 \pmod{p}$ and $0 \leqq i \leqq p - 1$. Obviously $i = 0$ is impossible so we have $1 \leqq i \leqq p - 1$. With each j we will associate the corresponding integer i. Since $ij \equiv ji \equiv 1 \pmod{p}$ we see that j is the integer associated with i. The integer 1 is associated with itself, and so is $p - 1$. Omitting these values for a moment we consider $2 \leqq j \leqq p - 2$. For these j we have $(j - 1, p) = (j + 1, p) = 1$ and hence $j^2 - 1 = (j - 1)(j + 1) \not\equiv 0 \pmod{p}$ by Theorem 1.8. Hence each of these j is associated with an $i \neq j$, $2 \leqq i \leqq p - 2$, and the associate of i is j itself. Thus the integers $2, 3, \cdots, p - 2$ can be paired off, j and its associate i, and $ji \equiv 1 \pmod{p}$. Multiplying all these pairs together we get $2 \cdot 3 \cdots (p - 2) \equiv 1 \pmod{p}$, and Wilson's theorem follows because $1 \cdot (p - 1) \equiv -1 \pmod{p}$.

Wilson's theorem and Fermat's theorem can be used to determine those primes p for which $x^2 \equiv -1 \pmod{p}$ has a solution. This is a special case of some results that we will take up later (see Chapter 3). However, it is interesting to see that this special case can be handled by fairly simple means.

Theorem 2.11 *Let* p *denote a prime. Then* $x^2 \equiv -1 \pmod{p}$ *has solutions if and only if* $p = 2$ *or* $p \equiv 1 \pmod{4}$.

Proof. If $p = 2$ we have the solution $x = 1$.

For any odd prime p we can write Wilson's theorem in the form

$$\left(1 \cdot 2 \cdots j \cdots \frac{p-1}{2}\right)\left(\frac{p+1}{2} \cdots (p - j) \cdots (p - 2)(p - 1)\right) \equiv -1 \pmod{p}.$$

The product on the left has been divided into two parts, each with the same number of factors. Pairing off j in the first half with $p - j$ in the second half, we can rewrite the congruence in the form

$$\prod_{j=1}^{(p-1)/2} j(p-j) \equiv -1 \pmod{p}.$$

But $j(p - j) \equiv -j^2 \pmod{p}$, and so we have, if $p \equiv 1 \pmod{4}$,

$$\prod_{j=1}^{(p-1)/2} j(p-j) \equiv \prod_{j=1}^{(p-1)/2} (-j^2) \equiv (-1)^{(p-1)/2}\left(\prod_{j=1}^{(p-1)/2} j\right)^2 \equiv \left(\prod_{j=1}^{(p-1)/2} j\right)^2 \pmod{p},$$

and so we have a solution, $\prod_{j=1}^{(p-1)/2} j$, of $x^2 \equiv -1 \pmod{p}$.

If $p \neq 2$ and $p \not\equiv 1 \pmod{4}$, then $p \equiv 3 \pmod{4}$. In this case, if for some integer x, $x^2 \equiv -1 \pmod{p}$, then we have $x^{p-1} \equiv (x^2)^{(p-1)/2} \equiv (-1)^{(p-1)/2} \equiv -1 \pmod{p}$ since $(p - 1)/2 \equiv 1 \pmod{2}$. But clearly $p \nmid x$, so we have $x^{p-1} \equiv 1 \pmod{p}$ by Theorem 2.7. This contradiction shows that $x^2 \equiv -1 \pmod{p}$ has no solution in this case.

PROBLEMS

1. List all integers x in the range $1 \leqq x \leqq 100$ that satisfy $x \equiv 7 \pmod{17}$.
2. Exhibit a complete residue system modulo 17 composed entirely of multiples of 3.
3. Exhibit a reduced residue system for the modulus 12; for 30.
4. If an integer x is even, observe that it must satisfy the congruence $x \equiv 0 \pmod{2}$. If an integer y is odd, what congruence does it satisfy? What congruence does an integer z of the form $6k + 1$ satisfy?
5. Write a single congruence that is equivalent to the pair of congruences $x \equiv 1 \pmod{4}$, $x \equiv 2 \pmod{3}$.
6. Prove that if p is a prime and $a^2 \equiv b^2 \pmod{p}$, then $p \mid (a + b)$ or $p \mid (a - b)$.
7. Show that if $f(x)$ is a polynomial with integral coefficients and if $f(a) \equiv k \pmod{m}$, then $f(a + tm) \equiv k \pmod{m}$ for every integer t.
8. Prove that any number which is a square must have one of the following for its units digit: 0, 1, 4, 5, 6, 9.
9. Prove that any fourth power must have one of 0, 1, 5, 6 for its units digit.
10. Evaluate $\phi(m)$ for $m = 1, 2, 3, \cdots, 12$.
11. Find the least positive integer x such that $13 \mid (x^2 + 1)$.
12. Prove that 19 is not a divisor of $4n^2 + 4$ for any integer n.
13. Exhibit a reduced residue system modulo 7 composed entirely of powers of 3.
14. Solve $3x \equiv 5 \pmod{11}$ by the method of Corollary 2.9.
15. Illustrate the proof of Theorem 2.10 for $p = 11$ and $p = 13$ by actually determining the pairs of associated integers.

16. The integers 12, 23, 34, 45, 56 are congruent to 1 modulo 11. To solve $5x \equiv 1 \pmod{11}$ we merely note that $45 = 5 \cdot 9$ and hence that $x = 9$ is a solution. Solve $ax \equiv 1 \pmod{11}$ for $a = 2, 3, \cdots, 10$.
17. Prove that $n^6 - 1$ is divisible by 7 if $(n, 7) = 1$.
18. Prove that $n^7 - n$ is divisible by 42, for any integer n.
19. Prove that $n^{12} - 1$ is divisible by 7 if $(n, 7) = 1$.
20. Prove that $n^{6k} - 1$ is divisible by 7 if $(n, 7) = 1$, k being any positive integer.
21. Prove that $n^{13} - n$ is divisible by 2, 3, 5, 7, and 13 for any integer n.
22. Prove that $n^{12} - a^{12}$ is divisible by 13 if n and a are prime to 13.
23. Prove that $n^{12} - a^{12}$ is divisible by 91 if n and a are prime to 91.
24. Prove that $\frac{1}{5}n^5 + \frac{1}{3}n^3 + \frac{7}{15}n$ is an integer for every integer n.
25. What is the last digit in the ordinary decimal representation of 3^{400}? *Suggestion:* $3^4 \equiv 1 \pmod 5$ by Fermat's theorem, and this with $3^4 \equiv 1 \pmod 2$ implies that $3^4 \equiv 1 \pmod{10}$. Hence $3^{4n} \equiv 1 \pmod{10}$ for any $n \geqq 1$.
26. What is the last digit in the ordinary decimal representation of 2^{400}?
27. What are the last two digits in the ordinary decimal representation of 3^{400}? *Suggestion:* use Theorem 2.8 to establish that $3^{20} \equiv 1 \pmod{25}$. In addition, $3^2 \equiv 1 \pmod 4$ so that $3^{20} \equiv 1 \pmod 4$, whence $3^{20} \equiv 1 \pmod{100}$.
28. Show that $-(m-1)/2, -(m-3)/2, \cdots, (m-3)/2, (m-1)/2$ is a complete residue system modulo m if m is odd, and that $-(m-2)/2, -(m-4)/2, \cdots, (m-2)/2, m/2$ is a complete residue system modulo m if m is even.
29. Show that $2, 4, 6, \cdots, 2m$ is a complete residue system modulo m if m is odd.
30. Show that $1^2, 2^2, \cdots, m^2$ is not a complete residue system modulo m if $m > 2$.
31. If n is composite, $n > 4$, prove that $(n-1)! \equiv 0 \pmod n$.
32. Show that an integer $m > 1$ is a prime if and only if m divides $(m-1)! + 1$.
33. For positive integers a, m, n with $m \neq n$, prove that

$$(a^{2m} + 1, a^{2n} + 1) = \begin{cases} 1, & \text{if } a \text{ is even,} \\ 2, & \text{if } a \text{ is odd.} \end{cases}$$

Suggestion: if p is a common divisor, $a^{2m} \equiv -1 \pmod p$. Raise this to the power 2^{n-m}, presuming $m < n$.
34. For m odd, prove that the sum of the elements of any complete residue system modulo m is congruent to zero modulo m; prove the analogous result for any reduced residue system for $m > 2$.
35. Find all sets of positive integers a, b, c satisfying all three congruences $a \equiv b \pmod c$, $b \equiv c \pmod a$, $c \equiv a \pmod b$. *Suggestion:* if a, b, c is such a set, so also is ka, kb, kc for any positive integer k. Hence it suffices to determine all "primitive" sets with the property $(a, b, c) = 1$. Also there is no loss in generality in assuming that $a \leqq b \leqq c$.
36. Find all triples a, b, c of non-zero integers such that $a \equiv b \pmod{|c|}$ $b \equiv c \pmod{|a|}$, $c \equiv a \pmod{|b|}$.

37. If p is an odd prime, prove that:

$$1^2 \cdot 3^2 \cdot 5^2 \cdots (p-2)^2 \equiv (-1)^{(p+1)/2} \pmod{p},$$

and

$$2^2 \cdot 4^2 \cdot 6^2 \cdots (p-1)^2 \equiv (-1)^{(p+1)/2} \pmod{p}.$$

38. Prove that $(a+b)^p \equiv a^p + b^p \pmod{p}$, where p is a prime.

39. If $r_1, r_2, \cdots, r_{p-1}$ is any reduced residue system modulo a prime p, prove that

$$\prod_{j=1}^{p-1} r_j \equiv -1 \pmod{p}.$$

40. If $r_1, r_2, \cdots, r_p$ and $r_1', r_2', \cdots, r_p'$ are any two complete residue systems modulo a prime $p > 2$, prove that the set $r_1r_1', r_2r_2', \cdots, r_pr_p'$ cannot be a complete residue system modulo p.

41. If p is any prime other than 2 or 5, prove that p divides infinitely many of the integers 9, 99, 999, 9999, $\cdots$. If p is any prime other than 2 or 5, prove that p divides infinitely many of the integers 1, 11, 111, 1111, $\cdots$.

42. If p is a prime, and if $h + k = p - 1$ with $h \geqq 0$ and $k \geqq 0$, prove that $h!k! + (-1)^h \equiv 0 \pmod{p}$.

43. For any prime p, if $a^p \equiv b^p \pmod{p}$, prove that $a^p \equiv b^p \pmod{p^2}$.

44. Given an integer n, prove that there is an integer m which to base ten contains only the digits 0 and 1 such that $n \mid m$. Prove that the same holds for digits 0 and 2, or 0 and 3, $\cdots$, or 0 and 9, but for no other pair of digits.

45. If n is composite, prove that $(n-1)! + 1$ is not a power of n.

46. If $1 \leqq k < n - 1$ prove that $(n-1)^2 \nmid (n^k - 1)$.

47. If p is a prime, prove that $(p-1)! + 1$ is a power of p if and only if $p = 2$, 3, or 5. *Suggestion:* if $p > 5$, $(p-1)!$ has factors 2, $p - 1$, and $(p-1)/2$, and so $(p-1)!$ is divisible by $(p-1)^2$.

48. Prove $\displaystyle\prod_{\substack{1 \leqq x \leqq n \\ (x,n)=(x+1,n)=1}} x \equiv 1 \pmod{n}$ if $n > 2$.

The symbol on the left denotes the product of all the positive integers x less than or equal to n such that both x and $x + 1$ are relatively prime to n.

49. Prove that there are infinitely many primes of the form $4n + 1$.

50. Prove that $(p-1)! \equiv p - 1 \pmod{1 + 2 + \cdots + (p-1)}$ if p is a prime.

51. For positive integers n let $\tau(n)$ denote the number of positive divisors of n, including n itself. For d such that $d \mid n$, $1 \leqq d \leqq \sqrt{n}$, pair d with n/d to prove that $\tau(n) < 2\sqrt{n}$.

2.2 Solutions of Congruences

In analogy with the solution of algebraic equations it is natural to consider the problem of solving a congruence. In the rest of this chapter we will let

$f(x)$ denote a polynomial with integral coefficients, and we will write $f(x) = a_0x^n + a_1x^{n-1} + \cdots + a_n$. If u is an integer such that $f(u) \equiv 0 \pmod{m}$ then we say that u is a solution of the congruence $f(x) \equiv 0 \pmod{m}$. Whether or not an integer is a solution of a congruence depends on the modulus m as well as on the polynomial $f(x)$. If the integer u is a solution of $f(x) \equiv 0 \pmod{m}$, and if $v \equiv u \pmod{m}$, Theorem 2.2 shows that v is also a solution. Because of this we will say that $x \equiv u \pmod{m}$ is a solution of $f(x) \equiv 0 \pmod{m}$, meaning that every integer congruent to u modulo m satisfies $f(x) \equiv 0 \pmod{m}$. For example, the congruence $x^2 - x + 4 \equiv 0 \pmod{10}$ has the solution 3. It also has the solution 8. We can say $x \equiv 3 \pmod{10}$ and $x \equiv 8 \pmod{10}$ are solutions. In this case, since $8 \equiv 3 \pmod{5}$, we can even say that $x \equiv 3 \pmod{5}$ is a solution. In the general case, if $f(x) \equiv 0 \pmod{m}$ has a solution u, it has infinitely many solutions—all integers v such that $v \equiv u \pmod{m}$. It is more reasonable to count the solutions in a different way. We will not count v as distinct from u if $v \equiv u \pmod{m}$. In the example 3 and 13 are not counted separately. However, 3 and 8 are both counted since $3 \not\equiv 8 \pmod{10}$.

Definition 2.4 *Let $r_1, r_2, \cdots, r_m$ denote a complete residue system modulo m. The number of solutions of $f(x) \equiv 0 \pmod{m}$ is the number of the r_i such that $f(r_i) \equiv 0 \pmod{m}$.*

It is clear from Theorem 2.2 that the number of solutions is independent of the choice of the complete residue system. Furthermore, the number of solutions cannot exceed the modulus m. If m is small it is a simple matter to just compute $f(r_i)$ for each of the r_i and thus to determine the number of solutions. In the above example the congruence has just two solutions. Some other examples are

$$x^2 + 1 \equiv 0 \pmod{7} \text{ has no solutions,}$$
$$x^2 + 1 \equiv 0 \pmod{5} \text{ has two solutions,}$$
$$x^2 - 1 \equiv 0 \pmod{8} \text{ has four solutions.}$$

Definition 2.5 *Let $f(x) = a_0x^n + a_1x^{n-1} + \cdots + a_n$. If $a_0 \not\equiv 0 \pmod{m}$ the degree of the congruence $f(x) \equiv 0 \pmod{m}$ is n. If $a_0 \equiv 0 \pmod{m}$, let j be the smallest positive integer such that $a_j \not\equiv 0 \pmod{m}$; then the degree of the congruence is $n - j$. If there is no such integer j, that is, if all the coefficients of $f(x)$ are multiples of m, no degree is assigned to the congruence.*

It should be noted that the degree of the congruence $f(x) \equiv 0 \pmod{m}$ is not the same thing as the degree of the polynomial $f(x)$. The degree of the congruence depends on the modulus; the degree of the polynomial does not. Thus if $g(x) = 6x^3 + 3x^2 + 1$, then $g(x) \equiv 0 \pmod{5}$ is of degree 3, and $g(x) \equiv 0 \pmod{2}$ is of degree 2, whereas $g(x)$ is of degree 3.

Theorem 2.12 *If $d \mid m$, $d > 0$, and if u is a solution of $f(x) \equiv 0 \pmod{m}$, then u is a solution of $f(x) \equiv 0 \pmod{d}$.*

Proof. This follows directly from Theorem 2.1*e*.

PROBLEMS

1. If $f(x) \equiv 0 \pmod{p}$ has exactly j solutions with p a prime, and $g(x) \equiv 0 \pmod{p}$ has no solutions, prove that $f(x)g(x) \equiv 0 \pmod{p}$ has exactly j solutions.
2. Denoting the number of solutions of $f(x) \equiv k \pmod{m}$ by $N(k)$, prove that $\sum_{k=1}^{m} N(k) = m$.
3. If a congruence $f(x) \equiv 0 \pmod{m}$ has m solutions, prove that any integer whatsoever is a solution. (In such a case the congruence is sometimes called an identical congruence.)
4. The fact that the product of any three consecutive integers is divisible by 3 leads to the identical congruence $x(x + 1)(x + 2) \equiv 0 \pmod{3}$. Generalize this, and write an identical congruence modulo m.

2.3 Congruences of Degree 1

Any congruence of degree 1 can be put in the form $ax \equiv b \pmod{m}$, $a \not\equiv 0 \pmod{m}$. From Corollary 2.9 we see that if $(a, m) = 1$, then $ax \equiv b \pmod{m}$ has exactly one solution, $x \equiv x_1 \pmod{m}$.

Let g denote (a, m). If $ax \equiv b \pmod{m}$ has a solution u, then $b \equiv au \pmod{m}$ and hence $b \equiv au \equiv 0 \pmod{g}$. Therefore, $ax \equiv b \pmod{m}$ has no solution if $g \nmid b$. However, if $g \mid b$ then, for u an integer, $au \equiv b \pmod{m}$ holds if and only if $(a/g)u \equiv (b/g) \pmod{m/g}$ by Theorem 2.3*a*. Now $(a/g, m/g) = 1$ and the congruence $(a/g)x \equiv (b/g) \pmod{m/g}$ has just one solution $x \equiv x_1 \pmod{m/g}$. In other words the solutions of $ax \equiv b \pmod{m}$ are the integers u such that $u \equiv x_1 \pmod{m/g}$, that is $u = x_1 + t(m/g)$, $t = 0, \pm 1, \pm 2, \cdots$. If t is given the values $0, 1, \cdots, g - 1$, then u takes on g values no two of which are congruent modulo m. If t is given any other value, the corresponding u will be congruent modulo m to one of these g values. Thus the solutions of $ax \equiv b \pmod{m}$ are $x \equiv x_1 + t(m/g) \pmod{m}$, $0 \leqq t \leqq g - 1$.

Theorem 2.13 *Let g denote (a, m). Then $ax \equiv b \pmod{m}$ has no solutions if $g \nmid b$. If $g \mid b$ it has g solutions $x \equiv (b/g)x_0 + t(m/g) \pmod{m}$, $t = 0, 1, \cdots, g - 1$ where x_0 is any solution of $(a/g)x \equiv 1 \pmod{m/g}$.*

Proof. This theorem follows from what has already been proved since $(a/g)x \equiv 1 \pmod{m/g}$ has a solution x_0 and then $x_1 = (b/g)x_0$ is a solution of $(a/g)x \equiv (b/g) \pmod{m/g}$.

For reasonably small numbers the solution of a congruence can often be obtained by inspection or by trying all integers of a complete residue system modulo m. However, if the numbers are large the actual numerical solution of a congruence of the form $ax \equiv b \pmod{m}$ can be rather lengthy. The hardest part is solving congruences $ax \equiv 1 \pmod{m}$ with $(a, m) = 1$. The solution as given in the proof of Corollary 2.9 is not usually very practical. A number of special methods of solution have been developed, but perhaps the best general method is to use Euclid's algorithm. Using Theorem 1.11 we determine $g = (a, m)$ and at the same time obtain integers u and v such that $au + mv = g$. Then we can take u for the x_0 in Theorem 2.13, and the rest is easy.

Another way to go about solving a congruence of degree 1 is to factor the modulus $m = \prod_{i=1}^{k} p_i^{e_i}$. Writing $m_i = p_i^{e_i}$ we note that the m_i are relatively prime in pairs and that $[m_1, m_2, \cdots, m_k] = m$. From Theorem 2.3*c* we see that the problem of solving $ax \equiv b \pmod{m}$ is equivalent to solving the set of congruences $ax \equiv b \pmod{m_i}$, $i = 1, 2, \cdots, k$, simultaneously. The individual congruences $ax \equiv b \pmod{m_i}$ may be easier to solve since their moduli m_i are smaller than m. Suppose the congruences $ax \equiv b \pmod{m_i}$ have the solutions $x \equiv u_i \pmod{m_i}$. There still remains the problem of finding the simultaneous solution x of the set of congruences. The proof of the next theorem will show how this can be done.

Theorem 2.14 *Chinese remainder theorem. Let $m_1, m_2, \cdots, m_r$ denote r positive integers that are relatively prime in pairs, and let $a_1, a_2, \cdots, a_r$ denote any r integers. Then the congruences $x \equiv a_i \pmod{m_i}$, $i = 1, 2, \cdots, r$, have common solutions. Any two solutions are congruent modulo $m_1 m_2 \cdots m_r$.*

Remark. If the moduli $m_1, m_2, \cdots, m_r$ are not relatively prime in pairs, there may not be any solution of the congruences. Necessary and sufficient conditions are given in Problem 14(*c*) in the next problem set.

Proof. Writing $m = m_1 m_2 \cdots m_r$ we see that m/m_j is an integer and that $(m/m_j, m_j) = 1$. Therefore, by Corollary 2.9, there are integers b_j such that $(m/m_j)b_j \equiv 1 \pmod{m_j}$. Clearly $(m/m_j)b_j \equiv 0 \pmod{m_i}$ if $i \neq j$. Now if we define x_0 as

$$x_0 = \sum_{j=1}^{r} \frac{m}{m_j} b_j a_j \tag{2.1}$$

we have

$$x_0 \equiv \sum_{j=1}^{r} \frac{m}{m_j} b_j a_j \equiv \frac{m}{m_i} b_i a_i \equiv a_i \pmod{m_i}$$

so that x_0 is a common solution of the original congruences.

If x_0 and x_1 are both common solutions of $x \equiv a_1 \pmod{m_i}$, $i = 1, 2, \cdots, r$, then $x_0 \equiv x_1 \pmod{m_i}$, for $i = 1, 2, \cdots, r$ and hence $x_0 \equiv x_1 \pmod{m}$ by Theorem 2.3*c*. This completes the proof.

The proof of this theorem provides us with an efficient method for solving a certain kind of problem. As an example let us find all integers that have remainders 1 or 2 when they are divided by each of 3, 4, and 5. In other words we are to find the common solutions of $x \equiv 1$ or $2 \pmod 3$, $x \equiv 1$ or $2 \pmod 4$, $x \equiv 1$ or $2 \pmod 5$. We have $m_1 = 3$, $m_2 = 4$, $m_3 = 5$, $m = 60$, and each a_i is 1 or 2. To find b_1 we solve $(60/3)b_1 \equiv 1 \pmod 3$; that is $20b_1 \equiv 1 \pmod 3$, which is the same as $-b_1 \equiv 1 \pmod 3$. We can take $b_1 = -1$ and then have $(m/m_1)b_1 = -20$. Similarly we obtain $b_2 = -1$ and $(m/m_2)b_2 = -15$. For b_3 we have $12b_3 \equiv 1 \pmod 5$, $2b_3 \equiv 1$, $4b_3 \equiv 2$, $-b_3 \equiv 2$, and can take $b_3 = -2$, $(m/m_3)b_3 = -24$. Using (2.1) we have merely to insert the values of the a_i in $x \equiv -20a_1 - 15a_2 - 24a_3 \pmod{60}$. Doing this we obtain the values given in the following table.

a_1	a_2	a_3	$x \pmod{60}$
1	1	1	$-20 - 15 - 24 \equiv -59 \equiv 1$
1	1	2	$-20 - 15 - 48 \equiv -83 \equiv -23$
1	2	1	$-20 - 30 - 24 \equiv -74 \equiv -14$
2	1	1	$-40 - 15 - 24 \equiv -79 \equiv -19$
2	2	1	$-40 - 30 - 24 \equiv -94 \equiv 26$
2	1	2	$-40 - 15 - 48 \equiv -103 \equiv 17$
1	2	2	$-20 - 30 - 48 \equiv -98 \equiv 22$
2	2	2	$-40 - 30 - 48 \equiv -118 \equiv 2$

The integers having remainders 1 or 2 when divided by 3, 4, 5 are given by $x \equiv 1, 2, 17, 22, 26, -14, -19, -23 \pmod{60}$.

PROBLEMS

1. Find all solutions of the congruences
(*a*) $20x \equiv 4 \pmod{30}$;
(*b*) $20x \equiv 30 \pmod 4$;
(*c*) $353x \equiv 254 \pmod{400}$.

2. How many solutions are there to each of the following congruences:
(*a*) $15x \equiv 25 \pmod{35}$;
(*b*) $15x \equiv 24 \pmod{35}$;
(*c*) $15x \equiv 0 \pmod{35}$?

3. Find the smallest positive integer (except $x = 1$) that satisfies the following congruences simultaneously: $x \equiv 1 \pmod 3$, $x \equiv 1 \pmod 5$, $x \equiv 1 \pmod 7$.
4. Find all integers that satisfy simultaneously: $x \equiv 2 \pmod 3$, $x \equiv 3 \pmod 5$, $x \equiv 5 \pmod 2$.
5. Solve the set of congruences: $x \equiv 1 \pmod 4$, $x \equiv 0 \pmod 3$, $x \equiv 5 \pmod 7$.
6. Find all integers that give the remainders 1, 2, 3 when divided by 3, 4, 5 respectively.
7. If a is selected at random from $1, 2, 3, \cdots, 14$, and b is selected at random from $1, 2, 3, \cdots, 15$, what is the probability that $ax \equiv b \pmod{15}$ has at least one solution? Exactly one solution?
8. Given any positive integer k, prove that there are k consecutive integers each divisible by a square > 1.
9. If x_2 is a solution of the congruence $ax \equiv b \pmod m$ (obtained perhaps by application of Euclid's algorithm to a and m), prove that $x \equiv x_2 + t(m/g) \pmod m$ gives all solutions as t runs through $0, 1, \cdots, g - 1$, where g is defined as $g = (a, m)$.
10. Suppose $(a, m) = 1$, and let x_1 denote a solution of $ax \equiv 1 \pmod m$. For $s = 1, 2, \cdots$, let $x_s = 1/a - (1/a)(1 - ax_1)^s$. Prove that x_s is an integer and that it is a solution of $ax \equiv 1 \pmod{m^s}$.
11. Suppose that $(a, m) = 1$. If $a = \pm 1$, the solution of $ax \equiv 1 \pmod{m^s}$ is obviously $x \equiv a \pmod{m^s}$. If $a = \pm 2$, then m is odd and $x \equiv \frac{1}{2}(1 - m^s)\frac{1}{2}a \pmod{m^s}$ is the solution of $ax \equiv 1 \pmod{m^s}$. For all other a use Problem 10 to show that the solution of $ax \equiv 1 \pmod{m^s}$ is $x \equiv k \pmod{m^s}$ where k is the nearest integer to $-(1/a)(1 - ax_1)^s$.
12. Solve $3x \equiv 1 \pmod{125}$ by Problem 11, taking $x_1 = 2$.
13. Let $m_1, m_2, \cdots, m_r$ be relatively prime in pairs. Assuming that each of the congruences $b_i x \equiv a_i \pmod{m_i}$, $i = 1, 2, \cdots, r$, is solvable, prove that the congruences have a simultaneous solution.
14. (*a*) Consider the set of congruences $x \equiv a_i \pmod{p^{e_i}}$, $i = 1, 2, \cdots, r$, with $e_1 \geqq e_2 \geqq \cdots \geqq e_r$. Prove that $x = a_1$ is a simultaneous solution of these congruences if $p^{e_i} \mid (a_1 - a_i)$ for $i = 2, 3, \cdots, r$.
(*b*) Let the canonical factoring of m be $p_1^{e_1}p_2^{e_2} \cdots p_k^{e_k}$. Prove that any simultaneous solution of the set of congruences $x \equiv a \pmod{p_i^{e_i}}$, $i = 1, 2, \cdots, k$, is a solution of $x \equiv a \pmod m$.
(*c*) Prove that the set of congruences $x \equiv a_i \pmod{m_i}$, $i = 1, 2, \cdots, n$, has a simultaneous solution if, and only if, $(m_i, m_j) \mid (a_i - a_j)$ holds for every pair of moduli, that is for every pair of subscripts i, j such that $1 \leqq i < j \leqq n$. Any two solutions are congruent modulo $[m_1, m_2, \cdots, m_n]$.
15. Let $(a, b) = 1$ and $c > 0$. Prove that there is an integer x such that $(a + bx, c) = 1$.
16. Consider a square divided up into n^2 equal squares. Number the columns of small squares $1, 2, \cdots, n$ from left to right. Similarly number the rows $1, 2, \cdots, n$ and let $\{c, r\}$ denote the small square in the cth column and rth row.

Let $a_0, b_0, a, b, \alpha, \beta$ be positive integers less than or equal to n and such that $(a, n) = (b, n) = (\alpha, n) = (\beta, n) = 1$. Write 1 in square $\{a_0, b_0\}$. Then count a columns to the right and b rows up from this square. If this takes you outside of the large square, count as if the large square were bent into a cylinder. Thus you will arrive at $\{a_0 + a, b_0 + b\}$ or $\{a_0 + a - n, b_0 + b\}$ or $\{a_0 + a, b_0 + b - n\}$ or $\{a_0 + a - n, b_0 + b - n\}$, whichever is actually one of the small squares. Write 2 in this square. Count a to the right and b up from 2 and insert 3 in the square. Continue until you have written in $1, 2, 3, \cdots, n$. Prove that m will have been written in $\{x_m, y_m\}$ where x_m and y_m are uniquely determined by

$$x_m \equiv a_0 + a(m - 1) \pmod{n}, \qquad 1 \leqq x_m \leqq n$$
$$y_m \equiv b_0 + b(m - 1) \pmod{n}, \qquad 1 \leqq y_m \leqq n, 1 \leqq m \leqq n.$$

Also show that all these squares $\{x_m, y_m\}$ are different but that continuing the process one more step would put $n + 1$ in $\{a_0, b_0\}$ which is already occupied by 1.

Now having reached $\{a_0, b_0\}$ again, count α to the right and β up and write $n + 1$ in this square which may or may not already be occupied. Then revert to the original process with step a, b to insert $n + 2, n + 3, \cdots, 2n$. Continue in this manner, using the extra step α, β just for $n + 1, 2n + 1, \cdots, (n - 1)n + 1$, and stopping when n^2 has been inserted.

For $1 \leqq m \leqq n^2$ prove that m is in $\{x_m, y_m\}$ where

$$x_m \equiv a_0 + a(m - 1) + \alpha\left[\frac{m-1}{n}\right] \pmod{n}, \qquad 1 \leqq x_m \leqq n,$$

$$y_m \equiv b_0 + b(m - 1) + \beta\left[\frac{m-1}{n}\right] \pmod{n}, \qquad 1 \leqq y_m \leqq n,$$

and $\left[\frac{m-1}{n}\right]$ is the quotient when n is divided into $m - 1$. Also prove that if $(a\beta - b\alpha, n) = 1$ then each square contains one and only one integer m, $1 \leqq m \leqq n^2$.

From now on assume $(a\beta - b\alpha, n) = 1$. Writing $m - 1 = qn + s$, $0 \leqq s \leqq n - 1$, show that the entries in the cth column are just the $m = qn + s + 1$ for which $0 \leqq q \leqq n - 1$, $0 \leqq s \leqq n - 1$, and $as \equiv c - a_0 - \alpha q \pmod{n}$. Prove that there is one and only one s for each q and that each s is distinct from all the others. Then show:

$$\text{Sum of } m \text{ in } c\text{th column} = \sum_{q=0}^{n-1} qn + \sum_{s=0}^{n-1} s + n = \frac{n(n^2 + 1)}{2}.$$

Prove the same for the sum in a row. Since $n(n^2 + 1)/2$ is independent of c, the sums in each row and in each column are the same, and the square array of integers is a so-called *magic square*.

Note that the initial square $\{a_0, b_0\}$ is subject to no conditions. The only essential conditions are that $a, b, \alpha, \beta, a\beta - b\alpha$ are relatively prime to n. Show that these conditions cannot be fulfilled if n is even. However, for odd n the values $a = b = \alpha = 1$, $\beta = 2$ always give a magic square.

2.4 The Function $\phi(n)$

We will return to the discussion of the solution of congruences in the next section. In this section we will use the Chinese remainder theorem to obtain an important property of the function $\phi(n)$ of Definition 2.3.

Theorem 2.15 *Let m and n denote any two positive, relatively prime integers. Then* $\phi(nm) = \phi(n)\phi(m)$.

Proof. Denote $\phi(m)$ by j and let $r_1, r_2, \cdots, r_j$ be a reduced residue system modulo m. Similarly write k for $\phi(n)$ and let $s_1, s_2, \cdots, s_k$ be a reduced residue system modulo n. If x is in a reduced residue system modulo mn, then $(x, m) = (x, n) = 1$, and hence $x \equiv r_h \pmod{m}$ and $x \equiv s_i \pmod{n}$ for some h and i. Conversely, if $x \equiv r_h \pmod{m}$ and $x \equiv s_i \pmod{n}$, then $(x, mn) = 1$. Thus a reduced residue system modulo mn can be obtained by determining all x such that $x \equiv r_h \pmod{m}$ and $x \equiv s_i \pmod{n}$ for some h and i. According to the Chinese remainder theorem, each pair h, i determines a single x modulo mn. Clearly, different pairs h, i yield different x modulo mn. There are jk of these pairs. Therefore a reduced residue system modulo mn contains $jk = \phi(m)\phi(n)$ numbers, and we have $\phi(mn) = \phi(m)\phi(n)$.

It is essential that m and n be relatively prime. In fact $\phi(2) = 1$ and $\phi(2^2) = 2 \neq \phi(2)\phi(2)$.

Theorem 2.16 *If* $n > 1$ *then* $\phi(n) = n \prod_{p|n} (1 - 1/p)$. *Also* $\phi(1) = 1$.

Remark. The symbol $\prod_{p|n}$ means the product over all primes that divide n. Thus if $n = p_1^{e_1}p_2^{e_2} \cdots p_r^{e_r}$ in canonical form, then $\prod_{p|n} (1 - 1/p)$ means $\prod_{j=1}^{r} (1 - 1/p_j)$. We will often use this notation as well as an analogous notation regarding sums. We will also write $\sum_{d|n}$ to mean the sum over all the positive divisors of n, prime or not. Furthermore we will sometimes use the convention that an empty sum is 0, an empty product is 1. Had we done so here we would not have had to treat $n = 1$ as a special case in the statement of the theorem.

Proof. It is obvious that $\phi(1) = 1$.

If $n > 1$ we can write $n = p_1^{e_1}p_2^{e_2} \cdots p_r^{e_r}$ in canonical form. Now $(p_j^{e_j}, p_{j+1}^{e_{j+1}}p_{j+2}^{e_{j+2}} \cdots p_r^{e_r}) = 1$ for $j = 1, 2, \cdots, r - 1$. Applying Theorem 2.15 repeatedly we obtain

$$\phi(n) = \prod_{j=1}^{r} \phi(p_j^{e_j}).$$

In order to compute $\phi(p^e)$, p a prime, we recall that $\phi(p^e)$ is the number of integers x such that $1 \leq x \leq p^e$, $(x, p^e) = 1$. There are p^e integers x between

1 and p^e, and we must count all of them except $p, 2p, 3p, \cdots, p^{e-1}p$. Therefore

$$\phi(p^e) = p^e - p^{e-1} = p^e\left(1 - \frac{1}{p}\right),$$

and hence

$$\phi(n) = \prod_{j=1}^{r} p_j^{e_j}\left(1 - \frac{1}{p_j}\right) = n\prod_{j=1}^{r}\left(1 - \frac{1}{p_j}\right) = n\prod_{p\mid n}\left(1 - \frac{1}{p}\right).$$

Another way of writing the result of Theorem 2.16 is

$$\phi(n) = p_1^{e_1-1}p_2^{e_2-1}\cdots p_r^{e_r-1}(p_1 - 1)(p_2 - 1)\cdots(p_r - 1).$$

Theorem 2.17 *For $n \geqq 1$ we have $\sum_{d\mid n}\phi(d) = n$.*

Proof. If $n = p^e$, p a prime, then

$$\begin{aligned}\sum_{d\mid n}\phi(d) &= \phi(1) + \phi(p) + \phi(p^2) + \cdots + \phi(p^e)\\ &= 1 + (p - 1) + p(p - 1) + \cdots + p^{e-1}(p - 1)\\ &= p^e = n.\end{aligned}$$

Therefore the theorem is true if n is a power of a prime. Now we proceed by induction. We suppose the theorem holds for integers with k or fewer distinct prime factors and consider any integer N with $k + 1$ distinct prime factors. Let p denote one of the prime factors of N, and let p^e be the highest power of p that divides N. Then $N = p^e n$, n has k distinct prime factors, and $(p, n) = 1$. Now as d ranges over the divisors of n, the set $d, pd, p^2d, \cdots, p^ed$ ranges over the divisors of N. Hence we have

$$\begin{aligned}\sum_{d\mid N}\phi(d) &= \sum_{d\mid n}\phi(d) + \sum_{d\mid n}\phi(pd) + \sum_{d\mid n}\phi(p^2d) + \cdots + \sum_{d\mid n}\phi(p^ed)\\ &= \sum_{d\mid n}\phi(d)\{1 + \phi(p) + \phi(p^2) + \cdots + \phi(p^e)\}\\ &= \sum_{d\mid n}\phi(d)\sum_{\delta\mid p^e}\phi(\delta) = np^e = N.\end{aligned}$$

In Chapter 4 we will obtain a different proof of this theorem. It will be independent of the results of this section, and we will find that the order can be reversed, that we can begin by proving Theorem 2.17 and obtain Theorem 2.16 from it. It is then an easy matter to obtain Theorem 2.15 from Theorem 2.16.

PROBLEMS

1. For what values of n is $\phi(n)$ odd?
2. Find the number of positive integers $\leqq 3600$ that are prime to 3600.
3. Find the number of positive integers $\leqq 3600$ that have a factor greater than 1 in common with 3600.
4. Find the number of positive integers $\leqq 7200$ that are prime to 3600.
5. Find the number of positive integers $\leqq 25200$ that are prime to 3600. (Observe that $25200 = 7 \times 3600$.)
6. If m and k are positive integers, prove that the number of positive integers $\leqq mk$ that are prime to m is $k\phi(m)$.
7. Show that $\phi(nm) = n\phi(m)$ if every prime that divides n also divides m.
8. If P denotes the product of the primes common to m and n, prove that $\phi(mn) = P\phi(m)\phi(n)/\phi(P)$. Hence if $(m, n) > 1$, prove $\phi(mn) > \phi(m)\phi(n)$.
9. If $\phi(m) = \phi(mn)$ and $n > 1$, prove that $n = 2$ and m is odd.
10. Characterize the set of positive integers n satisfying $\phi(2n) = \phi(n)$.
11. Characterize the set of positive integers satisfying $\phi(2n) > \phi(n)$.
12. Prove that there are infinitely many integers n so that $3 \nmid \phi(n)$.
13. Find all solutions x of $\phi(x) = 24$.
14. Prove that for a fixed integer n the equation $\phi(x) = n$ has only a finite number of solutions.
15. Find the smallest positive integer n so that $\phi(x) = n$ has no solution; exactly two solutions; exactly three solutions; exactly four solutions. (It has been conjectured that there is no integer n such that $\phi(x) = n$ has exactly one solution, but this is an unsolved problem.)
16. Prove that there is no solution of the equation $\phi(x) = 14$, and that 14 is the least positive even integer with this property. Apart from 14, what is the next smallest positive even integer n such that $\phi(x) = n$ has no solution?
17. Prove that for $n \geqq 2$ the sum of all positive integers less than n and prime to n is $n\phi(n)/2$.
18. If n has k distinct odd prime factors, prove that $2^k \mid \phi(n)$.
19. Define $f(n)$ as the sum of the positive integers less than n and prime to n. Prove that $f(m) = f(n)$ implies that $m = n$.
20. Let $\phi'(n)$ denote the number of integers x such that $1 \leqq x \leqq n$ and $(x, n) = (x + 1, n) = 1$. Prove

$$\phi'(n) = n \prod_{p \mid n} \left(1 - \frac{2}{p}\right).$$

21. (*a*) Let the canonical factorization of n be $n = \prod_{i=1}^{k} p_i^{\alpha_i}$. For every positive integer j, define

$$e_j(p_i) = \begin{cases} 1 \text{ if } p_i \mid j \\ 0 \text{ otherwise.} \end{cases}$$

Prove that $\sum_{j=1}^{n} e_j(p_1) = n/p_1$, and more generally

$$\sum_{j=1}^{n} e_j(p_1)e_j(p_2)\cdots e_j(p_j) = \frac{n}{p_1p_2\cdots p_j} \qquad \text{for } 1 \leqq r \leqq k.$$

(*b*) Prove that

$$\prod_{i=1}^{k} \{1 - e_j(p_i)\} = \begin{cases} 1 \text{ if } (j, n) = 1 \\ 0 \text{ otherwise,} \end{cases}$$

and hence that $\phi(n) = \sum_{j=1}^{n} \prod_{i=1}^{k} \{1 - e_j(p_i)\}$.

(*c*) Deduce that

$$\phi(n) = \sum_{j=1}^{n} \{1 - e_j(p_1) - e_j(p_2) - \cdots - e_j(p_k) + e_j(p_1)e_j(p_2)$$
$$+ e_j(p_1)e_j(p_3) + \cdots + e_j(p_{k-1})e_j(p_k) - e_j(p_1)e_j(p_2)e_j(p_3) - \text{etc.}\}$$

and so obtain an independent proof of Theorem 2.16.

22. If $d \mid n$ and $0 < d < n$, prove that $n - \phi(n) > d - \phi(d)$.

23. Prove the following generalization of Euler's theorem:

$$a^m \equiv a^{m-\phi(m)} \pmod{m}$$

for any integer a.

2.5 Congruences of Higher Degree

There is no general method for solving congruences. However, certain reductions can be made so that the problem finally becomes that of solving congruences with prime moduli. We can use the method of the Chinese remainder theorem in the first step of this reduction.

If $m = p_1^{e_1}p_2^{e_2}\cdots p_r^{e_r}$ then the congruence $f(x) \equiv 0 \pmod{m}$ is equivalent to the set of congruences $f(x) \equiv 0 \pmod{p_i^{e_i}}$, $i = 1, 2, \cdots, r$, in the sense that solutions of one are solutions of the other. If for some j, $1 \leqq j \leqq r$, the congruence $f(x) \equiv 0 \pmod{p_j^{e_j}}$ has no solution, then $f(x) \equiv 0 \pmod{m}$ has no solution. On the other hand, if all the congruences $f(x) \equiv 0 \pmod{p_i^{e_i}}$ have solutions, we can suppose that the ith congruence has exactly k_i solutions, say $a_i^{(1)}, a_i^{(2)}, \cdots, a_i^{(k_i)}$. No two of these are congruent modulo $p_i^{e_i}$, by Definition 2.4, and every solution of $f(x) \equiv 0 \pmod{p_i^{e_i}}$ is congruent to some $a_i^{(j)}$ modulo p_i.

Now an integer u is a root of $f(x) \equiv 0 \pmod{m}$ if and only if for each i there is a j_i such that $u \equiv a_i^{(j_i)} \pmod{p_i^{e_i}}$. Since the moduli $p_i^{e_i}$ are relatively prime in pairs, the Chinese remainder theorem is applicable. We determine integers b_i such that $mp_i^{-e_i}b_i \equiv 1 \pmod{p_i^{e_i}}$ and can then find u by means of (2.1):

$$u \equiv \sum_{i=1}^{r} \frac{m}{p_i^{e_i}} b_i a_i^{(j_i)} \pmod{m}. \tag{2.2}$$

When actually solving a problem it is usually best to compute the coefficients $mp_i^{-e_i}b_i$ first since they are independent of the choice of the j_i. It is then easy to insert the various values of the $a_i^{(j_i)}$ in (2.2), and the problem is solved.

There will be a different u modulo m for each choice of the integers $j_1, j_2, \cdots, j_r$, and each j_i can take on any of k_i values. Therefore the congruence $f(x) \equiv 0 \pmod{m}$ has $k_1k_2 \cdots k_r$ solutions. Since k_i is the number of solutions of $f(x) \equiv 0 \pmod{p_i^{e_i}}$, we have the following theorem.

Theorem 2.18 *Let $N(m)$ denote the number of solutions of the congruence $(x) \equiv 0 \pmod{m}$. Then $N(m) = \prod_{i=1}^{r} N(p_i^{e_i})$ if $m = p_1^{e_1}p_2^{e_2} \cdots p_r^{e_r}$ is the canonical factorization of m.*

Perhaps it should be remarked that the case in which $N(p_j^{e_j}) = 0$ for some j is not ruled out in Theorem 2.18.

Example. Solve $x^2 + x + 7 \equiv 0 \pmod{15}$.

Trying the values $x = 0, \pm 1, \pm 2$, we find that $x^2 + x + 7 \equiv 0 \pmod{5}$ has no solution. Since $15 = 3 \cdot 5$, the original congruence has no solution.

Example. Solve $x^2 + x + 7 \equiv 0 \pmod{189}$, given that $x \equiv 4, 13, -5 \pmod{27}$ are the solutions of $x^2 + x + 7 \equiv 0 \pmod{27}$ and that $x \equiv 0, -1 \pmod{7}$ are the solutions of $x^2 + x + 7 \equiv 0 \pmod{7}$.

In this example we have $m = 189 = 27 \cdot 7 = 3^3 \cdot 7$, $p_1^{e_1} = 27$, $p_2^{e_2} = 7$, $a_1^{(1)} = 4$, $a_1^{(2)} = 13$, $a_1^{(3)} = -5$, $a_2^{(1)} = 0$, $a_2^{(2)} = -1$. To find b_1 we multiply the congruence $7b_1 \equiv 1 \pmod{27}$ by 4 and obtain $b_1 \equiv 28b_1 \equiv 4 \pmod{27}$. For b_2 we have $27b_2 \equiv 1 \pmod{7}$ which gives us $b_2 \equiv -1 \pmod{7}$. We can now write (2.2) as

$$u \equiv 7 \cdot 4a_1^{(j_1)} + 27(-1)a_2^{(j_2)} \equiv 28a_1^{(j_1)} - 27a_2^{(j_2)} \pmod{189}$$

for the required roots u. Using the known values of the $a_i^{(j)}$ we quickly find $u \equiv -77$, $-14, -140, -50, 13, -113 \pmod{189}$.

If the numbers had been larger we might have had more trouble finding b_1 and b_2. In any case the methods mentioned in Section 2.3 can be used.

PROBLEMS

1. Solve the congruences:

$x^3 + 2x - 3 \equiv 0 \pmod{9}$;
$x^3 + 2x - 3 \equiv 0 \pmod{5}$;
$x^3 + 2x - 3 \equiv 0 \pmod{45}$.

2. Solve the congruence $x^3 + 4x + 8 \equiv 0 \pmod{15}$.
3. Solve the congruence $x^3 - 9x^2 + 23x - 15 \equiv 0 \pmod{503}$ by observing that 503 is a prime and that the polynomial factors into $(x - 1)(x - 3)(x - 5)$.
4. Solve the congruence $x^3 - 9x^2 + 23x - 15 \equiv 0 \pmod{143}$.

2.6 Prime Power Moduli

The problem of solving a congruence has now been reduced to that of solving a congruence whose modulus is a power of a single prime.

If r is a solution of $f(x) \equiv 0 \pmod{p^s}$, then $f(r) \equiv 0 \pmod{p^t}$ for $t = 1, 2, \cdots, s$. Let $x_s^{(1)}, x_s^{(2)}, \cdots, x_s^{(h_s)}$ be the solutions of $f(x) \equiv 0 \pmod{p^s}$. There may be no such solutions, or there may be many. Consider $s \geqq 2$. If there is a solution $x_s^{(i)}$ then there is a solution $x_{s-1}^{(j_i)}$ of $f(x) \equiv 0 \pmod{p^{s-1}}$ such that $x_s^{(i)} \equiv x_{s-1}^{(j_i)} \pmod{p^{s-1}}$. Therefore $x_s^{(i)} \equiv x_{s-1}^{(j_i)} + v_{s-1}p^{s-1} \pmod{p^s}$ for some integer v_{s-1}.

Remembering that $f(x)$ is a polynomial of degree n with integral coefficients, we see that $\frac{1}{1!}f'(x), \frac{1}{2!}f''(x), \cdots$ are polynomials with integral coefficients and that $f^{(t)}(x)$ is identically zero for $t > n$. Thus the Taylor's expansion of $f(x)$ is finite and we have

$$f(x + h) = f(x) + f'(x)h + \frac{1}{2}f''(x)h^2 + \cdots + \frac{1}{n!}f^{(n)}(x)h^n,$$

and then

$$0 \equiv f(x_s^{(i)}) \equiv f(x_{s-1}^{(j_i)} + v_{s-1}p^{s-1}) \equiv f(x_{s-1}^{(j_i)}) + f'(x_{s-1}^{(j_i)})v_{s-1}p^{s-1} \pmod{p^s}.$$

But $f(x_{s-1}^{(j_i)}) \equiv 0 \pmod{p^{s-1}}$ so we have

$$f'(x_{s-1}^{(j_i)})v_{s-1} \equiv -\frac{1}{p^{s-1}}f(x_{s-1}^{(j_i)}) \pmod{p}. \tag{2.3}$$

Conversely, if

$$f'(x_{s-1}^{(j)})v \equiv -\frac{1}{p^{s-1}}f(x_{s-1}^{(j)}) \pmod{p}, \tag{2.4}$$

then $f(x_{s-1}^{(j)} + vp^{s-1}) \equiv 0 \pmod{p^s}$. This shows us how to find all the solutions of $f(x) \equiv 0 \pmod{p^s}$, $s \geqq 2$, if we know those of $f(x) \equiv 0 \pmod{p^{s-1}}$. For each root $x_{s-1}^{(j)}$ we find all the solutions v of (2.4), and then the integers $x_{s-1}^{(j)} + vp^{s-1}$ will be solutions of $f(x) \equiv 0 \pmod{p^s}$. It can, of course, happen that there are no v corresponding to some x_{s-1}^{j}. In this case we have no solutions of $f(x) \equiv 0 \pmod{p^s}$ arising from this particular $x_{s-1}^{(j)}$.

We can say a little more about the solutions. In solving $f(x) \equiv 0 \pmod{p^s}$, $s \geqq 2$, we start with the solutions $x_1^{(j)}$ of $f(x) \equiv 0 \pmod p$. Fixing upon a particular $x_1^{(j_i)}$ we must first solve (2.3), with $s = 2$, for v_1. For each v_1 we have a root $x_2^{(k)} \equiv x_1^{(j_i)} + v_1 p \pmod{p^2}$ of $f(x) \equiv 0 \pmod{p^2}$. Using each one of these $x_2^{(k)}$ we must then solve (2.3) with $s = 3$, $j_i = k$, in order to find solutions of $f(x) \equiv 0 \pmod{p^3}$. But the congruence for v_2 has modulus p and $x_2^{(k)} \equiv x_1^{(j_i)} \pmod p$, and hence we can write it as

$$f'(x_1^{(j_i)})v_2 \equiv -\frac{1}{p^2} f(x_2^{(k)}) \pmod p.$$

This happens at each stage, and hence we may determine v_{s-1} from

$$(2.5) \qquad f'(x_1^{(j_i)})v_{s-1} \equiv -\frac{1}{p_{s-1}} f(x_{s-1}^{(k)}) \pmod p$$

for all the $x_{s-1}^{(k)}$ that ultimately arose from the solution $x_1^{(j_i)}$ of $f(x) \equiv 0 \pmod p$.

The congruence (2.5) is a linear congruence. If $f'(x_1^{(j_j)}) \not\equiv 0 \pmod p$, then there will be exactly one v_{s-1} for each of the $x_{s-1}^{(k)}$ arising ultimately from $x_1^{(j_i)}$. If $f'(x_1^{(j_i)}) \equiv 0 \pmod p$, then there will be p or no v_{s-1} according as $f(x_{s-1}^{(k)})/p^{s-1}$ is or is not congruent to 0 modulo p.

Example. Solve $x^2 + x + 7 \equiv 0 \pmod{27}$.

By trial we find that $x \equiv 1 \pmod 3$ is the only solution of $f(x) \equiv 0 \pmod 3$ for the present $f(x)$. Then $f'(x) = 2x + 1$ and $f'(1) \equiv 0 \pmod 3$. There is only one x_1, and (2.5) reduces to

$$0 \equiv -\frac{1}{3^{s-1}} f(x_{s-1}^{(k)}) \pmod 3,$$

which means that there are no v_{s-1} if $f(x_{s-1}^{(k)}) \not\equiv 0 \pmod{3^s}$ and that $v_{s-1} \equiv 0, 1, -1 \pmod 3$ if $f(x_{s-1}^{(k)}) \equiv 0 \pmod{3^s}$. We now find

$$\begin{array}{ll} x_1^{(1)} \equiv 1 \pmod 3, f(x_1^{(1)}) = 9, & v_1 \equiv 0, 1, -1 \pmod 3 \\ x_2^{(1)} \equiv 1 \pmod{3^2}, f(x_2^{(1)}) = 9, & \text{no } v_2 \\ x_2^{(2)} \equiv 4 \pmod{3^2}, f(x_2^{(2)}) = 27, & v_2 \equiv 0, 1, -1 \pmod 3 \\ x_2^{(3)} \equiv -2 \pmod{3^2}, f(x_2^{(3)}) = 9, & \text{no } v_2 \\ x_3^{(1)} \equiv 4 \pmod{3^3} & \\ x_3^{(2)} \equiv 13 \pmod{3^3} & \\ x_3^{(3)} \equiv -5 \pmod{3^3}. & \end{array}$$

Example. Solve $x^2 + x + 7 \equiv 0 \pmod{3^4}$.

Continuing from the previous example, we find

$$f(x_3^{(1)}) = 27, \qquad f(x_3^{(2)}) = 189, \qquad f(x_3^{(3)}) = 27.$$

The congruence has no solution since $27 \not\equiv 0$, $189 \not\equiv 0 \pmod{3^4}$.

Example. Solve $x^2 + x + 7 \equiv 0 \pmod{7^3}$.

The solutions of $f(x) \equiv 0 \pmod 7$ are $x \equiv 0, -1 \pmod 7$. Furthermore, $f'(0) = 1, f'(-1) = -1$. There will be just one $x_s^{(1)}$ corresponding to $x_1^{(1)} = 0$ and one $x_s^{(2)}$ corresponding to $x_1^{(2)} = -1$. Now (2.5) becomes

$$v_{s-1} \equiv -\frac{1}{7^{s-1}} f(x_{s-1}^{(1)}) \pmod 7 \qquad \text{corresponding to } x_1^{(1)} = 0$$

$$v_{s-1} \equiv \frac{1}{7^{s-1}} f(x_{s-1}^{(2)}) \pmod 7 \qquad \text{corresponding to } x_1^{(2)} = -1.$$

Then we find

$$x_1^{(1)} = 0, \quad f(x_1^{(1)}) = 7, \quad v_1 = -1, \quad x_2^{(1)} = -7, \quad f(x_2^{(1)}) = 49,$$
$$v_2 = -1, \quad x_3^{(1)} = -56,$$
$$x_1^{(2)} = -1, \quad f(x_1^{(2)}) = 7, \quad v_1 = 1, \quad x_2^{(2)} = 6, \quad f(x_2^{(2)}) = 49,$$
$$v_2 = 1, \quad x_3^{(2)} = 55.$$

The solutions of $x^2 + x + 7 \equiv 0 \pmod{7^3}$ are $x \equiv -56 \pmod{7^3}$ and $x \equiv 55 \pmod{7^3}$.

When solving numerical problems one often must determine whether or not one integer k divides another integer n. If $(k, 10) = 1$ and k is not too large, there is a rather simple way to do this. As a first example consider $k = 31, n = 23754$. Then $n - 4k = 23754 - 4 \cdot 30 - 4 = 10(2375 - 4 \cdot 3) = 10 \cdot 2363$. Since $(31, 10) = 1$ we see that $31 \mid 23754$ if and only if $31 \mid 2363$. We can repeat the argument to reduce 2363 still further. The entire process can be put in a more convenient form.

$$\begin{array}{r} 23754 \\ 12 \\ \hline 2363 \\ 9 \\ \hline 227 \\ 21 \\ \hline 1 \end{array} \qquad 31 \nmid 23754.$$

This process can be used for any k whose last digit is 1. We can write $k = 10j + 1$ and $n = 10a + b$. Then $n - bk = 10a + b - 10bj - b = 10(a - bj)$, and $k \mid n$ if and only if $k \mid (a - bj)$.

If the last digit of k is 9, we can write $k = 10j - 1$ and $n = 10a + b$, and we have $n + bk = 10a + b + 10bj - b = 10(a + bj)$. Then $k \mid n$ if and only if $k \mid (a + bj)$. If the last digit of k is 3, we can write $3k = 10j - 1$ and find $n + 3bk = 10(a + bj)$, and hence $k \mid n$ if and only if $k \mid (a + bj)$. Similarly, if the last digit of k is 7, we write $3k = 10j + 1$ and obtain $k \mid n$ if and only if $k \mid (a - bj)$.

Example.

$$\begin{array}{ll}
19 = 10 \cdot 2 - 1 & 3 \cdot 7 = 10 \cdot 2 + 1 \\
20513 & 8638 \\
\underline{6} & \underline{16} \\
2057 & 847 \\
\underline{14} & \underline{14} \\
219 & 70 \\
\underline{18} & \\
39 & \\
19 \nmid 20513 & 7 \mid 8638
\end{array}$$

PROBLEMS

1. The foregoing method determines whether k divides n, but in general the number we finally reach is not congruent to n modulo k. Consider the following scheme, exemplified for $n = 1234$.

$$\begin{array}{l}
1\,2\,3\,4 \\
3\,6\,9 \\
1\,0\,8 \\
3\,0 \\
9
\end{array}$$

Here we have written down 1234, then $3 \cdot 123$, $3 \cdot 36$, $3 \cdot 10$, $3 \cdot 3$, in turn. We drop off the right digit at each step and multiply what is left by 3. Now we have $n = 1234 \equiv 4 + 9 + 8 + 0 + 9 \equiv 30 \equiv 2 \pmod 7$ and also $n = 1234 \equiv 4 - 9 + 8 - 0 + 9 \equiv 12 \pmod{13}$. Show why this works for all positive integers n. What multiplier should be used instead of 3 if the modulus k is 9 or 11; if $k = 17$; if $k = 19$? For $k = 17$ and 19 the procedure is likely to be too long to be of any practical value. Find more satisfactory variations of the method. For example:

$$\begin{array}{l}
1\,7\,3\,4\,5\,6\,2 \\
8\,6\,7\,2\,5 \\
4\,3\,3\,5 \\
2\,1\,5 \\
1\,0
\end{array}$$

$1734562 \equiv 62 + 25 + 35 + 15 + 10 \equiv 147 \equiv 47 + 5 \equiv 52 \equiv 14 \pmod{19}$.

2. Show that for $k = 9$ the method in the text and the method of problem 1 are essentially the same and that they amount to the familiar process of "casting out nines."

3. Using the fact that $1001 = 7 \cdot 11 \cdot 13$, and assuming that you can recognize all multiples of 7 or 11 or 13 having no more than three digits, work out a scheme for testing for divisibility by 7 or 11 or 13 simultaneously.

4. Prove $(y + vp^{s-1})^j \equiv y^j + jy^{j-1}vp^{s-1} \pmod{p^s}$ if $s \geqq 2$.

Prove

$$f(y + vp^{s-1}) - f(y) \equiv \sum_{i=0}^{n-1} (n - i)a_i y^{n-i-1} vp^{s-1} \pmod{p^s} \text{ if } s \geqq 2$$

and

$$f(x) = \sum_{i=0}^{n} a_i x^{n-i}.$$

This can be used to replace the use of Taylor's expansion at the beginning of this section.

5. Apply the method of this section to solve $ax - 1 \equiv 0 \pmod{p^s}$, $(a, p) = 1$. How are these solutions related to those given by Problem 11 of section 2.3 with m replaced by p^s?

6. Solve $x^5 + x^4 + 1 \equiv 0 \pmod{3^4}$.

7. Solve $x^3 + x + 57 \equiv 0 \pmod{5^3}$.

8. Solve $x^2 + 5x + 24 \equiv 0 \pmod{36}$.

9. Solve $x^3 + 10x^2 + x + 3 \equiv 0 \pmod{3^3}$.

10. Solve $x^3 + x^2 - 4 \equiv 0 \pmod{7^3}$.

11. Solve $x^3 + x^2 - 5 \equiv 0 \pmod{7^3}$.

2.7 Prime Modulus

We have now reduced the problem of solving $f(x) \equiv 0 \pmod{m}$ to its last stage, congruences with prime moduli. It is here that we will not be able to find a general method. However, there are some general facts concerning the solutions, and we will find that we are led to some new and interesting matters.

As before, we write $f(x) = a_0 x^n + a_1 x^{n-1} + \cdots + a_n$ and we assume that p is a prime and $a_0 \not\equiv 0 \pmod{p}$.

Theorem 2.19 *If the degree n of $f(x) \equiv 0 \pmod{p}$ is greater than or equal to p, then either every integer is a solution of $f(x) \equiv 0 \pmod{p}$ or there is a polynomial $g(x)$ having integral coefficients, with leading coefficient 1, and such that $g(x) \equiv 0 \pmod{p}$ is of degree less that p and the solutions of $g(x) \equiv 0 \pmod{p}$ are precisely those of $f(x) \equiv 0 \pmod{p}$.*

Proof. Dividing $f(x)$ by $x^p - x$ we obtain $f(x) = q(x)(x^p - x) + r(x)$ where $q(x)$ is a polynomial with integral coefficients and $r(x)$ is either zero or a polynomial with integral coefficients and degree less than p. Fermat's theorem shows that $u^p - u \equiv 0 \pmod{p}$, and hence $f(u) \equiv r(u) \pmod{p}$ for every integer u. Therefore if $r(x)$ is zero, or if every coefficient in $r(x)$ is divisible by p, then every integer is a solution of $f(x) \equiv 0 \pmod{p}$. The only other possibility is $r(x) = b_m x^m + b_{m-1} x^{m-1} + \cdots + b_0$, $m < p$, with at least one coefficient not divisible by p. Let b_k be the coefficient with largest subscript k such that $(p, b_k) = 1$. Then there is an integer b such that $bb_k \equiv 1 \pmod{p}$, and clearly $r(x) \equiv 0 \pmod{p}$ and $br(x) \equiv 0 \pmod{p}$ have the same solutions. The requirements of the theorem are satisfied if we define $g(x)$ as

$$x^k + b(b_{k-1}x^{k-1} + b_{k-2}x^{k-2} + \cdots + b_0).$$

In the statement of Theorem 2.19, $g(x)$ is described as having the property that $g(x) \equiv 0 \pmod{p}$ and $f(x) \equiv 0 \pmod{p}$ have the same solutions. From the proof of the theorem we see that $g(u) \equiv bf(u) \pmod{p}$ for every integer u. However, we do not say that the polynomials $g(x)$ and $bf(x)$ are congruent modulo p; we will use this last statement to mean that each coefficient in $g(x)$ is congruent modulo p to the corresponding coefficient in $bf(x)$.

Theorem 2.20 *The congruence $f(x) \equiv 0 \pmod{p}$ of degree n has at most n solutions.*

Proof. The proof is by induction on the degree of $f(x) \equiv 0 \pmod{p}$. I $n = 0$, the polynomial $f(x)$ is just a_0 with $a_0 \not\equiv 0 \pmod{p}$, and hence the congruence has no solution. If $n = 1$, the congruence has exactly one solution by Theorem 2.13. Assuming the truth of the theorem for all congruences of degree $<n$, suppose that there are more than n solutions of the congruence $f(x) \equiv 0 \pmod{p}$ of degree n. Let the leading term of $f(x)$ be a_0x^n and let $u_1, u_2, \cdots, u_n, u_{n+1}$ be solutions of the congruence, with $u_i \not\equiv u_j \pmod{p}$ for $i \neq j$. We define $g(x)$ by the equation

$$g(x) = f(x) - a_0(x - u_1)(x - u_2) \cdots (x - u_n),$$

noting the cancellation of a_0x^n on the right. Then either $g(x)$ is identically zero, or it is a polynomial of degree k, $0 \leqq k < n$.

We wish to prove that $g(x)$ is either identically zero or is a polynomial having all its coefficients divisible by p. If this were not so the congruence $g(x) \equiv 0 \pmod{p}$ would have a degree, say h, and we see that $h \leqq k < n$. But $g(x) \equiv 0 \pmod{p}$ has n solutions $u_1, u_2, \cdots, u_n$, and this is impossible by our induction hypothesis.

Now what we have proved about $g(x)$ shows that $g(u) \equiv 0 \pmod{p}$ for all integers u and hence that $f(u) \equiv a_0(u - u_1)(u - u_2) \cdots (u - u_n) \pmod{p}$ for all integers u. In particular,

$$a_0(u_{n+1} - u_1)(u_{n+1} - u_2) \cdots (u_{n+1} - u_n) \equiv f(u_{n+1}) \equiv 0 \pmod{p}.$$

But this contradicts Theorem 1.15, and hence the assumption that $f(x) \equiv 0 \pmod{p}$ has more than n solutions is false.

Corollary 2.21 *If $b_0x^n + b_1x^{n-1} + \cdots + b_n \equiv 0 \pmod{p}$ has more than n solutions then all the coefficients b_j are divisible by p.*

Theorem 2.22 *The congruence $f(x) \equiv 0 \pmod{p}$ of degree n, with leading coefficient $a_0 = 1$, has n solutions if and only if $f(x)$ is a factor of $x^p - x$ modulo p, that is, if and only if $x^p - x = f(x)q(x) + ps(x)$ where $q(x)$ and $s(x)$ have integral coefficients, and where either $s(x)$ is a polynomial of degree less than n or $s(x)$ is zero.*

Proof. If $f(x) \equiv 0 \pmod{p}$ has n solutions, then $n \leqq p$. Dividing $x^p - x$ by $f(x)$, we find $x^p - x = f(x)q(x) + r(x)$ where $r(x)$ is zero or $r(x)$ has degree less than n. For every solution u of $f(x) \equiv 0 \pmod{p}$ we have $u^p - u \equiv 0 \pmod{p}$, and hence $r(u) \equiv 0 \pmod{p}$. Therefore if $r(x)$ is not zero, it is a polynomial of degree less than n having n solutions. According to Corollary 2.21 all the coefficients of $r(x)$ are divisible by p, and we can write $r(x) = ps(x)$.

Conversely, if $x^p - x = f(x)q(x) + ps(x)$, then $f(u)q(u) \equiv u^p - u - ps(u) \equiv 0 \pmod{p}$ for every integer u. Therefore $f(x)q(x) \equiv 0 \pmod{p}$ has p solutions. But $q(x)$ is of degree p–n with leading coefficient 1, and hence by Theorem 2.20 the congruence $q(x) \equiv 0 \pmod{p}$ has at most p–n solutions, $v_1, v_2, \cdots, v_k$, say, with $k \leqq p$–n. If u is any of the other p–k residues modulo p, then $(q(u), p)) = 1$ and $f(u)q(u) \equiv 0 \pmod{p}$, and we have $f(u) \equiv 0 \pmod{p}$ by Theorem 1.15. Hence $f(x) \equiv 0 \pmod{p}$ has at least $p - k \geqq p - (p - n) = n$ solutions. This, with Theorem 2.20, shows that $f(x) \equiv 0 \pmod{p}$ has exactly n solutions.

The restriction $a_0 = 1$ in this theorem is needed so that we may divide $x^p - x$ by $f(x)$ and obtain a polynomial $q(x)$ with integral coefficients. However, it is not very much of a restriction. We can always find an integer a such that $aa_0 \equiv 1 \pmod{p}$. Then $af(x) - (aa_0 - 1)x^n \equiv 0 \pmod{p}$ has the same solutions as $f(x) \equiv 0 \pmod{p}$, and $af(x) - (aa_0 - 1)x^n$ has its leading coefficient 1.

PROBLEMS

1. Reduce the following congruences to equivalent congruences of degree $\leqq 6$:
(*a*) $x^{11} + x^8 + 5 \equiv 0 \pmod{7}$;
(*b*) $x^{20} + x^{13} + x^7 + x \equiv 2 \pmod{7}$;
(*c*) $x^{15} - x^{10} + 4x - 3 \equiv 0 \pmod{7}$.

2. Prove that $2x^3 + 5x^2 + 6x + 1 \equiv 0 \pmod{7}$ has three solutions by use of Theorem 2.22.

3. Prove that $x^{14} + 12x^2 \equiv 0 \pmod{13}$ has thirteen solutions and so it is an identical congruence.

4. Prove that if $f(x) \equiv 0 \pmod{p}$ has j solutions $x \equiv a_1$, $x \equiv a_2, \cdots, x \equiv a_j \pmod{p}$, there is a polynomial $q(x)$ such that $f(x) \equiv (x - a_1)(x - a_2) \cdots (x - a_j)q(x) \pmod{p}$. *Suggestion:* begin by showing that there is a $q_1(x)$ such that $f(x) \equiv (x - a_1)q_1(x) \pmod{p}$ and that $q_1(x) \equiv 0 \pmod{p}$ has solutions $x \equiv a_2, x \equiv a_3, \cdots, x \equiv a_j \pmod{p}$. Then use induction.

5. With the assumptions and notation of the previous problem, prove that if the degree of $f(x)$ is j, then $q(x)$ is a constant and can be taken as the leading coefficient of $f(x)$.

6. Prove that Fermat's theorem implies that

$$x^{p-1} - 1 \equiv (x-1)(x-2)\cdots(x-p+1) \pmod{p}$$

and

$$x^p - x \equiv x(x-1)(x-2)\cdots(x-p+1) \pmod{p}.$$

7. By comparing coefficients of x in the previous problem, give another proof of Wilson's theorem.

8. Let m be composite. Prove that Theorem 2.20 is false if "mod p" is replaced by "mod m."

2.8 Congruences of Degree Two, Prime Modulus

If $f(x) \equiv 0 \pmod{p}$ is of degree two, then $f(x) = ax^2 + bx + c$, and a is relatively prime to p. We will suppose $p > 2$ since the case $p = 2$ offers no difficulties. Then p is odd, and $4af(x) = (2ax + b)^2 + 4ac - b^2$. Hence u is a solution of $f(x) \equiv 0 \pmod{p}$ if and only if $2au + b \equiv v \pmod{p}$, where v is a solution of $v^2 \equiv b^2 - 4ac \pmod{p}$. Furthermore, since $(2a, p) = 1$, for each solution v there is one, and only one, u modulo p such that $2au + b \equiv v \pmod{p}$. Clearly different v modulo p yield different u modulo p. Thus the problem of solving the congruence of degree two is reduced to that of solving a congruence of the form $x^2 \equiv a \pmod{p}$.

In Chapter 3 we will consider the congruence $x^2 \equiv a \pmod{p}$ in detail. For the present we will merely obtain some general results concerning the more general congruence $x^n \equiv a \pmod{p}$ and certain related concepts.

PROBLEM

1. Reduce the following congruences to the form $x^2 \equiv a \pmod{p}$:
(*a*) $4x^2 + 2x + 1 \equiv 0 \pmod{5}$; (*b*) $3x^2 - x + 5 \equiv 0 \pmod{7}$;
(*c*) $2x^2 + 7x - 10 \equiv 0 \pmod{11}$; (*d*) $x^2 + x - 1 \equiv 0 \pmod{13}$.

2.9 Power Residues

Definition 2.6 *If $x^n \equiv a \pmod{p}$ has a solution, then a is called an nth power residue modulo p.*

Definition 2.7 *Let m denote a positive integer and a any integer such that $(a, m) = 1$. Let h denote the smallest positive integer such that $a^h \equiv 1 \pmod{m}$. We say that a belongs to the exponent h modulo m.*

Since $a^{\phi(m)} \equiv 1 \pmod{m}$ by Euler's theorem, we see that every a relatively prime to m belongs to some exponent $h \leqq \phi(m)$ modulo m. Dividing $\phi(m)$ by h we obtain $\phi(m) = qh + r$, $0 \leqq r < h$. But then $a^r \equiv a^{r+qh} \equiv a^{\phi(m)} \equiv 1 \pmod{m}$. Since h is the least positive integer such that $a^h \equiv 1 \pmod{m}$, and since $0 \leqq r < h$, we see that r cannot be positive. Therefore $r = 0$, and we have the first assertion in the following theorem.

Theorem 2.23 *If a belongs to the exponent h modulo m then $h \mid \phi(m)$. Furthermore $a^j \equiv a^k \pmod{m}$ if and only if $h \mid (j - k)$.*

Proof. There is no loss in generality in assuming $j > k$, and since $(a, m) = 1$, the congruence $a^j \equiv a^k \pmod{m}$ is equivalent to $a^{j-k} \equiv 1 \pmod{m}$. Thus the second assertion in the theorem follows as in the proof of the first assertion.

Theorem 2.24 *If a belongs to the exponent h modulo m then a^k belongs to the exponent $h/(h, k)$ modulo m.*

Proof. According to Theorem 2.23, $(a^k)^j \equiv 1 \pmod{m}$ if and only if $h \mid kj$. But $h \mid kj$ if and only if $\{h/(h, k)\} \mid \{k/(h, k)\}j$ and hence if and only if $\{h/(h, k)\} \mid j$. Therefore the least positive integer j such that $(a^k)^j \equiv 1 \pmod{m}$ is $j = h/(h, k)$.

Definition 2.8 *If a belongs to the exponent $\phi(m)$ modulo m, then a is called a primitive root modulo m.*

Theorem 2.25 *If p is a prime, then there exist $\phi(p - 1)$ primitive roots modulo p. The only integers having primitive roots are p^e, $2p^e$, 1, 2, and 4, with p an odd prime.*

Proof. Each integer a, $1 \leqq a \leqq p - 1$, belongs to some exponent h modulo p with $h \mid (p - 1)$. If a belongs to the exponent h, then $(a^k)^h \equiv 1 \pmod{p}$ for all k, and $1, a, a^2, \cdots, a^{h-1}$ are distinct modulo p. Therefore, from Theorem 2.20, these h numbers are all the solutions of $x^h \equiv 1 \pmod{p}$. By Theorem 2.24 just $\phi(h)$ of these numbers belong to the exponent h modulo p. The others belong to smaller exponents. Also, any integer a that belongs to the exponent h modulo p is a solution of $x^h \equiv 1 \pmod{p}$. Therefore for each h that divides $p - 1$ there will be either $\phi(h)$ or no integers a, $1 \leqq a \leqq p - 1$, such that a belongs to the exponent h modulo p. Let $\psi(h)$ denote the number of the integers a that belong to the exponent h modulo p. Then $\psi(h) \leqq \phi(h)$ for each h that divides $p - 1$ and $\sum_{h \mid p-1} \psi(h) = p - 1$. But $\sum_{h \mid p-1} \phi(h) = p - 1$ according to Theorem 2.17, so we have $\sum_{h \mid p-1} (\psi(h) - \phi(h)) = 0$ and $\psi(h) - \phi(h) \leqq 0$. This implies $\psi(h) = \phi(h)$ for all h that divide $p - 1$, and in particular $\psi(p - 1) = \phi(p - 1) > 0$. This proves the first part of the theorem.

It is easily seen that $\phi(n)$ is even for $n > 2$. Let $m = 2^f \prod_{i=1}^{k} p_i^{e_i}$ where the p_i are distinct odd primes, $f \geqq 0$, $e_i > 0$, and $k \geqq 1$. If $(a, m) = 1$ we have $a^{\phi(p_1^{e_1})} \equiv 1 \pmod{p_1^{e_1}}$ and $a^{\phi(m/p_1^{e_1})} \equiv 1 \pmod{m/p_1^{e_1}}$. Suppose that $k \geqq 2$ or $f \geqq 2$. Then both $\phi(p_1^{e_1})$ and $\phi(m/p_1^{e_1})$ are even, and therefore $a^{\frac{1}{2}\phi(p_1^{e_1})\phi(m/p_1^{e_1})}$ is congruent to 1 modulo $p_1^{e_1}$ and modulo $m/p_1^{e_1}$, hence modulo m. Therefore we do not have a primitive root if $k \geqq 2$, or if $k = 1$ and $f \geqq 2$.

Next we prove that 2^n has no primitive root if $n \geqq 3$. First note that for any odd integer a, say $a = 2b + 1$, then $a^2 = 4b(b+1) + 1 = 1 + 8c$, $a^4 = (1 + 8c)^2 = 1 + 16d$, $a^8 = (1 + 16d)^2 = 1 + 32r$. Continuing by induction we get for $n \geqq 3$,

$$a^{2^{n-2}} = 1 + 2^n g,\ a^{2^{n-2}} \equiv 1 \pmod{2^n}.$$

But $\phi(2^n) = 2^{n-1} > 2^{n-2}$, and hence 2^n has no primitive root if $n \geqq 3$. This together with the preceding paragraph establishes that the only positive integers that can possibly have primitive roots are p^e, $2p^e$, 1, 2, and 4, with p an odd prime. Finally we prove that there is a primitive root in each of these cases.

Consider $m = p^e$ and let a be a primitive root modulo p. Let $b = a + pt$. Then by the binomial theorem,

$$b^{p-1} \equiv a^{p-1} + (p-1)a^{p-2}pt \pmod{p^2}$$

and we can choose t to make $b^{p-1} = 1 + n_1 p$ with $n_1 \not\equiv 0 \pmod{p}$. Noting that $(1 + np^{j-1})^p \equiv 1 + np^j \pmod{p^{2j-1}}$ we use induction to see that $b^{p^{j-1}(p-1)} = 1 + n_j p^j$ with $n_j \equiv n_{j-1} \pmod{p^{j-1}}$. Then $n_j \equiv n_1 \not\equiv 0 \pmod{p}$. For $e \geqq 2$ let b belong to the exponent h modulo p^e. Then $h \mid p^{e-1}(p-1)$ and hence $h = p^s d$, $s \leqq e - 1$, $d \mid (p-1)$, which implies $b^{p^s(p-1)} \equiv 1 \pmod{p^e}$, $1 + n_{s+1}p^{s+1} \equiv 1 \pmod{p^e}$, and therefore $s \geqq e - 1$, $s = e - 1$. We also have $b^d \equiv b^{p^s d} \equiv 1 \pmod{p}$, which implies that $(p-1) \mid d$ by Theorem 2.23 since $b \equiv a \pmod{p}$ and a belongs to the exponent $p - 1$. Then we have $d = p - 1$, $h = \phi(p^e)$ and b is a primitive root modulo p^e. Notice that b is independent of e.

Now consider $m = 2p^e$ and let a be a primitive root modulo p^e. Let $b = a$ or $a + p^e$, whichever is odd. Then $b^h \equiv 1 \pmod{2}$ for all h, and $b^h \equiv a^h \equiv 1 \pmod{p^e}$ if and only if $p^{e-1}(p-1) \mid h$. This implies that b is a primitive root modulo $2p^e$.

Finally, we note that 3 is a primitive root modulo 4, and 1 is a primitive root modulo 2 and 1.

Theorem 2.26 *Suppose that m has a primitive root g. Then* $g^j \equiv g^k \pmod{m}$ *if and only if* $j \equiv k \pmod{\phi(m)}$; *in particular,* $g^j \equiv 1 \pmod{m}$ *if and only if* $\phi(m) \mid j$. *The set* $g, g^2, \cdots, g^{\phi(m)}$ *forms a reduced residue system modulo m, so that if a is any integer satisfying* $(a, m) = 1$, *there is one and only one* g^j *in the set such that* $g^j \equiv a \pmod{m}$.

Proof. The first part of the theorem is a special case of Theorem 2.23. It follows that $g, g^2, \cdots, g^{\phi(m)}$ are incongruent in pairs modulo m, and so this set forms a reduced residue system modulo m.

The exponent j such that $g^j \equiv a \pmod{m}$ is called the *index* of a. The index depends on m and g as well as on a. Indices behave very much like logarithms, and they are sometimes useful as an aid to computation as well as being of theoretical interest.

Theorem 2.27 *If p is a prime and $(a, p) = 1$, then the congruence $x^n \equiv a \pmod{p}$ has $(n, p - 1)$ solutions or no solutions according as*

$$a^{(p-1)/(n,p-1)} \equiv 1 \pmod{p} \quad \text{or} \quad a^{(p-1)/(n,p-1)} \not\equiv 1 \pmod{p}.$$

Proof. Let b denote $(n, p - 1)$. If $x^n \equiv a \pmod{p}$ has a solution u, then

$$a^{(p-1)/b} \equiv u^{n(p-1)/b} \equiv u^{(p-1)(n/b)} \equiv 1 \pmod{p}.$$

Therefore $x^n \equiv a \pmod{p}$ has no solution if $a^{(p-1)/b} \not\equiv 1 \pmod{p}$.

Conversely, suppose that $a^{(p-1)/b} \equiv 1 \pmod{p}$. By Theorems 2.25 and 2.26, there is a primitive root g modulo p and an exponent j such that $g^j \equiv a \pmod{p}$. Thus we have

$$g^{j(p-1)/b} \equiv a^{(p-1)/b} \equiv 1 \pmod{p}$$

and this implies that $j(p - 1)/b \equiv 0 \pmod{p - 1}$ by Theorem 2.26, so that $b \mid j$. Now any solution of $x^n \equiv a \pmod{p}$, if any exist, can also be written as a power of g, say g^y, modulo p. Hence the solutions in x, if any, of $x^n \equiv a \pmod{p}$ correspond to the solutions in y of $g^{yn} \equiv g^j \pmod{p}$. This congruence, by Theorem 2.23, has solutions if and only if $yn \equiv j \pmod{p - 1}$ has solutions, which it does by Theorem 2.13 since $b \mid j$. Moreover there are $(n, p - 1)$ solutions by Theorem 2.13, and thus we have $(n, p - 1)$ solutions of $x^n \equiv a \pmod{p}$.

Corollary 2.28 *If p is an odd prime and $(a, p) = 1$ then $x^2 \equiv a \pmod{p}$ has two or no solutions according as $a^{(p-1)/2} \equiv 1$ or $-1 \pmod{p}$.*

Proof. From Fermat's theorem we have

$$(a^{(p-1)/2} - 1)(a^{(p-1)/2} + 1) \equiv a^{p-1} - 1 \equiv 0 \pmod{p},$$

and hence $a^{(p-1)/2} \equiv \pm 1 \pmod{p}$.

This result is not useful for numerical calculations unless p is very small. For example, if we ask whether the congruence $x^2 \equiv 19 \pmod{61}$ has solutions, Corollary 2.28 leads into a very lengthy calculation. In Chapter 3 better methods for answering such questions are developed.

PROBLEMS

1. Find a primitive root of the prime 3; the prime 5; the prime 7; the prime 11; the prime 13.
2. Find a primitive root of 23.
3. How many primitive roots does the prime 13 have?
4. To what exponents do each of 1, 2, 3, 4, 5, 6 belong modulo 7? To what exponents do they belong modulo 11?
5. Let p be an odd prime. Prove that a belongs to the exponent 2 modulo p if and only if $a \equiv -1 \pmod{p}$.
6. If a belongs to the exponent h modulo m, prove that no two of $a, a^2, a^3, \cdots, a^h$ are congruent modulo m.
7. If p is an odd prime, how many solutions are there to $x^{p-1} \equiv 1 \pmod{p}$; to $x^{p-1} \equiv 2 \pmod{p}$?
8. Prove that 3 is a primitive root of 17 by observing that the powers of 3 are congruent to 3, 9, 10, 13, 5, 15, 11, 16, 14, 8, 7, 4, 12, 2, 6, 1 modulo 17. Then use Theorem 2.27 to decide how many solutions each of the following congruences has:

(*a*) $x^{12} \equiv 16 \pmod{17}$ (*b*) $x^{48} \equiv 9 \pmod{17}$
(*c*) $x^{20} \equiv 13 \pmod{17}$ (*d*) $x^{11} \equiv 9 \pmod{17}$.

Suggestion: $16 \equiv 3^8$, $13 \equiv 3^4 \pmod{17}$.
9. Using the data of the previous problem, decide which of the congruences $x^2 \equiv 1$, $x^2 \equiv 2$, $x^2 \equiv 3, \cdots, x^2 \equiv 16 \pmod{17}$, have solutions.
10. Prove that if p is a prime and $(a, p) = 1$ and $(n, p - 1) = 1$, then $x^n \equiv a \pmod{p}$ has exactly one solution.
11. Prove that if g is a primitive root modulo a prime p, and if $(k, p - 1) = 1$, then g^k is also a primitive root.
12. Prove that if a belongs to the exponent 3 modulo a prime p, then $1 + a + a^2 \equiv 0 \pmod{p}$, and $1 + a$ belongs to the exponent 6.
13. Prove that if a belongs to the exponent h modulo a prime p, and if h is even, then $a^{h/2} \equiv -1 \pmod{p}$.
14. Prove that if a belongs to an exponent h, and b to an exponent k, modulo m, then ab belongs to an exponent which is a divisor of hk. Furthermore, if $(h, k) = 1$, then ab belongs to the exponent hk modulo m. *Suggestion:* if ab belongs to the exponent r, then

$$1 \equiv (ab)^r \equiv (ab)^{hr} \equiv (a^h)^r b^{hr} \equiv b^{hr} \pmod{m},$$

so that $k \mid hr$.
15. Given that $ab \equiv 1 \pmod{m}$, and that a belongs to the exponent h modulo m, prove that b belongs to the exponent h. Then prove that if a prime $p > 3$, the product of all the primitive roots of p is congruent to 1 modulo p.
16. Let a and $n > 1$ be any integers such that

$$a^{n-1} \equiv 1 \pmod{n} \quad \text{but} \quad a^x \not\equiv 1 \pmod{n}$$

for every proper divisor x of $n - 1$. Prove that n is a prime.

17. For any prime p and any integer a such that $(a, p) = 1$, say that a is a cubic residue of p if $x^3 \equiv a \pmod{p}$ has at least one solution. Prove that if p is of the form $3k + 2$, then all integers in a reduced residue system modulo p are cubic residues, whereas if p is of the form $3k + 1$, only one-third of the members of a reduced residue system are cubic residues.
18. Prove Wilson's theorem using primitive roots.
19. Let p be an odd prime. Let $r_1, r_2, \cdots, r_{p-1}$ be the integers $1, 2, \cdots, p - 1$ in any order. Prove that at least two of the numbers $1 \cdot r_1, 2 \cdot r_2, \cdots, (p - 1)r_{p-1}$ are congruent modulo p.

2.10 Number Theory from an Algebraic Viewpoint

In this and the next section we discuss some of the forms in which the elementary concepts of number theory turn up in algebra. The theory of numbers provides a rich source of examples of the structures of abstract algebra. We shall treat briefly three of these structures: groups, rings, and fields.

Before giving the technical definition of a group, let us explain some of the language used. Operations like addition and multiplication are called "binary operations" because two elements are added, or multiplied, to produce a third element. The subtraction of pairs of elements, $a - b$, is likewise a binary operation. So also is exponentiation, a^b, in which the element a is raised to the bth power. Now, a group consists of a set of elements together with a binary operation on those elements, such that certain properties hold. The number theoretic groups with which we deal will have integers or sets of integers as elements, and the operation will be either addition or multiplication. However, a general group can have elements of any sort and any kind of binary operation, just so long as it satisfies the conditions that we shall impose shortly.

We begin with a general binary operation denoted by $\oplus$, and we shall presume that this binary operation is single-valued. This means that for each pair a, b of elements, $a \oplus b$ has a unique value or is not defined. A set of elements is said to be "closed" with respect to an operation $\oplus$, or closed "under" the operation, if $a \oplus b$ is defined and is an element of the set for every pair of elements a, b of the set. For example, the natural numbers $1, 2, 3, \cdots$ are closed under addition but are not closed under subtraction. An element e is said to be an "identity element" of a set with respect to the operation $\oplus$ if the property

$$a \oplus e = e \oplus a = a$$

holds for every element a in the set. In case the elements of the set are numbers, then e is the zero element, $e = 0$, if $\oplus$ is ordinary addition, whereas e is the unity element, $e = 1$, if $\oplus$ is ordinary multiplication. Assuming the existence of an identity element e, an element a is said to have an "inverse," written a^{-1}, if the property

$$a \oplus a^{-1} = a^{-1} \oplus a = e$$

holds. If the elements are numbers and $\oplus$ is ordinary addition, we usually write $a + b$ for $a \oplus b$ and $-a$ for the inverse a^{-1} because the additive inverse is the negative of the number a. On the other hand, if the operation $\oplus$ is ordinary multiplication, we write $a \cdot b$ for $a \oplus b$. In this case the notation a^{-1} is the customary one in elementary algebra for the multiplicative inverse. Here, and throughout this section, the word "number" means any sort of number, integral, rational, real, or complex.

Definition 2.9 *A group G is a set of elements a, b, c, $\cdots$ together with a single-valued binary operation $\oplus$ such that*

(1) *the set is closed under the operation;*
(2) *the associative law holds, namely,*
$a \oplus (b \oplus c) = (a \oplus b) \oplus c$ *for all elements a, b, c in G;*
(3) *the set has a unique identity element, e;*
(4) *each element in G has a unique inverse in G.*

A group G is called "abelian" or "commutative" if $a \oplus b = b \oplus a$ for every pair of elements a, b in G. A "finite group" is one with a finite number of elements; otherwise it is an "infinite group." If a group is finite, the number of its elements is called the "order" of the group.

Properties 1, 2, 3, and 4 are not the minimum possible postulates for a group. For example, in postulate 4 we could have required merely that each element a have a left inverse, that is an inverse a' such that $a' \oplus a = e$, and then we could prove the other half of postulate 4 as a consequence. However, to avoid too lengthy a discussion of group theory, we leave such refinements to the books on algebra.

The set of all integers $0, \pm 1, \pm 2, \cdots$ is a group under addition; in fact it is an abelian group. But the integers are not a group under multiplication because of the absence of inverses for all elements except ± 1.

Another example of a group is obtained by considering congruences modulo m. In case $m = 6$, to give a concrete example, we are familiar with such simple congruences as

$$3 + 4 \equiv 1 \pmod{6}, \qquad 5 + 5 \equiv 4 \pmod{6}.$$

We get "the additive group modulo 6" by taking a complete residue system, say 0, 1, 2, 3, 4, 5, and replacing congruence modulo 6 by equality:

$$3 + 4 = 1, \qquad 5 + 5 = 4.$$

The complete addition table for this system is:

$\oplus$	0	1	2	3	4	5
0	0	1	2	3	4	5
1	1	2	3	4	5	0
2	2	3	4	5	0	1
3	3	4	5	0	1	2
4	4	5	0	1	2	3
5	5	0	1	2	3	4

Of course, any complete residue system modulo 6 would do just as well; thus 1, 2, 3, 4, 5, 6, or 7, −2, 17, 30, 8, 3, could serve as the elements, provided we perform additions modulo 6. If we were to use the system 7, −2, 17, 30, 8, 3, the addition table would look quite different. However, the two groups are essentially the same; we have just renamed the elements: 0 is now called 30, 1 is 7, and so on. We say that the two groups are "isomorphic," and we do not consider isomorphic groups as being different. Thus we speak of "the" additive group modulo 6, not "an" additive group modulo 6.

Definition 2.10 *Two groups, G with operation $\oplus$ and G' with operation $\odot$, are said to be isomorphic if there is a one-to-one correspondence between the elements of G and those of G', such that if a in G corresponds to a' in G', and b in G corresponds to b' in G', then $a \oplus b$ in G corresponds to $a' \odot b'$ in G'.*

Another way of thinking of the additive group modulo 6 is in terms of the so-called residue classes. Put two integers a and b into the same residue class modulo 6 if $a \equiv b \pmod 6$, and the result is to separate all integers into six residue classes:

$$
\begin{aligned}
C_0&: \cdots, -18, -12, -6, 0, 6, 12, 18, \cdots \\
C_1&: \cdots, -17, -11, -5, 1, 7, 13, 19, \cdots \\
C_2&: \cdots, -16, -10, -4, 2, 8, 14, 20, \cdots \\
C_3&: \cdots, -15, -9, -3, 3, 9, 15, 21, \cdots \\
C_4&: \cdots, -14, -8, -2, 4, 10, 16, 22, \cdots \\
C_5&: \cdots, -13, -7, -1, 5, 11, 17, 23, \cdots
\end{aligned}
$$

If any element in class C_2 is added to any element in class C_3, the sum is an element in class C_5, so it is reasonable to write $C_2 + C_3 = C_5$. Similarly we observe that $C_3 + C_4 = C_1$, $C_5 + C_3 = C_2$, etc., and so we could make up an addition table for these classes. But the addition table so constructed would be simply a repetition of the addition table of the elements 0, 1, 2, 3, 4, 5 modulo 6. Thus the six classes $C_0, C_1, C_2, C_3, C_4, C_5$ form a group under this addition that is isomorphic to the additive group modulo 6. This residue class formulation of the additive group modulo 6 has the advantage that such a peculiar equation as $5 + 5 = 4$ (in which the symbols have a different meaning than in elementary arithmetic) is replaced by the more reasonable form $C_5 + C_5 = C_4$.

Theorem 2.29 *Any complete residue system modulo m forms a group under addition modulo m. Two complete residue systems modulo m constitute isomorphic groups under addition, and so we speak of "the" additive group modulo m.*

Proof. Let us begin with the complete residue system $0, 1, 2, \cdots, m - 1$ modulo m. This system is closed under addition modulo m, and the associative property of addition is inherited from the corresponding property for all integers, that is $a + (b + c) = (a + b) + c$ implies $a + (b + c) \equiv (a + b) + c \pmod{m}$. The identity element is 0, and it is unique. Finally, the additive inverse of 0 is 0, and the additive inverse of any other element a is $m - a$. These inverses are unique.

Passing from the system $0, 1, \cdots, m - 1$ to any complete residue system $r_0, r_1, \cdots, r_{m-1}$, we observe that all the above observations hold with a replaced by r_a, $a = 0, 1, \cdots, m - 1$, so that we have essentially the same group with new notation.

PROBLEMS

1. Which of the following are groups?
(*a*) the even integers under addition;
(*b*) the odd integers under addition;
(*c*) the integers under subtraction;
(*d*) the even integers under multiplication;
(*e*) all integers which are multiples of 7, under addition;
(*f*) all rational numbers under addition (recall that a rational number is one of the form a/b where a and b are integers, with $b \neq 0$);
(*g*) the same set as in (*f*), but under multiplication;
(*h*) the set as in (*f*) with the zero element deleted, under multiplication;
(*i*) all rational numbers a/b having $b = 1$ or $b = 2$, under addition;
(*j*) all rational numbers a/b having $b = 1$, $b = 2$, or $b = 3$, under addition.

2. Let G have as elements the four pairs $(1, 1)$, $(1, -1)$, $(-1, 1)$, $(-1, -1)$, and let $(a, b) \oplus (c, d) = (ac, bd)$. Prove that G is a group.
3. Using the complete residue system $7, -2, 17, 30, 8, 3$, write out the addition table for the additive group modulo 6. Rewrite this table replacing 7 by 1, 30 by 0, etc. Verify that this table gives the same values for $a \oplus b$ as the one in the text.
4. Prove that the set of elements e, a, b, c, with the following table for the binary operation,

$\oplus$	e	a	b	c
e	e	a	b	c
a	a	e	c	b
b	b	c	a	e
c	c	b	e	a

is a group. Prove that this group is isomorphic to the additive group modulo 4.
5. Prove that the set of elements e, u, v, w, with the following table for the binary operation,

$\oplus$	e	u	v	w
e	e	u	v	w
u	u	e	w	v
v	v	w	e	u
w	w	v	u	e

is a group. Prove that this group is not isomorphic to the additive group modulo 4, but that it is isomorphic to the group described in Problem 2.
6. Prove that the set of elements 1, 2, 3, 4, under the operation of multiplication modulo 5, is a group which is isomorphic to the group in Problem 4.
7. Prove that the set of complex numbers $+1$, -1, $+i$, $-i$, where $i^2 = -1$, is a group under multiplication and that it is isomorphic to the group in Problem 4.
8. Prove that the isomorphism property is "transitive," that is, if a group G_1 is isomorphic to G_2, and if G_2 is isomorphic to G_3, then G_1 is isomorphic to G_3.

9. Prove that the elements 1, 3, 5, 7 under multiplication modulo 8 form a group which is isomorphic to the group in Problem 5.

10. Prove that there are essentially only two groups of order 4, that is that any group of order 4 is isomorphic to one of the groups in Problems 4 and 5.

11. For any positive integer $m > 1$, separate all integers into classes $C_0, C_1, \cdots, C_{m-1}$, putting integers r and s into the same class if $r \equiv s \pmod{m}$, thus

$$C_0: \cdots, -2m, -m, 0, m, 2m, \cdots$$
$$C_1: \cdots, -2m+1, -m+1, 1, m+1, 2m+1, \cdots$$
etc.

Prove that if any two integers, one from class C_a and one from class C_b, are added, the sum is always an integer in a unique class, namely, either C_{a+b} or C_{a+b-m} according as $a + b < m$ or $a + b \geqq m$. Define the sum $C_a + C_b = C_{a+b}$ or $C_a + C_b = C_{a+b-m}$ accordingly, and prove that these classes form a group under this addition. Prove that this group is isomorphic to the additive group modulo m.

2.11 Multiplicative Groups, Rings, and Fields

Theorem 2.30 *Let $m > 1$ be a positive integer. Any reduced residue system modulo m is a group under multiplication modulo m. The group is of order $\phi(m)$. Any two such groups are isomorphic, and so we speak of "the multiplicative group modulo m."*

Proof. Let us consider any reduced residue system $r_1, r_2, \cdots, r_n$ where $n = \phi(m)$. This set is closed under multiplication modulo m by Theorem 1.8. The associative property of multiplication is inherited from the corresponding property for integers, because $a(bc) = (ab)c$ implies that $a(bc) \equiv (ab)c \pmod{m}$. The reduced residue system contains one element, say r_j, such that $r_j \equiv 1 \pmod{m}$, and this is clearly the unique identity element of the group. Finally, for each r_i, the congruence $xr_i \equiv r_j \pmod{m}$ has a solution by Theorem 2.13, and this solution is unique within the reduced residue system $r_1, r_2, \cdots, r_n$. Two different reduced residue systems modulo m are congruent, element by element, modulo m, and so we have an isomorphism between the two groups.

Notation. We have been using the symbol $\oplus$ for the binary operation of the group, and we have found that in particular groups $\oplus$ may represent addition or multiplication or some other operation. In dealing with general groups it is convenient to drop the symbol $\oplus$, just as the dot representing ordinary multiplication is usually omitted in algebra. We will write ab for $a \oplus b$, abc for $a \oplus (b \oplus c) = (a \oplus b) \oplus c$, a^2 for $a \oplus a$, a^3 for $a \oplus (a \oplus a)$, etc. Also, $abcd$ can be written for $(a \oplus b \oplus c) \oplus d = (a \oplus b) \oplus (c \oplus d)$ etc., as can be seen by applying induction to the associative law. We will even use the word multiplication for the operation $\oplus$, but it must be remembered that we do not mean the ordinary multiplication of arithmetic. In fact we are dealing with general groups so that a is not a number, it is just an abstract element of a group. It is convenient to write a^0 for e, a^{-2} for $(a^{-1})^2$, a^{-3} for $(a^{-1})^3$, etc. It is not difficult to show that the usual laws of exponents are valid under this definition.

Theorem 2.31 *In any group G, $ab = ac$ implies $b = c$. If a is any element of a finite group G with identity element e, then there is a (unique) smallest positive integer r such that $a^r = e$.*

Proof. The first part of the theorem is established by multiplying $ab = ac$ on the left by a^{-1}, thus $a^{-1}(ab) = a^{-1}(ac)$, $(a^{-1}a)b = (a^{-1}a)c$, $eb = ec$, $b = c$. To prove the second part, consider the series of elements obtained by repeated multiplication by a,

$$e, a, a^2, a^3, a^4, \cdots.$$

Since the group is finite, and since the members of this series are elements of the group, there must occur a repetition of the form $a^s = a^t$ with, say, $s < t$. But this equation can be written in the form $a^s e = a^s a^{t-s}$, whence $a^{t-s} = e$. Thus there is some positive integer, $t - s$, such that $a^{t-s} = e$ and the smallest positive exponent with this property is the value of r in the theorem.

Definition 2.11 *Let G be any group, finite or infinite, and a an element of G. If $a^s = e$ for some positive integer s, then a is said to be of finite order. If a is of finite order, the order of a is the smallest positive integer r such that $a^r = e$. If there is no positive integer s such that $a^s = e$, then a is said to be of infinite order. A group G is said to be cyclic if it contains an element a such that the powers of a*

$$\cdots, a^{-3}, a^{-2}, a^{-1}, a^0 = e, a, a^2, a^3, \cdots$$

comprise the whole group; such an element a is said to generate the group and is called a generator.

Theorem 2.31 shows that all the elements of a finite group are of finite order. Every group, finite or infinite, contains at least the single element e that is of finite order. There are infinite groups consisting entirely of elements of finite order.

If a cyclic group is finite, and has generator a, then the group consists of $e, a, a^2, a^3, \cdots, a^{r-1}$, where r is the order of the element a. All other powers of a are superfluous because they merely repeat these.

Theorem 2.32 *The order of an element of a finite group G is a divisor of the order of the group. If the order of the group is denoted by n, then $a^n = e$.*

Proof. Let the element a have order r. It is readily seen that

$$(A) \qquad e, a, a^2, a^3, \cdots, a^{r-1}$$

are r distinct elements of G. If these r elements do not exhaust the group, there is some other element, say b_2. Then we can prove that

$$(B) \qquad b_2, b_2a, b_2a^2, b_2a^3, \cdots, b_2a^{r-1}$$

are r distinct elements, all different from the r elements of A. For in the first place if $b_2a^s = b_2a^t$, then $a^s = a^t$ by Theorem 2.31. And on the other hand, if $b_2a^s = a^t$, then $b_2 = a^{t-s}$, so that b_2 would be among the powers of a.

If G is not exhausted by the sets A and B, then there is another element b_3 which gives rise to r new elements

$$b_3, b_3a, b_3a^2, b_3a^3, \cdots, b_3a^{r-1},$$

all different from the elements in A and B, by a similar argument. This process of obtaining new elements $b_2, b_3, \cdots$ must terminate since G is finite. So if the last batch of new elements is, say

$$b_k, b_ka, b_ka^2, b_ka^3, \cdots, b_ka^{r-1},$$

then the order of the group G is kr, and the first part of the theorem is proved. To prove the second part, we observe that $n = kr$ and $a^r = e$ by Theorem 2.31, whence $a^n = e$.

It can be noted that Theorem 2.32 implies the theorems of Fermat and Euler, where the set of integers relatively prime to the modulus m is taken as the group. In making this implication, the reader will see the necessity of "translating" the language and notation of group theory into that of number theory. In the same way we note that the language of Definition 2.7, that "a belongs to the exponent h modulo m," is translated into group theoretic language as "the element a of the multiplicative group modulo m has order h." Also the "primitive root modulo m" of Definition 2.8 is called a "generator" of the multiplicative group modulo m in group theory.

Definition 2.12 *A ring is a set of at least two elements with two binary operations, $\oplus$ and $\odot$, such that it is a commutative group under $\oplus$, is closed under $\odot$, and such that $\odot$ is associative and distributive with respect to $\oplus$. The identity element with respect to $\oplus$ is called the "zero" of the ring. If all the elements of a ring, other than the zero, form a commutative group under $\odot$, then it is called a field.*

It is customary to call $\oplus$ addition and $\odot$ multiplication and to write $a + b$ for $a \oplus b$, ab for $a \odot b$. The conditions on $\odot$ for a ring are then $a(bc) = (ab)c$, $a(b + c) = ab + ac$, $(b + c)a = ba + ca$. In general, the elements $a, b, c, \cdots$ are not numbers, and the operations of addition and multiplication are not the ordinary ones of arithmetic. However, the only rings and fields that will be considered here will have numbers for elements, and the operations will either be ordinary addition and multiplication, or addition and multiplication modulo m.

Theorem 2.33 *The set Z_m of elements* $0, 1, 2, \cdots, m - 1$, *with addition and multiplication defined modulo m, is a ring for any integer $m > 1$. Such a ring is a field if and only if m is a prime.*

Proof. We have already seen in Theorem 2.29 that any complete residue system modulo m is a group under addition modulo m. This group is commutative, and the associative and distributive properties of multiplication modulo m are inherited from the corresponding properties for ordinary multiplication. Therefore Z_m is a ring.

Next, by Theorem 2.30 any reduced residue system modulo m is a group under multiplication modulo m. If m is a prime p, the reduced residue system of Z_p is $1, 2, \cdots, p - 1$, that is, all the elements of Z_p other than 0. Since 0 is the zero of the ring, Z_p is a field. On the other hand if m is not a prime, then m is of the form ab with $1 < a \leqq b < m$. Then the elements of Z_m other than 0 do not form a group under multiplication modulo m because there is no inverse for the element a, no solution of $ax \equiv 1 \pmod{m}$. Thus Z_m is not a field.

Some questions can be settled very readily by using the fields Z_p. For example consider the following problem: prove that for any prime $p > 3$ the sum

$$\frac{1}{1^2} + \frac{1}{2^2} + \frac{1}{3^2} + \cdots + \frac{1}{(p - 1)^2},$$

if written as a rational number a/b has the property that $p \mid a$. In the field Z_p the term $1/j^2$ in the above sum is j^{-2} or x^2 where x is the least positive integer such that $xj \equiv 1 \pmod{p}$. Hence in Z_p the problem can be put in the form, prove that the sum $1^{-2} + 2^{-2} + \cdots + (p - 1)^{-2}$ is the zero element of the field. But the inverses of $1, 2, 3, \cdots, p - 1$ are just the same elements again in some order, so we can write

$$1^{-2} + 2^{-2} + \cdots + (p - 1)^{-2} = 1^2 + 2^2 + \cdots + (p - 1)^2.$$

For this final sum there is a well-known formula for the sum of the squares of the natural numbers giving $p(p - 1)(2p - 1)/6$. But this is zero in Z_p, because of the factor p, except in the cases $p = 2$ and $p = 3$ where division by 6 is meaningless.

PROBLEMS

1. Prove that the multiplicative group modulo 9 is isomorphic to the additive group modulo 6.
2. Prove that the additive group modulo m is cyclic with 1 as generator. Prove that any one of $\phi(m)$ elements could serve as generator.
3. Prove that any two cyclic groups of order m are isomorphic.
4. Prove that the group of all integers under addition is an infinite cyclic group.
5. If a is an element of order r of a group G, prove that $a^k = e$ if and only if $r \mid k$.
6. What is the smallest positive integer m such that the multiplicative group modulo m is not cyclic?
7. A subgroup S of a group G is a subset of elements of G which form a group under the same binary operation. If G is finite, prove that the order of a subgroup S is a divisor of the order of G.
8. Prove Theorem 2.32, for the case in which the group is commutative, in a manner analogous to the proof of Theorem 2.8.
9. Prove Theorem 2.8 by the method used in the proof of Theorem 2.32.
10. Let G consist of all possible sequences $(a_1, a_2, a_3, \cdots)$ with each $a_i = 1$ or -1. Let $(a_1, a_2, a_3, \cdots) \oplus (b_1, b_2, b_3, \cdots) = (a_1b_1, a_2b_2, a_3b_3, \cdots)$. Show that G is an infinite group all of whose elements are of finite order.
11. Let G consist of a, b, c, d, e, f and let $\oplus$ be defined by the following table.

$\oplus$	e	a	b	c	d	f
e	e	a	b	c	d	f
a	a	e	d	f	b	c
b	b	f	e	d	c	a
c	c	d	f	e	a	b
d	d	c	a	b	f	e
f	f	b	c	a	e	d

Show that G is a noncommutative group.
12. Prove that the multiplicative group modulo p is cyclic if p is a prime.
13. Exhibit the addition and multiplication tables for the elements of the field of residues modulo 7.
14. Prove that the set of all integers under ordinary addition and multiplication is a ring but not a field.

15. Prove that the set of all even integers under ordinary addition and multiplication is a ring.

16. Prove that the set 0, 3, 6, 9 is a ring under addition and multiplication modulo 12.

17. Prove that in any field $a0 = 0a = 0$ for every element a.

18. Let a be a divisor of m, say $m = aq$ with $1 < a < m$. Prove that the set of elements $0, a, 2a, 3a, \cdots, (q-1)a$, with addition and multiplication modulo m, forms a ring. Under what circumstances is it a field?

19. Prove that the set of all rational numbers forms a field.

20. Prove that the set of all rational functions $f(x)/g(x)$, where $f(x)$ and $g(x)$ are polynomials with integral coefficients, and $g(x) \not\equiv 0$, forms a field.

21. If x, y, z, are any numbers, real or complex, the cancellation law states that $xy = xz$ implies $y = z$ if $x \neq 0$. There is a "weak" cancellation law that states that $x^2y = x^2z$ implies $xy = xz$ whether x is zero or not. Consider the set of all integers modolo m with multiplication modulo m, $m > 1$. Show that the cancellation law holds if and only if m is a prime, and that the "weak" law holds if and only if m is a square-free, that is m is a product of distinct primes.

22. Consider the system of all integers modulo m under multiplication modulo m, $m > 1$. Show that:

(*a*) It is not a group.

(*b*) It is associative.

(*c*) It is commutative.

(*d*) It is closed.

(*e*) It has a unique unity—an element u such that $ux = x$ for all x.

(*f*) It has a unique zero—an element z such that $zx = z$ for all x.

23. For $m = 30$ find all idempotent elements (x such that $x^2 = x$). Also find the elements w such that $wx = z$, the zero element, for some $x \neq z$.

24. An integral domain is a ring with the following additional properties: (i) there is a unique identity element with respect to multiplication; (ii) multiplication is commutative; (iii) if $ab = ac$ and $a \neq 0$, then $b = c$. Prove that any field is an integral domain. Which of the following are integral domains?

(*a*) the set of all integers;

(*b*) the set Z_m of Theorem 2.33;

(*c*) the set $F[x]$ of all polynomials in a variable or indeterminate x, with coefficients in a field F.

25. Let m be a positive integer and consider the set of all the divisors of m. For numbers in this set define two operations $\odot$ and $\oplus$ as $a \odot b = (a, b)$, $a \oplus b = [a, b]$, g.c.d. and l.c.m. Prove that $\odot$ and $\oplus$ are associative and commutative. Prove the distributive law $a \odot (b \oplus c) = (a \odot b) \oplus (a \odot c)$ and its dual $a \oplus (b \odot c) = (a \oplus b) \odot (a \oplus c)$. Show that $a \odot a = a \oplus a = a$. Also prove $1 \odot a = 1$ and $1 \oplus a = a$, so that 1 behaves like an ordinary zero. and $m \odot a = a$, and $m \oplus a = m$. Define a relation $\olessthan$ as $a \olessthan b$ if $a \odot b = a$, Prove $a \olessthan a$, that $\olessthan$ is transitive, and that $a \olessthan b$ if and only if $a \oplus b = b$.

Prove that if m is not divisible by any square other than 1, then corresponding to each divisor a there is a divisor a' such that $a \odot a' = 1$, $a \oplus a' = m$. (These algebras with square-free m are examples of Boolean algebras.)

26. Prove that for any prime $p > 2$ the sum

$$\frac{1}{1^3} + \frac{1}{2^3} + \cdots + \frac{1}{(p-1)^3},$$

if written as a rational number a/b, has the property that $p \mid a$. *Suggestion:* Interpret the sum in Z_p, and use the result

$$1^3 + 2^3 + \cdots + n^3 = n^2(n+1)^2/4.$$

NOTES ON CHAPTER 2

In **§2.1** it was noted that (i) $a \equiv a \pmod{m}$, (ii) $a \equiv b \pmod{m}$ if and only if $b \equiv a \pmod{m}$, and (iii) $a \equiv b \pmod{m}$ and $b \equiv c \pmod{m}$ imply $a \equiv c \pmod{m}$. Thus the congruence relation has the (i) reflexive property, (ii) the symmetric property, and (iii) the transitive property, and so the congruence relation is a so-called *equivalence relation.*

If $(a, m) = 1$ the congruence $ax \equiv b \pmod{m}$ could be solved as $x \equiv ba^{-1} \pmod{m}$ or $x = ba^{-1}$ where a^{-1} is an integer satisfying $aa^{-1} \equiv 1 \pmod{m}$. However, we do not use the symbol a^{-1} in this way, reserving it to mean $1/a$.

A generalization by Gauss of Fermat's Theorem 2.7 can be found in the Special Topics, page 262 ff. Also on page 263 is given a proof of the first part of Theorem 2.25 by a group theoretic approach.

For interesting discussions of magic squares see Chapter 4 of the book by Harold M. Stark, listed in the General References on page 270, and the paper by D. N. Lehmer, "On the congruences connected with certain magic squares," *Trans. Amer. Math. Soc.*, **31**, 529–551 (1929).

3

Quadratic Reciprocity

3.1 Quadratic Residues

Definition 3.1 *For all a such that $(a, m) = 1$, a is called a quadratic residue modulo m if the congruence $x^2 \equiv a \pmod{m}$ has a solution. If it has no solution, then a is called a quadratic nonresidue modulo m.*

Since $a + m$ is a quadratic residue or nonresidue modulo m according as a is or is not, we consider as distinct residues or nonresidues only those that are distinct modulo m. The quadratic residues modulo 5 are 1 and 4, whereas 2 and 3 are the nonresidues.

Definition 3.2 *If p denotes an odd prime and $(a, p) = 1$, the Legendre symbol $\left(\frac{a}{p}\right)$ is defined to be 1 if a is a quadratic residue, -1 if a is a quadratic nonresidue modulo p.*

Theorem 3.1 *Let p be an odd prime and let a and b denote integers relatively prime to p. Then*

(a) $\left(\frac{a}{p}\right) \equiv a^{(p-1)/2} \pmod{p}$,

(b) $\left(\frac{a}{p}\right)\left(\frac{b}{p}\right) = \left(\frac{ab}{p}\right)$,

(c) $a \equiv b \pmod{p}$ *implies that* $\left(\frac{a}{p}\right) = \left(\frac{b}{p}\right)$,

(d) $\left(\frac{a^2}{p}\right) = 1, \quad \left(\frac{a^2b}{p}\right) = \left(\frac{b}{p}\right), \quad \left(\frac{1}{p}\right) = 1, \quad \left(\frac{-1}{p}\right) = (-1)^{(p-1)/2}$.

Remark. If $\left(\frac{a}{p}\right) = 1$ then $a^{(p-1)/2} \equiv 1 \pmod{p}$ and Corollary 2.28 shows that $x^2 \equiv a \pmod{p}$ has exactly two solutions; alternatively, this conclusion can be drawn from Theorem 2.20, because if $x^2 \equiv a \pmod{p}$ has one solution x_0 then $-x_0$ is a distinct solution because p is odd.

Proof. Part (*a*) of the theorem follows from Corollary 2.28. The remaining parts are all simple consequences of part (*a*). Part (*a*) can also be proved without the use of Corollary 2.28 as follows: If $\left(\frac{a}{p}\right) = 1$, then $x^2 \equiv a \pmod{p}$ has a solution, say x_0. Then, by Theorem 2.7, $a^{(p-1)/2} \equiv x_0^{p-1} \equiv 1 \equiv \left(\frac{a}{p}\right)$ $\pmod{p}$.

On the other hand, if $\left(\frac{a}{p}\right) \equiv -1$, then $x^2 \equiv a \pmod{p}$ has no solution, and we proceed as in the proof of Theorem 2.10. To each j satisfying $1 \leqq j \leqq p - 1$ we associate the unique integer i such that $ji \equiv a \pmod{p}$, $0 \leqq i \leqq p - 1$. We see that $i = 0$ is impossible and that the associate of i is j. Since $x^2 \equiv a \pmod{p}$ has no solution, no integer j is associated with itself. Thus the integers $1, 2, \cdots, p - 1$, can be paired off, j and its associate i, and $ji \equiv a \pmod{p}$. There are $(p-1)/2$ pairs. Multiplying all these pairs together we get $(p - 1)! \equiv a^{(p-1)/2} \pmod{p}$. Using Theorem 2.10 we obtain $a^{(p-1)/2} \equiv -1 \equiv \left(\frac{a}{p}\right) \pmod{p}$.

Theorem 3.2 *Lemma of Gauss. Let p be an odd prime and let $(a, p) = 1$. Consider the integers $a, 2a, 3a, \cdots, \{(p-1)/2\}a$ and their least non-negative residues modulo p. If n denotes the number of these residues that exceed $\frac{p}{2}$, then $\left(\frac{a}{p}\right) = (-1)^n$.*

Proof. Let $r_1, r_2, \cdots, r_n$ denote the residues that exceed $p/2$, and let $s_1, s_2, \cdots, s_k$ denote the remaining residues. The r_i and s_i are all distinct, and none is zero. Furthermore $n + k = (p - 1)/2$. Now $0 < p - r_i < p/2$, $i = 1, 2, \cdots, n$, and the numbers $p - r_i$ are distinct. Also no $p - r_i$ is an s_j for if $p - r_i = s_j$ then $r_i \equiv \rho a$, $s_j \equiv \sigma a$, for some ρ, σ, $1 \leqq \rho \leqq (p-1)/2$, $1 \leqq \sigma \leqq (p-1)/2$, and $p - \rho a \equiv \sigma a \pmod{p}$. Since $(a, p) = 1$ this implies $a(\rho + \sigma) \equiv 0$, $\rho + \sigma \equiv 0 \pmod{p}$, which is impossible. Thus $p - r_1, p - r_2, \cdots, p - r_n, s_1, s_2, \cdots, s_k$ are all distinct, are all at least 1 and less than $p/2$, and they are $n + k = (p - 1)/2$ in number. That is, they are just the integers $1, 2, \cdots, (p - 1)/2$ in some order. Multiplying them together we have

$$(p - r_1)(p - r_2) \cdots (p - r_n)s_1 s_2 \cdots s_k = 1 \cdot 2 \cdots \frac{p-1}{2}$$

and then

$$(-r_1)(-r_2) \cdots (-r_n)s_1 s_2 \cdots s_k \equiv 1 \cdot 2 \cdots \frac{p-1}{2} \pmod{p},$$

$$(-1)^n r_1 r_2 \cdots r_n s_1 s_2 \cdots s_k \equiv 1 \cdot 2 \cdots \frac{p-1}{2} \pmod{p},$$

$$(-1)^n a \cdot 2a \cdot 3a \cdots \frac{p-1}{2} a \equiv 1 \cdot 2 \cdots \frac{p-1}{2} \pmod{p}.$$

We can cancel the factors $2, 3, \cdots, (p-1)/2$ to obtain $(-1)^n a^{(p-1)/2} \equiv 1 \pmod{p}$ which gives us $(-1)^n \equiv a^{(p-1)/2} \equiv \left(\frac{a}{p}\right) \pmod{p}$ by Theorem 3.1*a*.

Definition 3.3 *For real x, the symbol $[x]$ denotes the greatest integer less than or equal to x.*

For example, $[15/2] = 7$, $[-15/2] = -8$, $[-15] = -15$.

Theorem 3.3 *If p is an odd prime and $(a, 2p) = 1$, then*

$$\left(\frac{a}{p}\right) = (-1)^t \quad \textit{where} \quad t = \sum_{j=1}^{(p-1)/2} \left[\frac{ja}{p}\right]; \quad \textit{also} \quad \left(\frac{2}{p}\right) = (-1)^{(p^2-1)/8}.$$

Proof. We use the same notation as in the proof of Theorem 3.2. The r_i and s_i are just the least positive remainders obtained on dividing the integers ja by p, $j = 1, 2, \cdots, (p-1)/2$. The quotient in this division is easily seen to be $q = [ja/p]$. Then for $(a, p) = 1$, whether a is odd or even, we have

$$\sum_{j=1}^{(p-1)/2} ja = \sum_{j=1}^{(p-1)/2} p\left[\frac{ja}{p}\right] + \sum_{j=1}^{n} r_j + \sum_{j=1}^{k} s_j$$

and

$$\sum_{j=1}^{(p-1)/2} j = \sum_{j=1}^{n} (p - r_j) + \sum_{j=1}^{k} s_j = np - \sum_{j=1}^{n} r_j + \sum_{j=1}^{k} s_j$$

and hence by subtraction,

$$(a-1)\sum_{j=1}^{(p-1)/2} j = p\left(\sum_{j=1}^{(p-1)/2}\left[\frac{ja}{p}\right] - n\right) + 2\sum_{j=1}^{n} r_j.$$

But

$$\sum_{j=1}^{(p-1)/2} j = \frac{p^2 - 1}{8}$$

so we have

$$(a-1)\frac{p^2-1}{8} \equiv \sum_{j=1}^{(p-1)/2} \left[\frac{ja}{p}\right] - n \pmod{2}.$$

If a is odd, this implies $n \equiv \sum_{j=1}^{(p-1)/2} \left[\frac{ja}{p}\right] \pmod{2}$. If $a = 2$ it implies $n \equiv (p^2-1)/8 \pmod{2}$ since $[2j/p] = 0$ for $1 \leqq j \leqq (p-1)/2$. Our theorem now follows by Theorem 3.2.

Although Theorem 3.1(*a*), Theorem 3.2, and the first part of Theorem 3.3 are of considerable importance in theoretical considerations, they are too cumbersome to use for calculations unless p is very small. However, the other parts of Theorems 3.1 and 3.3 are useful in numerical cases. The second part of Theorem 3.3 involves $(-1)^{(p^2-1)/8}$, and this can be easily computed if p is reduced modulo 8. For example if $p = 59$ then $p \equiv 3 \pmod{8}$ and $(-1)^{(p^2-1)/8} = (-1)^{(3^2-1)/8}$. Finally, we point out that the problem of numerical evaluation of $\left(\frac{a}{p}\right)$, apart from the cases $a = \pm 1, \pm 2$, is treated in the next section.

PROBLEMS

1. Find $[3/2]$, $[-3/2]$, $[\pi]$, $[-7]$, $[x]$ for $0 \leqq x < 1$.
2. With reference to the notation of Theorem 1.2 prove that $q = [b/a]$.
3. Prove that 3 is a quadratic residue of 13, but a quadratic nonresidue of 7.
4. Find the values of $\left(\frac{a}{p}\right)$ in each of the 12 cases, $a = -1, 2, -2, 3$ and $p = 11, 13, 17$.
5. Prove that the quadratic residues of 11 are 1, 3, 4, 5, 9, and list all solutions of each of the ten congruences $x^2 \equiv a \pmod{11}$ and $x^2 \equiv a \pmod{11^2}$ where $a = 1, 3, 4, 5, 9$.
6. List the quadratic residues of each of the primes 7, 13, 17, 29, 37.
7. Which of the following congruences have solutions? How many?
(*a*) $x^2 \equiv 2 \pmod{61}$ (*b*) $x^2 \equiv 2 \pmod{59}$
(*c*) $x^2 \equiv -2 \pmod{61}$ (*d*) $x^2 \equiv -2 \pmod{59}$
(*e*) $x^2 \equiv 2 \pmod{122}$ (*f*) $x^2 \equiv 2 \pmod{118}$
(*g*) $x^2 \equiv -2 \pmod{122}$ (*h*) $x^2 \equiv -2 \pmod{118}$.
8. How many solutions are there to each of the congruences:
(*a*) $x^2 \equiv -1 \pmod{61}$ (*b*) $x^2 \equiv -1 \pmod{59}$
(*c*) $x^2 \equiv -1 \pmod{365}$ (*d*) $x^2 \equiv -1 \pmod{3599}$
(*e*) $x^2 \equiv -1 \pmod{122}$ (*f*) $x^2 \equiv -1 \pmod{244}$?

9. Let p be a prime, and let $(a, p) = (b, p) = 1$. Prove that if $x^2 \equiv a \pmod{p}$ and $x^2 \equiv b \pmod{p}$ are not solvable, then $x^2 \equiv ab \pmod{p}$ is solvable.

10. Prove that if p is an odd prime then $x^2 \equiv 2 \pmod{p}$ has solutions if and only if $p \equiv 1$ or $7 \pmod{8}$.

11. Denote quadratic residues by r, nonresidues by n. Prove that r_1r_2 and n_1n_2 are residues and that rn is a nonresidue for a prime p. Show that there are $(p - 1)/2$ quadratic residues and $(p - 1)/2$ nonresidues for an odd prime p.

12. Let g be a primitive root of an odd prime p. Prove that the quadratic residues modulo p are congruent to $g^2, g^4, g^6, \cdots, g^{p-1}$, and the nonresidues are congruent to $g, g^3, g^5, g^7, \cdots, g^{p-2}$.

13. Prove that if r is a quadratic residue modulo $m > 2$, then $r^{\phi(m)/2} \equiv 1 \pmod{m}$. *Suggestion:* use the fact that there is some integer a such that $r \equiv a^2 \pmod{m}$.

14. Show that if a is a quadratic residue modulo m, and $ab \equiv 1 \pmod{m}$, then b is also a quadratic residue. Then prove that the product of the quadratic residues modulo p is congruent to $+1$ or -1 according as the prime p is of the form $4k + 3$ or $4k + 1$.

15. Prove that if p is a prime having the form $4k + 3$, and if m is the number of quadratic residues less than $p/2$, then $1 \cdot 3 \cdot 5 \cdots (p - 2) \equiv (-1)^{m+k+1} \pmod{p}$, and $2 \cdot 4 \cdot 6 \cdots (p - 1) \equiv (-1)^{m+k} \pmod{p}$.

16. Prove that the quadratic residues modulo p are congruent to $1^2, 2^2, 3^2, \cdots, \{(p - 1)/2\}^2$, where p is an odd prime. Hence prove that if $p > 3$, the sum of the quadratic residues is divisible by p.

17. For all primes p prove that $x^8 \equiv 16 \pmod{p}$ is solvable. *Suggestion:* use Theorem 2.27.

18. Let p be an odd prime. Prove that if there is an integer x such that

$p \mid (x^2 + 1)$ then $p \equiv 1 \pmod{4}$;

$p \mid (x^2 - 2)$ then $p \equiv 1$ or $7 \pmod{8}$;

$p \mid (x^2 + 2)$ then $p \equiv 1$ or $3 \pmod{8}$;

$p \mid (x^4 + 1)$ then $p \equiv 1 \pmod{8}$.

Show that there are infinitely many primes of each of the forms $8n + 1$, $8n + 3$, $8n + 5$, $8n + 7$. *Suggestion:* use Theorem 2.27 for the case $p \mid (x^4 + 1)$.

19. Let p be an odd prime. Prove that every primitive root of p is a quadratic nonresidue. Prove that every quadratic nonresidue is a primitive root if and only if p is of the form $2^{2^n} + 1$ where n is a non-negative integer, i.e., if and only if $p = 3$ or p is a Fermat number.

3.2 Quadratic Reciprocity

Theorem 3.4 *The Gaussian reciprocity law. If p and q are distinct odd primes, then*

$$\left(\frac{p}{q}\right)\left(\frac{q}{p}\right) = (-1)^{\{(p-1)/2\}\{(q-1)/2\}}.$$

Proof. Let S be the set of all pairs of integers (x, y) satisfying $1 \leqq x \leqq (p-1)/2$, $1 \leqq y \leqq (q-1)/2$. The set S has $(p-1)(q-1)/4$ members. Separate this set into two mutually exclusive subsets S_1 and S_2 according as $qx > py$ or $qx < py$. Note that there are no pairs (x, y) in S such that $qx = py$. The set S_1 can be described as the set of all pairs (x, y) such that $1 \leqq x \leqq (p-1)/2$, $1 \leqq y < qx/p$. The number of pairs in S_1 is then seen to be $\sum_{x=1}^{(p-1)/2} [qx/p]$. Similarly S_2 consists of the pairs (x, y) such that $1 \leqq y \leqq (q-1)/2$, $1 \leqq x < py/q$, and the number of pairs in S_2 is $\sum_{y=1}^{(q-1)/2} [py/q]$. Thus we have

$$\sum_{j=1}^{(p-1)/2}\left[\frac{qj}{p}\right] + \sum_{j=1}^{(q-1)/2}\left[\frac{pj}{q}\right] = \frac{p-1}{2}\frac{q-1}{2}$$

and hence

$$\left(\frac{p}{q}\right)\left(\frac{q}{p}\right) = (-1)^{\{(p-1)/2\}\{(q-1)/2\}}$$

by Theorem 3.3.

This theorem, together with Theorem 3.1 and the second part of Theorem 3.3, makes the computation of $\left(\frac{a}{p}\right)$ fairly simple. For example, we have

$$\left(\frac{-42}{61}\right) = \left(\frac{-1}{61}\right)\left(\frac{2}{61}\right)\left(\frac{3}{61}\right)\left(\frac{7}{61}\right),$$

$$\left(\frac{-1}{61}\right) = (-1)^{60/2} = 1,$$

$$\left(\frac{2}{61}\right) = (-1)^{(61^2-1)/8} = -1,$$

$$\left(\frac{3}{61}\right) = \left(\frac{61}{3}\right)(-1)^{(2/2)(60/2)} = \left(\frac{1}{3}\right) = 1,$$

$$\left(\frac{7}{61}\right) = \left(\frac{61}{7}\right)(-1)^{(6/2)(60/2)} = \left(\frac{5}{7}\right) = \left(\frac{7}{5}\right)(-1)^{(4/2)(6/2)} = \left(\frac{2}{5}\right)$$
$$= (-1)^{24/8} = -1.$$

Hence $\left(\frac{-42}{61}\right) = 1$. This computation demonstrates a number of different sorts of steps; it was chosen for this purpose and is not the shortest possible. A shorter way is

$$\left(\frac{-42}{61}\right) = \left(\frac{19}{61}\right) = \left(\frac{61}{19}\right)\cdot 1 = \left(\frac{4}{19}\right) = 1.$$

One could also obtain the value of $\left(\frac{-42}{61}\right)$ by use of Theorem 3.2 or the first part of Theorem 3.3, but the computation would be considerably longer.

There is another kind of problem that is of some importance. As an example, let us find all odd primes p such that 3 is a quadratic residue modulo p. We have

$$\left(\frac{3}{p}\right) = \left(\frac{p}{3}\right)(-1)^{(p-1)/2},$$

$$\left(\frac{p}{3}\right) = \begin{cases} \left(\frac{1}{3}\right) = 1 & \text{if } p \equiv 1 \pmod 3 \\ \left(\frac{2}{3}\right) = -1 & \text{if } p \equiv 2 \pmod 3, \end{cases}$$

and

$$(-1)^{(p-1)/2} = \begin{cases} 1 & \text{if } p \equiv 1 \pmod 4 \\ -1 & \text{if } p \equiv 3 \pmod 4. \end{cases}$$

Thus $\left(\frac{3}{p}\right) = 1$ if and only if $p \equiv 1 \pmod 3$, $p \equiv 1 \pmod 4$, or $p \equiv 2 \pmod 3$, $p \equiv 3 \pmod 4$; that is $p \equiv 1$ or $11 \pmod{12}$.

Results of this type are sometimes useful if one is trying to determine whether or not a certain number is a prime. Consider the number 9997. We might notice that $9997 = 100^2 - 3$, so that $3 \equiv 100^2 \pmod p$ if $p \mid 9997$. That is, either $p = 3$ or $\left(\frac{3}{p}\right) = 1$. Since $3 \nmid 9997$ and $p \neq 2$ we have $\left(\frac{3}{p}\right) = 1$, and hence $p \equiv 1$ or $11 \pmod{12}$. We need test only p such that $p \leqq \sqrt{9997}$. If we list the numbers $1, 13, 25, \cdots, 97$, and $11, 23, 35, \cdots, 95$, we can delete 1 and all composite numbers, and we find just eleven primes that must be tested. It will be found that 13 divides 9997 and that $9997 = 13 \cdot 769$. Is 769 a prime? If $p \mid 769$ then $p \mid 9997$ and it will be in our list. The only p in our list such that $p \leqq \sqrt{769}$ are 13, 11, and 23. None of these divides 769, and hence 769 is a prime.

Just as we determined which primes have 3 as a quadratic residue, so for any odd prime p we can analyze which primes have p as a residue. This is done in effect in the following result.

Theorem 3.5 *Let p be an odd prime. For any odd prime $q > p$ let r be determined as follows. First if p is of the form $4n + 1$, define r as the least positive remainder when q is divided by p; thus $q = kp + r$, $0 < r < p$. Next if p is of the form $4n + 3$, there is a unique r defined by the relations $q = 4kp \pm r$, $0 < r < 4p$, $r \equiv 1 \pmod 4$. Then in both cases $\left(\frac{p}{q}\right) = \left(\frac{r}{p}\right)$.*

Proof. If $p = 4n + 1$, by Theorems 3.4 and 3.1*c* we see that $\left(\frac{p}{q}\right) = \left(\frac{q}{p}\right) = \left(\frac{r}{p}\right)$. In case $p = 4n + 3$, we first prove that r exists to satisfy the conditions stated. Let r_0 be the least positive remainder when q is divided by $4p$, so $0 < r_0 < 4p$. If $r_0 \equiv 1 \pmod 4$, take $r = r_0$; if $r_0 \equiv 3 \pmod 4$ take $r = 4p - r_0$. The uniqueness of r is readily established.

If $q = 4kp + r$, then $q \equiv r \equiv 1 \pmod 4$ and again $\left(\frac{p}{q}\right) = \left(\frac{q}{p}\right) = \left(\frac{r}{p}\right)$. If $q = 4kp - r$ then $q \equiv -r = 3 \pmod 4$ and by Theorems 3.4, 3.1*c*, and 3.1*d* we have

$$\left(\frac{p}{q}\right) = -\left(\frac{q}{p}\right) = -\left(\frac{-r}{p}\right) = -\left(\frac{-1}{p}\right)\left(\frac{r}{p}\right) = \left(\frac{r}{p}\right).$$

For example, suppose we want to determine all odd primes q that have 11 as a quadratic residue. A complete set of quadratic residues r of 11 satisfying $0 < r < 44$ and $r \equiv 1 \pmod 4$ is 1, 5, 9, 25, 37. Hence by Theorem 3.5 the odd primes q having 11 as a quadratic residue are precisely those primes of the form $44k \pm r$ where $r = 1, 5, 9, 25$, or 37.

PROBLEMS

1. In the example preceding Theorem 3.5, why was it unnecessary to test any primes $p > \sqrt{9997}$?
2. Prove that if p and q are distinct primes of the form $4k + 3$, and if $x^2 \equiv p \pmod q$ has no solutions, then $x^2 \equiv q \pmod p$ has two solutions.
3. Prove that if a prime p is a quadratic residue of an odd prime q, and p is of the form $4k + 1$, then q is a quadratic residue of p.
4. Which of the following congruences are solvable?
(*a*) $x^2 \equiv 5 \pmod{227}$ (*b*) $x^2 \equiv 5 \pmod{229}$
(*c*) $x^2 \equiv -5 \pmod{227}$ (*d*) $x^2 \equiv -5 \pmod{229}$
(*e*) $x^2 \equiv 7 \pmod{1009}$ (*f*) $x^2 \equiv -7 \pmod{1009}$
(Note that 227, 229, and 1009 are primes.)
5. Find the values of $\left(\frac{p}{q}\right)$ in the nine cases obtained from all combinations of $p = 7, 11, 13$ and $q = 227, 229, 1009$.
6. Decide whether $x^2 \equiv 150 \pmod{1009}$ is solvable or not.
7. Find all primes p such that $x^2 \equiv 13 \pmod p$ has a solution.
8. Find all primes p such that $\left(\frac{10}{p}\right) = 1$.

9. Find all primes p such that $\left(\frac{5}{p}\right) = -1$.
10. Of which primes is -2 a quadratic residue?
11. If a is a quadratic nonresidue of each of the odd primes p and q, is $x^2 \equiv a \pmod{pq}$ solvable?
12. In the proof of Theorem 3.4 consider the pairs (x, y) as points in a plane. Let O, A, B, C denote the points $(0, 0)$, $(p/2, 0)(p/2, q/2)$, $(0, q/2)$, respectively, and draw the lines OA, OB, OC, AB, and BC. Repeat the proof of Theorem 3.4 using geometrical language—pairs of points, etc.
13. Prove that there are infinitely many primes of each of the forms $3n + 1$ and $3n - 1$. *Suggestion:* first determine the primes p such that $\left(\frac{-3}{p}\right) = 1$.

3.3 The Jacobi Symbol

Definition 3.4 *Let* $(P, Q) = 1$, $Q > 0$, *Q odd, so that* $Q = q_1 q_2 \cdots q_s$ *where the q_i are odd primes, not necessarily distinct. Then the Jacobi symbol* $\left(\frac{P}{Q}\right)$ *is defined by*

$$\left(\frac{P}{Q}\right) = \prod_{j=1}^{s} \left(\frac{P}{q_j}\right)$$

where $\left(\frac{P}{q_j}\right)$ *is the Legendre symbol.*

If Q is an odd prime, the Jacobi symbol and Legendre symbol are indistinguishable. However, this can cause no confusion since their values are the same in this case. Clearly $\left(\frac{P}{Q}\right) = \pm 1$ but *it is not true* that $\left(\frac{P}{Q}\right) = 1$ implies that P is a quadratic residue modulo Q. For example, $\left(\frac{2}{9}\right) = 1$ but $x^2 \equiv 2 \pmod 9$ has no solution. A number a is a quadratic residue modulo Q only if $(a, Q) = 1$ and a is a quadratic residue modulo all primes p that divide Q. If $\left(\frac{a}{Q}\right) = -1$, then a is not a quadratic residue.

Theorem 3.6 *Suppose that Q and Q' are odd and positive and that* $(PP', QQ') = 1$. *Then*

(*a*) $\left(\frac{P}{Q}\right)\left(\frac{P}{Q'}\right) = \left(\frac{P}{QQ'}\right)$,

(*b*) $\left(\frac{P}{Q}\right)\left(\frac{P'}{Q}\right) = \left(\frac{PP'}{Q}\right)$,

(c) $\left(\frac{P^2}{Q}\right) = \left(\frac{P}{Q^2}\right) = 1,$

(d) $\left(\frac{P'P^2}{Q'Q^2}\right) = \left(\frac{P'}{Q'}\right),$

(e) $P' \equiv P \pmod{Q}$ *implies* $\left(\frac{P'}{Q}\right) = \left(\frac{P}{Q}\right).$

Proof. First (*a*) is obvious from the definition of $\left(\frac{P}{Q}\right)$ and (*b*) follows from the definition and Theorem 3.1*b*. Then (*c*) follows from (*b*) and (*a*) and so also does (*d*). To prove (*e*) we write $Q = q_1q_2 \cdots q_s$. Then $P' \equiv P \pmod{q_j}$ so that $\left(\frac{P'}{q_j}\right) = \left(\frac{P}{q_j}\right)$ by Theorem 3.1*c*, and then we have (*e*) from Definition 3.4.

Theorem 3.7 *If Q is odd and $Q > 0$, then*

$$\left(\frac{-1}{Q}\right) = (-1)^{(Q-1)/2} \quad \text{and} \quad \left(\frac{2}{Q}\right) = (-1)^{(Q^2-1)/8}$$

Proof. We have

$$\left(\frac{-1}{Q}\right) = \prod_{j=1}^{s} \left(\frac{-1}{q_j}\right) = \prod_{j=1}^{s} (-1)^{(q_j-1)/2} = (-1)^{\sum_{j=1}^{s} (q_j-1)/2}.$$

If a and b are odd, then

$$\frac{ab-1}{2} - \left(\frac{a-1}{2} + \frac{b-1}{2}\right) = \frac{(a-1)(b-1)}{2} \equiv 0 \pmod{2}$$

and hence

$$\frac{a-1}{2} + \frac{b-1}{2} \equiv \frac{ab-1}{2} \pmod{2}.$$

Applying this repeatedly we obtain

$$(3\cdot1) \qquad \sum_{j=1}^{s} \frac{q_j-1}{2} \equiv \frac{1}{2}\left(\prod_{j=1}^{s} q_j - 1\right) \equiv \frac{Q-1}{2} \pmod{2},$$

and thus $\left(\frac{-1}{Q}\right) = (-1)^{(Q-1)/2}$.

Similarly, if a and b are odd, then

$$\frac{a^2b^2-1}{8} - \left(\frac{a^2-1}{8} + \frac{b^2-1}{8}\right) = \frac{(a^2-1)(b^2-1)}{8} \equiv 0 \pmod{8},$$

so we have

$$\frac{a^2-1}{8}+\frac{b^2-1}{8}\equiv\frac{a^2b^2-1}{8} \pmod 2,$$

$$\sum_{j=1}^{s}\frac{q_j^2-1}{8}\equiv\frac{Q^2-1}{8} \pmod 2,$$

and hence,

$$\left(\frac{2}{Q}\right)=\prod_{j=1}^{s}\left(\frac{2}{q_j}\right)=(-1)^{\sum_{j=1}^{s}(q_j{}^2-1)/8}=(-1)^{(Q^2-1)/8}.$$

Theorem 3.8 *If P and Q are odd and positive and if $(P, Q) = 1$, then*

$$\left(\frac{P}{Q}\right)\left(\frac{Q}{P}\right)=(-1)^{\{(P-1)/2\}\{(Q-1)/2\}}.$$

Proof. Writing $P=\prod_{i=1}^{r}p_i$ as well as $Q=\prod_{j=1}^{s}q_j$, we have

$$\left(\frac{P}{Q}\right)=\prod_{j=1}^{s}\left(\frac{P}{q_j}\right)=\prod_{j=1}^{s}\prod_{i=1}^{r}\left(\frac{p_i}{q_j}\right)=\prod_{j=1}^{s}\prod_{i=1}^{r}\left(\frac{q_j}{p_i}\right)(-1)^{\{(p_i-1)/2\}\{(q_j-1)/2\}}$$

$$=\left(\frac{Q}{P}\right)(-1)^{\sum_{j=1}^{s}\sum_{i=1}^{r}\{(p_i-1)/2\}\{(q_j-1)/2\}}$$

where we have used Theorem 3.4. But

$$\sum_{j=1}^{s}\sum_{i=1}^{r}\frac{p_i-1}{2}\frac{q_j-1}{2}=\sum_{i=1}^{r}\frac{p_i-1}{2}\sum_{j=1}^{s}\frac{q_j-1}{2}$$

and

$$\sum_{i=1}^{r}\frac{p_i-1}{2}\equiv\frac{P-1}{2},\qquad\sum_{j=1}^{s}\frac{q_j-1}{2}\equiv\frac{Q-1}{2} \pmod 2$$

as in (3.1) in the proof of Theorem 3.7. Therefore we have

$$\left(\frac{P}{Q}\right)=\left(\frac{Q}{P}\right)(-1)^{\{(P-1)/2\}\{(Q-1)/2\}}$$

which proves the theorem.

The theorem we have just proved shows that the Jacobi symbol obeys the law of reciprocity. It is worthwhile to consider what has been done. In this chapter we have been interested in quadratic residues. The definition of the Legendre symbol is a natural one to make. We then proved the useful and celebrated law of reciprocity for this symbol. The Jacobi symbol is an extension of the Legendre symbol, defining $\left(\frac{P}{Q}\right)$ for composite Q. However, at first it might have seemed more natural to define $\left(\frac{P}{Q}\right)$ to be 1 for quadratic residues P and -1 for nonresidues modulo Q. Had this been done, there would have been no reciprocity law ($P = 5$, $Q = 9$ is an example). What we have done is this: we have dropped the connection with quadratic residues in favor of the law of reciprocity. This does not mean that the Jacobi symbol cannot be used in computations like those in Section 3.2. In fact, the Jacobi symbol plays an important role in such calculations. In Section 3.2 we used the reciprocity law to invert the symbol $\left(\frac{p}{q}\right)$ to $\left(\frac{q}{p}\right)$, but we could do it only if q was a prime. In order to compute $\left(\frac{a}{p}\right)$ we had to factor a and consider a product of Legendre symbols. Now however, using Jacobi symbols we do not need to factor a if it is odd and positive. We compute $\left(\frac{a}{p}\right)$ as a Jacobi symbol and then know the quadratic character of a modulo p if p is a prime.

For example:

$$\left(\frac{105}{317}\right) = \left(\frac{317}{105}\right) = \left(\frac{2}{105}\right) = 1,$$

and hence 105 is a quadratic residue modulo the prime number 317.

PROBLEMS

1. Evaluate: $\left(\frac{-23}{83}\right)$; $\left(\frac{51}{71}\right)$; $\left(\frac{71}{73}\right)$; $\left(\frac{-35}{97}\right)$.

2. Which of the following congruences are solvable:
(*a*) $x^2 \equiv 10 \pmod{127}$;
(*b*) $x^2 \equiv 73 \pmod{173}$;
(*c*) $x^2 \equiv 137 \pmod{401}$?

3. Which of the following congruences are solvable:
(*a*) $x^2 \equiv 11 \pmod{61}$; (*b*) $x^2 \equiv 42 \pmod{97}$;
(*c*) $x^2 \equiv -43 \pmod{79}$; (*d*) $x^2 - 31 \equiv 0 \pmod{103}$?

4. Show that if p and q are odd primes one of which is of the form $4k + 1$, then $\left(\frac{p}{q}\right) = \left(\frac{q}{p}\right)$.

5. Prove that $\sum_{j=1}^{p-1} \left(\frac{j}{p}\right) = 0$, p an odd prime.

6. If p is an odd prime and $(a, p) = 1$, prove that $ax^2 + bx + c \equiv 0 \pmod{p}$ has two, one, or no solutions according as $b^2 - 4ac$ is a quadratic residue, is congruent to zero, or is a quadratic nonresidue modulo p.

7. Use Wilson's theorem to prove that if p is a prime of the form $4n + 3$, then

$$1 \cdot 2 \cdot 3 \cdots \frac{p-1}{2} \equiv (-1)^m \pmod{p}$$

where m is the number of quadratic nonresidues among the factors on the left side.

8. (*a*) Let p be an odd prime with a quadratic residue a. Prove that $x^2 \equiv a \pmod{p^2}$ has exactly two solutions, by writing $x = x_1 + py$, where x_1 is a solution of $x^2 \equiv a \pmod{p}$.

(*b*) Generalize by the use of mathematical induction, and establish that $x^2 \equiv a \pmod{p^k}$ has exactly two solutions.

9. Let the prime divisors of the odd integer m be $p_1, p_2, \cdots, p_n$, and let $(a, m) = 1$. Prove that $x^2 \equiv a \pmod{m}$ has a solution if and only if $\left(\frac{a}{p_i}\right) = 1$, for $i = 1, 2, \cdots, n$.

10. For which primes p do there exist integers x and y with $(x, p) = 1$, $(y, p) = 1$, such that $x^2 + y^2 \equiv 0 \pmod{p}$?

11. For which prime powers p^a do there exist integers x and y with $(x, p) = 1$, $(y, p) = 1$, such that $x^2 + y^2 \equiv 0 \pmod{p^a}$?

12. For which positive integers n do there exist integers x and y with $(x, n) = 1$, $(y, n) = 1$, such that $x^2 + y^2 \equiv 0 \pmod{n}$?

13. Let the integers $1, 2, \cdots, p - 1$ modulo p, p an odd prime, be divided into two nonempty sets S_1 and S_2 so that the product of two elements in the same set is in S_1, whereas the product of an element of S_1 and an element of S_2 is in S_2. Prove that S_1 consists of the quadratic residues, S_2 of the nonresidues, modulo p. *Suggestion:* use a primitive root modulo p.

14. Let k be odd. Prove: if $a \geqq n$, then $x^2 \equiv 2^a k \pmod{2^n}$ has at least one solution. If $a < n$, then the congruence has a solution if and only if a is even and $x^2 \equiv k \pmod{2^{n-a}}$ has a solution.

15. Let k be odd. Prove that $x^2 \equiv k \pmod{2}$ has exactly one solution. Furthermore, $x^2 \equiv k \pmod{2^2}$ is solvable if and only if $k \equiv 1 \pmod{4}$, in which case there are two solutions.

16. Let k be odd, and let $n \geqq 3$. Prove that $x^2 \equiv k \pmod{2^n}$ is solvable if and only if $k \equiv 1 \pmod{8}$. *Suggestion:* use mathematical induction. Assuming that $x^2 \equiv k \pmod{2^n}$ has a solution u, show that an integer t can be found so that $(u + 2^{n-1}t)^2 \equiv k \pmod{2^{n+1}}$.

17. Assume that $n \geqq 3$ and $k \equiv 1 \pmod{8}$. Prove that any solution u of $x^2 \equiv k \pmod{2^n}$ gives rise to three other solutions, $-u$, $u + 2^{n-1}$, and $-u + 2^{n-1}$. Prove that these four solutions are incongruent modulo 2^n.

18. Prove that if u and v are any odd numbers, then one of $u - v$ and $u + v$ is of the form $4m + 2$.

19. Let $n \geqq 3$ and $k \equiv 1 \pmod{8}$. Prove that if u and v are two incongruent solutions of $x^2 \equiv k \pmod{2^n}$, then v has one of the three forms $-u$, $u + 2^{n-1}$, $-u + 2^{n-1}$ modulo 2^n. Hence the congruence has exactly four solutions. *Suggestion:* analyze $u^2 \equiv v^2 \pmod{2^n}$, and hence $(u - v)(u + v) \equiv 0 \pmod{2^n}$, in the light of the preceding problem.

20. Consider the congruence $x^2 \equiv a \pmod{p^s}$ with p a prime, $s \geqq 1$, $a = p^t b$, $(b, p) = 1$. Prove: if $t \geqq s$ the congruence is solvable. If $t < s$ the congruence is solvable if and only if t is even and $x^2 \equiv b \pmod{p^{s-t}}$ is solvable.

21. Consider the congruence $x^2 \equiv a \pmod m$. For each prime factor p of m, let p^{s_p} denote the highest power of p that divides m, and p^{t_p} the highest power that divides a, so that $s_p \geqq 1$, $t_p \geqq 0$. Write c_p for a/pt_p. Prove that the congruence is solvable if and only if

(1) for each prime factor p of m such that $t_p < s_p$, the integer t_p is even and $\left(\frac{c_p}{p}\right) = 1$;

(2) in case m is even and $t_2 < s_2$, then t_2 is even and $c_2 \equiv 1 \pmod{2^r}$ where $r = \min(3, s_2 - t_2)$.

22. Let p be any odd prime. Let $f(a)$ denote the number of solutions x, y of $x^2 - y^2 \equiv a \pmod p$, where two solutions x_1, y_1 and x_2, y_2 are counted separately unless $x_1 \equiv x_2$ and $y_1 \equiv y_2 \pmod p$. Prove that $f(a) = p - 1$ unless $p \mid a$, in which case the result is $f(a) = 2p - 1$.

For the next few problems we extend the range of meaning of the symbol $\left(\frac{a}{p}\right)$ by defining it to be 0 whenever the prime $p \mid a$.

23. Prove that

$$\sum_{m=1}^{p} \left(\frac{am + b}{p}\right) = 0,$$

assuming $a \not\equiv 0 \pmod p$. Also prove that $\left(\frac{ab}{p}\right) = \left(\frac{a}{p}\right)\left(\frac{b}{p}\right)$, and that $\left(\frac{a}{p}\right) = \left(\frac{b}{p}\right)$ if $a \equiv b \pmod p$.

24. For any odd prime p consider the sequence $\left(\frac{1}{p}\right), \left(\frac{2}{p}\right), \cdots, \left(\frac{p-1}{p}\right)$. Any consecutive pair of terms of the sequence is of one of the four types, 1, 1 or -1, -1 or 1, -1 or -1, 1. Denote the number of occurrences of each of these types by $N(1, 1)$, $N(-1, -1)$, $N(1, -1)$, and $N(-1, 1)$ respectively. Prove that

$$2N(1, 1) - 2N(-1, -1) = \sum_{x=1}^{p-2} \left\{\left(\frac{x}{p}\right) + \left(\frac{x+1}{p}\right)\right\},$$

$$2N(1, -1) - 2N(-1, 1) = \sum_{x=1}^{p-2} \left\{\left(\frac{x}{p}\right) - \left(\frac{x+1}{p}\right)\right\}.$$

25. Prove that $\sum_{x=1}^{p} \left(\frac{x}{p}\right)\left(\frac{x+1}{p}\right) = -1$ if p is any odd prime. *Suggestion:* define $s(a, p) = \sum_{x=1}^{p} \left(\frac{x}{p}\right)\left(\frac{x+a}{p}\right)$ and prove that $s(a, p) = a(1, p)$ if $p \nmid a$. Then evaluate $\sum_{a=1}^{p-1} s(a, p)$.

26. Using the notation of the two preceding problems show that $-s(1, p)$ is the excess of the number of changes of sign in the sequence $\left(\frac{1}{p}\right), \left(\frac{2}{p}\right), \cdots, \left(\frac{p-1}{p}\right)$ over the number of times that the sign does not change. Hence prove that

$$N(1, -1) + N(-1, 1) - N(1, 1) - N(-1, -1) = +1.$$

Then establish in case $p \equiv 1 \pmod 4$ that

$$N(1, 1) + 1 = N(1, -1) = N(-1, 1) = N(-1, -1) = (p - 1)/4,$$

and in case $p \equiv 3 \pmod 4$ that

$$N(1, -1) - 1 = N(1, 1) = N(-1, 1) = N(-1, -1) = (p - 3)/4.$$

27. Prove that if p is an odd prime then $\sum_{m=1}^{p} \left(\frac{m^2 - b}{p}\right) = -1$ unless $p \mid b$, in which case the sum has value $p - 1$.

NOTES ON CHAPTER 3

Theorem 3.5 is a variation of a result by Peter Hagis, *Amer. Math. Monthly*, **77**, 397 (1970).

On page 264 in Special Topics there is a discussion of the calendar problem of finding the day of the week given the date, using a simple congruence involving the greatest integer function of Definition 3.3.

4

Some Functions of Number Theory

4.1 Greatest Integer Function

The function $[x]$ was introduced in Definition 3.3. It is defined for all real x and it assumes integral values only. Many of its properties are included in the following theorem.

Theorem 4.1 *Let x and y be real numbers. Then we have*

(*a*) $[x] \leqq x < [x] + 1,\ x - 1 < [x] \leqq x,\ 0 \leqq x - [x] < 1.$

(*b*) $[x] = \sum_{1 \leqq i \leqq x} 1$ if $x \geqq 0$.

(*c*) $[x + m] = [x] + m$ *if m is an integer.*

(*d*) $[x] + [y] \leqq [x + y] \leqq [x] + [y] + 1.$

(*e*) $[x] + [-x] = \begin{cases} 0 & \text{if } x \text{ is an integer,} \\ -1 & \text{otherwise.} \end{cases}$

(*f*) $\left[\frac{[x]}{m}\right] = \left[\frac{x}{m}\right]$ *if m is a positive integer.*

(*g*) $x - [x]$ *is the fractional part of x.*

(*h*) $-[-x]$ *is the least integer $\geqq x$.*

(*i*) $[x + \frac{1}{2}]$ *is the nearest integer to x. If two integers are equally near to x, it is the larger of the two.*

(*j*) $-[-x + \frac{1}{2}]$ *is the nearest integer to* x. *If two integers are equally near to* x, *it is the smaller of the two.*

(*k*) *If* n *and* a *are positive integers,* $[n/a]$ *is the number of integers among* $1, 2, 3, \cdots, n$ *that are divisible by* a.

Proof. The first part of (*a*) is just the definition of $[x]$ in algebraic form. The two other parts are rearrangements of the first part.

In (*b*) the sum is vacuous if $x < 1$. We adopt the convention that a vacuous sum is zero. Then, for $x \geqq 0$, the sum counts the number of positive integers i that are less than or equal to x. This number is evidently just $[x]$.

Part (*c*) is obvious from the definition of $[x]$.

To prove (*d*) we write $x = n + \nu$, $y = m + \mu$, where n and m are integers and $0 \leqq \nu < 1$, $0 \leqq \mu < 1$. Then

$$\begin{aligned}[x] + [y] = n + m \leqq [n + \nu + m + \mu] &= [x + y] \\ &= n + m + [\nu + \mu] \leqq n + m + 1 = [x] + [y] + 1.\end{aligned}$$

Again writing $x = n + \nu$, we also have $-x = -n - 1 + 1 - \nu$, $0 < 1 - \nu \leqq 1$. Then

$$\begin{aligned}[x] + [-x] &= n + [-n - 1 + 1 - \nu] \\ &= n - n - 1 + [1 - \nu] = \begin{cases} 0 & \text{if } \nu = 0 \\ -1 & \text{if } \nu > 0, \end{cases}\end{aligned}$$

and we have (*e*).

To prove (*f*) we write $x = n + \nu$, $n = qm + r$, $0 \leqq \nu < 1$, $0 \leqq r \leqq m - 1$, and have

$$\left[\frac{x}{m}\right] = \left[\frac{qm + r + \nu}{m}\right] = q + \left[\frac{r + \nu}{m}\right] = q,$$

since $0 \leqq r + \nu < m$. Then (*f*) follows because

$$\left[\frac{[x]}{m}\right] = \left[\frac{n}{m}\right] = \left[q + \frac{r}{m}\right] = q.$$

Part (*g*) is nothing more than a definition of the words "fractional part of x."

Replacing x by $-x$ in (*a*) we get $-x - 1 < [-x] \leqq -x$ and hence $x \leqq -[-x] < x + 1$, which proves (*h*).

To prove (*i*) we let n be the nearest integer to x, taking the larger one if two are equally distant. Then $n = x + \theta$, $-\frac{1}{2} < \theta \leqq \frac{1}{2}$, and $[x + \frac{1}{2}] = n + [-\theta + \frac{1}{2}] = n$, since $0 \leqq -\theta + \frac{1}{2} < 1$.

The proof of (*j*) is similar to that of (*i*).

To prove part (k) we note that if $a, 2a, 3a, \cdots, ja$ are all the positive integers $\leqq n$ that are divisible by a, then we must prove that $[n/a] = j$. But we see that $(j + 1)a$ exceeds n, and so

$$ja \leqq n < (j + 1)a, \qquad j \leqq n/a < j + 1, \qquad [n/a] = j.$$

Theorem 4.2 *Let p denote a prime. Then the largest exponent e such that $p^e \mid n!$ is*

$$e = \sum_{i=1}^{\infty} \left[\frac{n}{p^i}\right].$$

Proof. If $p^i > n$, then $[n/p^i] = 0$. Therefore the sum terminates; it is not really an infinite series. The theorem is easily proved by mathematical induction. It is true for $1!$. Assume it is true for $(n - 1)!$ and let j denote the largest integer such that $p^j \mid n$. Since $n! = n \cdot (n - 1)!$, we must prove that $\sum [n/p^i] - \sum [(n - 1)/p^i] = j$. But

$$\left[\frac{n}{p^i}\right] - \left[\frac{n-1}{p^i}\right] = \begin{cases} 1 & \text{if } p^i \mid n \\ 0 & \text{if } p^i \nmid n \end{cases}$$

and hence

$$\sum \left[\frac{n}{p^i}\right] - \sum \left[\frac{n-1}{p^i}\right] = j.$$

The preceding proof is short, but it is rather artificial. A different proof can be based on a simple, but interesting, observation. If $a_1, a_2, \cdots, a_n$ are non-negative integers let $f(1)$ denote the number of them that are greater than or equal to 1, $f(2)$ the number greater than or equal to 2, etc. Then

$$a_1 + a_2 + \cdots + a_n = f(1) + f(2) + f(3) + \cdots.$$

Now, for $1 \leqq j \leqq n$, we let a_j be the largest integer such that $p^{a_j} \mid j$. Then $f(1)$ counts the number of integers $\leqq n$ that are divisible by p, $f(2)$ the number divisible by p^2, etc. Hence $f(k)$ counts the integers $p^k, 2p^k, 3p^k, \cdots, [n/p^k]p^k$, so that $f(k) = [n/p^k]$. Thus we see that

$$e = a_1 + a_2 + \cdots + a_n = \sum_{i=1}^{\infty} f(i) = \sum_{i=1}^{\infty} \left[\frac{n}{p^i}\right].$$

Formula (f) of Theorem 4.1 shortens the work of computing e in Theorem 4.2. For example, if we wish to find the highest power of 7 that divides $1000!$ we compute

$$[1000/7] = 142, \qquad [142/7] = 20, \qquad [20/7] = 2, \qquad [2/7] = 0.$$

Adding we find that $7^{164} \mid 1000!$, $7^{165} \nmid 1000!$.

The applications of Theorem 4.2 are not restricted to numerical problems. As an example, let us prove that

$$\frac{n!}{a_1!a_2!\cdots a_r!}$$

is an integer if $a_i \geqq 0$, $a_1 + a_2 + \cdots + a_r = n$. To do this we merely have to show that every prime divides the numerator to at least as high a power as it divides the denominator. Using Theorem 4.2 we need only prove

$$\sum \left[\frac{n}{p^i}\right] \geqq \sum \left[\frac{a_1}{p^i}\right] + \sum \left[\frac{a_2}{p^i}\right] + \cdots + \sum \left[\frac{a_r}{p^i}\right].$$

But repeated use of Theorem 4.1*d* gives us

$$\left[\frac{a_1}{p^i}\right] + \left[\frac{a_2}{p^i}\right] + \cdots + \left[\frac{a_r}{p^i}\right] \leqq \left[\frac{a_1 + a_2 + \cdots + a_r}{p^i}\right] = \left[\frac{n}{p^i}\right].$$

Summing this over i we have our desired result.

A special case of what has just been proved is that *the product of any k consecutive integers is divisible by $k!$*. To see this we first observe that if one of the k consecutive integers is 0, the result is immediate because 0 is divisible by any nonzero integer. Thus we need consider only k consecutive positive integers, or k consecutive negative integers, and the latter case can be subsumed under the former. Taking the largest of the k consecutive positive integers as n, we look at the product

$$n(n-1)(n-2)\cdots(n-k+1) \quad \text{or} \quad \{n!\}/(n-k)!.$$

Taking $r = 2$, $a_1 = k$, $a_2 = n - k$ in what was proved above we see that $\{n!\}/\{a_1!a_2!\}$ or $n!/\{k!(n-k)!\}$ is an integer, and so $k!$ is a divisor of $n!/(n-k)!$.

Another way to draw this conclusion is to use the fact that $n!/\{k!(n-k)!\}$ is a binomial coefficient, specifically the coefficient of $x^k y^{n-k}$ in the expansion of $(x + y)^n$, and all such coefficients are integers.

A slightly more complicated example is to prove that

$$\frac{(ab)!}{a!(b!)^a}$$

is an integer. We must show that

$$\sum \left[\frac{ab}{p^i}\right] - \sum \left[\frac{a}{p^i}\right] - a \sum \left[\frac{b}{p^i}\right] \geqq 0$$

for every prime p. Let r and s denote the integers such that $p^r \leqq a < p^{r+1}$ and $p^s \leqq b < p^{s+1}$. Then

$$\sum\left[\frac{ab}{p^i}\right] - \sum\left[\frac{a}{p^i}\right] - a\sum\left[\frac{b}{p^i}\right]$$

$$= \sum_{i=1}^{s}\left[\frac{ab}{p^i}\right] + \sum_{i=s+1}^{r+s}\left[\frac{ab}{p^i}\right] + \sum_{i=r+s+1}^{\infty}\left[\frac{ab}{p^i}\right] - \sum_{i=1}^{r}\left[\frac{a}{p^i}\right] - \sum_{i=1}^{s} a\left[\frac{b}{p^i}\right]$$

$$= \sum_{i=1}^{s}\left(\left[\frac{ab}{p^i}\right] - a\left[\frac{b}{p^i}\right]\right) + \sum_{i=1}^{r}\left(\left[\frac{ab}{p^{s+i}}\right] - \left[\frac{a}{p^i}\right]\right) + \sum_{i=r+s+1}^{\infty}\left[\frac{ab}{p^i}\right]$$

$$\geqq \sum_{i=1}^{s}\left(\left[\frac{ab}{p^i}\right] - a\left[\frac{b}{p^i}\right]\right) + \sum_{i=1}^{r}\left(\left[\frac{ap^s}{p^{s+i}}\right] - \left[\frac{a}{p^i}\right]\right)$$

$$= \sum_{i=1}^{s}\left(\left[\frac{ab}{p^i}\right] - a\left[\frac{b}{p^i}\right]\right) \geqq 0$$

since $[ab/p^i] \geqq a[b/p^i]$ by repeated use of Theorem 4.1*d*.

PROBLEMS

1. What is the highest power of 2 dividing 533!? The highest power of 3? The highest power of 6? The highest power of 12? The highest power of 70?

2. If 100! were written out in the ordinary decimal notation without the factorial sign, how many zeros would there be in a row at the right end?

3. For what real numbers x is it true that

(*a*) $[x] + [x] = [2x]$;

(*b*) $[x + 3] = 3 + [x]$;

(*c*) $[x + 3] = 3 + x$;

(*d*) $[x + \frac{1}{2}] + [x - \frac{1}{2}] = [2x]$;

(*e*) $[9x] = 9$?

4. Given that $[x + y] = [x] + [y]$ and $[-x - y] = [-x] + [-y]$, prove that x or y is an integer.

5. Find formulas for the highest exponent e of the prime p such that p^e divides (*a*) the product $2 \cdot 4 \cdot 6 \cdots (2n)$ of the first n even numbers; (*b*) the product of the first n odd numbers.

6. For any real number x prove that $[x] + [x + \frac{1}{2}] = [2x]$.

7. For any positive real numbers x and y prove that $[x] \cdot [y] \leqq [xy]$.

8. For any positive real numbers x and y prove that

$$[x - y] \leqq [x] - [y] \leqq [x - y] + 1.$$

9. Prove that $(2n)!/(n!)^2$ is even if n is a positive integer.

10. Let m be any real number not zero or a positive integer. Prove that an x exists so that the equation of Theorem 4.1*f* is false.

11. If p and q are distinct primes, prove that the divisors of p^2q^3 coincide with the terms of $(1 + p + q^2)(1 + q + q^2 + q^3)$ when the latter is multiplied out.
12. For any integers a and $m \geqq 2$, prove that $a - m[a/m]$ is the least non-negative residue of a modulo m. Write a similar expression for the least positive residue of a modulo m.
13 If a and b are positive integers such that $(a, b) = 1$, and ρ is a real number such that $a\rho$ and $b\rho$ are integers, prove that ρ is an integer. Hence prove that $\rho = n!/(a!b!)$ is an integer if $(a, b) = 1$ and $a + b = n + 1$. Generalize this to prove that

$$\rho = \frac{n!}{a_1!a_2! \cdots a_r!}$$

is an integer if $(a_1, a_2, \cdots, a_r) = 1$ and $a_1 + a_2 + \cdots + a_r = n + 1$.
14. Consider an integer $n \geqq 1$ and the integers i, $1 \leqq i \leqq n$. For each $k = 0, 1, 2, \cdots$ find the number of i's that are divisible by 2^k but not by 2^{k+1}. Thus prove

$$\sum_{j=1}^{\infty} \left[\frac{n}{2^j} + \frac{1}{2}\right] = n,$$

and hence that we get the correct value for the sum $n/2 + n/4 + n/8 + \cdots$ if we replace each term by its nearest integer, using the larger one if two exist.
15. If n is any positive integer and ξ any real number, prove that

$$[\xi] + \left[\xi + \frac{1}{n}\right] + \cdots + \left[\xi + \frac{n-1}{n}\right] = [n\xi].$$

16. Prove that $[2\alpha] + [2\beta] \geqq [\alpha] + [\beta] + [\alpha + \beta]$ holds for every pair of real numbers, but that $[3\alpha] + [3\beta] \geqq [\alpha] + [\beta] + [2\alpha + 2\beta]$ does not.
17. For every positive integer n, prove that $n!(n - 1)!$ is a divisor of $(2n - 2)!$.
18. If $(m, n) = 1$, prove that

$$\sum_{x=1}^{n-1} \left[\frac{mx}{n}\right] = \frac{(m-1)(n-1)}{2}.$$

19. If $m \geqq 1$, prove that $[(1 + \sqrt{3})^{2m+1}]$ is divisible by 2^{m+1} but not by 2^{m+2}.
20. Let θ be real, and $0 < \theta < 1$. Define

$$g_n = \begin{cases} 0 \text{ if } [n\theta] = [(n-1)\theta], \\ 1 \text{ otherwise.} \end{cases}$$

Prove that
$$\lim_{n\to\infty} \frac{g_1 + g_2 + \cdots + g_n}{n} = \theta.$$

21. Let n be an odd integer >5. If n factors into the product of two integers, $n = uv$, with $u > v$ and $u - v \leqq \sqrt[4]{64n}$, prove that the roots of $x^2 - 2[\sqrt{n} + 1]x + n = 0$ are integers. *Suggestion:* use the identity $\{(u + v)/2\}^2 - \{(u - v)/2\}^2 = uv$ to get bounds on the integer $(u + v)/2$.

22. Let α be a positive irrational number. Prove that the two sequences,

$[1 + \alpha], [2 + 2\alpha], \cdots, [n + n\alpha], \cdots$, and

$$[1 + \alpha^{-1}], [2 + 2\alpha^{-1}], \cdots, [n + n\alpha^{-1}], \cdots,$$

together contain every positive integer exactly once. Prove that this is false if α is rational.

23. Let S be the set of integers given by $[\alpha x]$ and $[\beta x]$ for $x = 1, 2, \cdots$. Prove that S consists of every positive integer, each appearing exactly once, if and only if α and β are positive irrational numbers such that $\frac{1}{\alpha} + \frac{1}{\beta} = 1$.

24. For positive real numbers α, β, γ define $f(\alpha, \beta, \gamma)$ as the sum of all positive terms of the series

$$\left[\frac{\gamma - \alpha}{\beta}\right] + \left[\frac{\gamma - 2\alpha}{\beta}\right] + \left[\frac{\gamma - 3\alpha}{\beta}\right] + \left[\frac{\gamma - 4\alpha}{\beta}\right] + \cdots.$$

(If there are no positive terms, define $f(\alpha, \beta, \gamma) = 0$.) Prove that $f(\alpha, \beta, \gamma) = f(\beta, \alpha, \gamma)$. *Suggestion:* $f(\alpha, \beta, \gamma)$ is related to the number of solutions of $\alpha x + \beta y \leqq \gamma$ in positive integer pairs x, y.

25. Prove that if p is a prime and $0 \leqq n \leqq p^k$ then

$$\binom{p^k}{n} \equiv \begin{cases} 1 \pmod p \text{ if } n = 0 \text{ or } p^k, \\ 0 \pmod p \text{ if } 1 \leqq n \leqq p^k - 1, \end{cases}$$

where $\binom{p^k}{n}$ is the binomial coefficient.

26. Prove that if p is a prime and $0 \leqq n \leqq p^k$ then

$$\binom{p^k}{n} \equiv \begin{cases} 1 \pmod{p^2} \text{ if } n = 0 \text{ or } p^k, \\ pa'(-1)^{a-1} \pmod{p^2} \text{ if } n = ap^{k-1}, 1 \leqq a \leqq p - 1, \\ \qquad aa' \equiv 1 \pmod p, \\ 0 \pmod{p^2} \text{ otherwise.} \end{cases}$$

27. Prove that if p is a prime and $m = \sum_{j=0}^{r} a_j p^j$, $n = \sum_{j=0}^{r} b_j p^j$, $0 \leqq a_j \leqq p - 1$, $0 \leqq b_j \leqq p - 1$, then

$$\binom{m}{n} \equiv \binom{a_0}{b_0}\binom{a_1}{b_1}\binom{a_2}{b_2} \cdots \binom{a_r}{b_r} \pmod p.$$

Suggestion: consider $(1 + x)^m \pmod p$.

4.2 Arithmetic Functions

Functions such as $\phi(n)$ of Theorem 2.5 that are defined for all positive integers n are called *arithmetic functions*, or *number theoretic functions*, or *numerical functions*. Specifically, an *arithmetic function* f is one whose domain is the positive integers and whose range is a subset of the complex numbers.

Definition 4.1 *For positive integers n we make the following definitions.*

$\tau(n)$ is the number of positive divisors of n.

$\sigma(n)$ is the sum of the positive divisors of n.

$\sigma_k(n)$ is the sum of the kth *powers of the positive divisors of n.*

For example, $\tau(6) = 4$, $\sigma(6) = 12$, $\sigma_2(6) = 50$. These are all arithmetic functions. The value of k can be any real number, positive, negative, or zero. The functions $\tau(n)$ and $\sigma(n)$ are merely special cases of $\sigma_k(n)$, because $\tau(n) = \sigma_0(n)$, $\sigma(n) = \sigma_1(n)$. It is convenient to use the symbols $\sum_{d|n} f(d)$ and $\prod_{d|n} f(d)$ for the sum and product of $f(d)$ over all positive divisors d of n. Thus we write

$$\tau(n) = \sum_{d|n} 1, \qquad \sigma(n) = \sum_{d|n} d, \qquad \sigma_k(n) = \sum_{d|n} d^k.$$

Theorem 4.3 *If $n = p_1^{e_1} p_2^{e_2} \cdots p_r^{e_r}$ then*

$$\tau(n) = (e_1 + 1)(e_2 + 1) \cdots (e_r + 1). \quad \textit{Also } \tau(1) = 1.$$

Proof. A positive integer d divides n if, and only if, $d = p_1^{f_1} p_2^{f_2} \cdots p_r^{f_r}$ with $0 \leqq f_i \leqq e_i$ for $i = 1, 2, \cdots, r$. Thus there are just $(e_1 + 1)(e_2 + 1) \cdots (e_r + 1)$ such d.

If $(m, n) = 1$ it follows from Theorem 4.3 that $\tau(mn) = \tau(m)\tau(n)$.

Definition 4.2 *If $f(n)$ is an arithmetic function not identically zero such that $f(mn) = f(m)f(n)$ for every pair of positive integers m, n satisfying $(m, n) = 1$, then $f(n)$ is said to be multiplicative. If $f(mn) = f(m)f(n)$ whether m and n are relatively prime or not, $f(n)$ is said to be totally multiplicative or completely multiplicative.*

If f is a multiplicative function, $f(n) = f(n)f(1)$ for every positive integer n, and since $f(n)$ is not zero for all values of n, we see that $f(1) = 1$.

From the definition of a multiplicative function f it follows by mathematical induction that if $m_1, m_2, \cdots, m_r$ are positive integers which are relative prime in pairs, then

$$f(m_1 m_2 \cdots m_r) = f(m_1)f(m_2) \cdots f(m_r).$$

In particular, this result would hold if the integers $m_1, m_2, \cdots m_r$ are prime powers of distinct primes. Since every positive integer >1 can be factored into a product of prime powers of distinct primes, it follows that if f is a multiplicative function and we know the value of $f(p^\alpha)$ for every prime p and every positive integer α, then the value of $f(n)$ for every positive integer n can be readily determined by multiplication. For example $f(3600) = f(2^4)f(3^2)f(5^2)$. Similarly, if g is a totally multiplicative function and we know the value of $g(p)$ for every prime p, then the value of $g(n)$ for every positive integer n can be readily determined. For example $g(3600) = \{g(2)\}^4 \{g(3)\}^2 \{g(5)\}^2$.

These basic properties can be stated in another way. First, if f and g are multiplicative functions such that $f(p^j) = g(p^j)$ for all primes p and all positive integers j, then $f(n) = g(n)$ for all positive integers n, so that $f = g$. Second, if f and g are totally multiplicative functions such that $f(p) = g(p)$ for all primes p, then $f = g$.

Theorem 4.4 *Let $f(n)$ be a multiplicative function and let $F(n) = \sum_{d|n} f(d)$. Then $F(n)$ is multiplicative.*

Proof. Suppose $(m, n) = 1$. If $m = 1$ or $n = 1$ then $F(m) = 1$ or $F(n) = 1$, so that $F(mn) = F(m)F(n)$ is clear in these cases. Otherwise m and n have canonical representations

$$n = p_1^{\alpha_1} p_2^{\alpha_2} \cdots p_r^{\alpha_r}, \qquad m = q_1^{\beta_1} q_2^{\beta_2} \cdots q_s^{\beta_s},$$

with positive exponents α_i and β_i and where $p_1, p_2, \cdots, p_r, q_1, q_2, \cdots, q_s$ are distinct primes. The positive divisors d_1 of n are just the numbers $d_1 = p_1^{\gamma_1} p_2^{\gamma_2} \cdots p_r^{\gamma_r}$ for all possible choices of the γ's satisfying $0 \leqq \gamma_i \leqq \alpha_i$. Similarly the positive divisors d_2 of m are given by $d_2 = q_1^{\delta_1} q_2^{\delta_2} \cdots q_s^{\delta_s}$, $0 \leqq \delta_i \leqq \beta_i$. Therefore as d_1 runs through all positive divisors of n, and d_2 runs through all positive divisors of m, their product $d_1 d_2$ runs through the values $d = d_1 d_2 = p_1^{\gamma_1} p_2^{\gamma_2} \cdots p_r^{\gamma_r} q_1^{\delta_1} q_2^{\delta_2} \cdots q_s^{\delta_s}$, $0 \leqq \gamma_i \leqq \alpha_i$, $0 \leqq \delta_i \leqq \beta_i$; but these are just all the positive divisors of $nm = p_1^{\alpha_1} p_2^{\alpha_2} \cdots p_r^{\alpha_r} q_1^{\beta_1} q_2^{\beta_2} \cdots q_s^{\beta_s}$. In other words

$$\sum_{d_1|n} \sum_{d_2|m} f(d_1 d_2) = \sum_{d|nm} f(d).$$

It is clear that $(d_1, d_2) = 1$, and hence we have

$$F(nm) = \sum_{d|nm} f(d) = \sum_{d_1|n} \sum_{d_2|m} f(d_1 d_2) = \sum_{d_1|n} \sum_{d_2|m} f(d_1) f(d_2)$$

$$= \sum_{d_1|n} f(d_1) \sum_{d_2|m} f(d_2) = F(n)F(m).$$

We could have used this theorem and Definition 4.1 to prove that $\tau(n)$ is multiplicative. Since $\tau(n) = \sum_{d|n} 1$ is of the form $\sum_{d|n} f(d)$, and since the function $f(n) = 1$ is multiplicative, Theorem 4.4 can be used, and we see that $\tau(n)$ is multiplicative. Then Theorem 4.3 would have been easy to prove. If p_i is a prime, then $\tau(p_i^{e_i}) = e_i + 1$, since $p_i^{e_i}$ has the $e_i + 1$ positive divisors $1, p_i, p_i^2, \cdots, p_i^{e_i}$ and no more. Then, since $\tau(n)$ is multiplicative, we have

$$\tau(p_1^{e_1} p_2^{e_2} \cdots p_r^{e_r}) = \tau(p_1^{e_1})\tau(p_2^{e_2}) \cdots \tau(p_r^{e_r}) = (e_1 + 1)(e_2 + 1) \cdots (e_r + 1).$$

This exemplifies a useful method for handling certain arithmetic functions. We will use it to find a formula for $\sigma(n)$ in the following theorem. However, it should be pointed out that $\sigma(n)$ can also be found quite simply in the same manner as we first obtained the formula for $\tau(n)$.

Theorem 4.5 *If* $n = p_1^{e_1}p_2^{e_2}\cdots, p_r^{e_r}$, *then*

$$\sigma(n) = \prod_{i=1}^{r} \frac{p_i^{e_i+1} - 1}{p_i - 1}, \qquad \sigma(1) = 1.$$

Proof. By definition $\sigma(n) = \sum_{d|n} d$, so we can apply Theorem 4.4 with $f(n) = n$, $F(n) = \sigma(n)$. Thus $\sigma(n)$ is multiplicative and $\sigma(n) = \prod_{i=1}^{r} \sigma(p_i^{e_i})$. But the positive divisors of $p_i^{e_i}$ are just $1, p_i, p_i^2, \cdots, p_i^{e_i}$ whose sum is $(p_i^{e_i+1} - 1)/(p_i - 1)$.

PROBLEMS

1. Find the smallest integer x for which $\phi(x) = 6$.
2. Find the smallest integer x for which $\tau(x) = 6$.
3. Find the smallest positive integer n so that $\sigma(x) = n$ has no solutions; exactly one solution; exactly two solutions; exactly three solutions.
4. Find the smallest positive integer m for which there is another positive integer $n \neq m$ such that $\sigma(m) = \sigma(n)$.
5. Prove that $\prod_{d|n} d = n^{\tau(n)/2}$.
6. Prove that $\sum_{d|n} d = \sum_{d|n} n/d$, and more generally than $\sum_{d|n} f(d) = \sum_{d|n} f(n/d)$.
7. Prove that $\sigma_{-k}(n) = n^{-k}\sigma_k(n)$.
8. Find a formula for $\sigma_k(n)$.
9. If $f(n)$ and $g(n)$ are multiplicative functions, and $g(n) \neq 0$ for every n, show that the functions $F(n) = f(n)g(n)$ and $G(n) = f(n)/g(n)$ are also multiplicative.
10. Give an example to show that if $f(n)$ is totally multiplicative, $F(n)$ need not also be totally multiplicative, where $F(n)$ is defined as $\sum_{d|n} f(d)$.
11. Prove that the number of positive irreducible fractions $\leqq 1$ with denominator $\leqq n$ is $\phi(1) + \phi(2) + \phi(3) + \cdots + \phi(n)$.
12. Prove that the number of divisors of n is odd if and only if n is a perfect square. If the integer $k \geqq 1$, prove that $\sigma_k(n)$ is odd if and only if n is a square or double a square.
13. Given any positive integer $n > 1$, prove that there are infinitely many integers x satisfying $\tau(x) = n$.
14. Given any positive integer n, prove that there is only a finite number of integers x satisfying $\sigma(x) = n$.
15. Prove that if $(a, b) > 1$ then $\sigma_k(ab) < \sigma_k(a)\sigma_k(b)$ and $\tau(ab) < \tau(a)\tau(b)$.
16. We say (following Euclid) that m is a perfect number if $\sigma(m) = 2m$, that is, if m is the sum of all its positive divisors other than itself. If $2^n - 1$ is a

prime p, prove that $2^{n-1}p$ is a perfect number. Use this result to find three perfect numbers.

17. Prove that an integer q is a prime if and only if $\sigma(q) = q + 1$.

18. Show that if $\sigma(q) = q + k$ where $k \mid q$ and $k < q$, then $k = 1$.

19. Prove that every even perfect number has the form given in Problem 16. *Suggestion:* assume that $2^{n-1}q$ is a perfect number, where $n > 1$ and q is odd. Write $\sigma(q) = q + k$ and so deduce from $\sigma(2^{n-1}q) = 2^n q$ that $q = k(2^n - 1)$. Thus $k \mid q$ and $k < q$.

20. For any integer $n \geqq 2$ define $\nu(n)$ as $(-1)^j$, where j is the total number of prime factors of n. For example, if $n = 16$, then $j = 4$; if $n = 72$, $j = 5$. Also define $\nu(1) = 1$. Prove that $\nu(n)$ is a totally multiplicative function and that

$$\sum_{d \mid n} \nu(d) = \begin{cases} 1 \text{ if } n \text{ is a perfect square,} \\ 0 \text{ otherwise.} \end{cases}$$

21. If $d \mid n$ and $\delta \mid (n/d)$, then $d \mid (n/\delta)$. Prove that the set of ordered pairs (d, δ) where d ranges over all positive divisors of a fixed integer n and, for each value of d, δ ranges over all positive divisors of n/d, is a symmetric set in the sense that if (a, b) is in the set, so is (b, a).

22. Prove that the set of pairs in the preceding problem is the same as the set of pairs (d, δ) over all positive d and δ such that $d\delta \mid n$.

23. Consider the set of ordered pairs (d, γ) where d ranges over all positive divisors of a fixed integer n, and for each such d, γ ranges over all positive divisors of d. Prove that this is the same as the set of ordered pairs $(\beta\gamma, \gamma)$ where γ ranges over all divisors of n and, for each such γ, β runs through the positive divisors of n/γ.

4.3 The Moebius Inversion Formula

Definition 4.3 *The Moebius function $\mu(n)$ is defined by*

$$\mu(n) = \begin{cases} 1 & \text{if } n = 1 \\ 0 & \text{if } a^2 \mid n \text{ for some } a > 1 \\ (-1)^r & \text{if } n = p_1 p_2 \cdots p_r,\ p_i \text{ distinct primes.} \end{cases}$$

Theorem 4.6 *The function $\mu(n)$ is multiplicative and*

$$\sum_{d \mid n} \mu(d) = \begin{cases} 1 & \text{if } \ n = 1 \\ 0 & \text{if } \ n > 1. \end{cases}$$

Proof. It is clear from the definition that $\mu(n)$ is multiplicative. If $F(n) = \sum_{d \mid n} \mu(d)$, then $F(n)$ is multiplicative by Theorem 4.4. Since $F(1) = \mu(1) = 1$ and $F(p^e) = \sum_{f=0}^{e} \mu(p^f) = 1 + (-1) = 0$, we have the desired result.

Theorem 4.7 *Moebius inversion formula. If $F(n) = \sum_{d \mid n} f(d)$ for every positive integer n, then $f(n) = \sum_{d \mid n} \mu(d) F(n/d)$.*

Proof. We have

$$\sum_{d\mid n}\mu(d)F\left(\frac{n}{d}\right)=\sum_{d\mid n}\mu(d)\sum_{\delta\mid(n/d)}f(\delta)=\sum_{\delta\mid n}\sum_{d\mid(n/\delta)}\mu(d)f(\delta)$$
$$=\sum_{\delta\mid n}f(\delta)\sum_{d\mid(n/\delta)}\mu(d)=f(n)$$

by Theorem 4.6.

Theorem 4.8 *If* $f(n)=\sum_{d\mid n}\mu(d)F(n/d)$ *for every positive integer* n, *then* $F(n)=\sum_{d\mid n}f(d)$.

Proof. Again by Theorem 4.6 we find

$$\sum_{d\mid n}f(d)=\sum_{d\mid n}\sum_{\delta\mid d}\mu(\delta)F\left(\frac{d}{\delta}\right)=\sum_{d\mid n}\sum_{\gamma\mid d}\mu\left(\frac{d}{\gamma}\right)F(\gamma)$$
$$=\sum_{\gamma\mid n}\sum_{\beta\gamma\mid n}\mu\left(\frac{\beta\gamma}{\gamma}\right)F(\gamma)=\sum_{\gamma\mid n}F(\gamma)\sum_{\beta\mid(n/\gamma)}\mu(\beta)$$
$$=F(n).$$

It should be noted that Theorem 4.7 and its converse, Theorem 4.8, do not require that $f(n)$ or $F(n)$ be multiplicative. A generalization of these two results is given in Theorem 4.14 at the end of the next section.

The last two theorems are often very useful. As an example we will obtain the results of Section 2.4 concerning Euler's ϕ-function in a different way. In Theorem 2.5 we saw that $\phi(n)$ is the number of positive integers less than or equal to n that are relatively prime to n. Let S denote the set of integers $1, 2, \cdots, n$, that is, the set of integers i satisfying $1\leqq i\leqq n$. We separate S into subsets S_d, where $d\mid n$, by putting i into S_d if $(i, n)=d$. Then each element of S is in exactly one S_d. Moreover, i is in S_d if and only if i is of the form jd with $1\leqq j\leqq n/d$ and $(j, n/d)=1$. Therefore there are exactly $\phi(n/d)$ elements in S_d. Since there are n elements in S, we have $n=\sum_{d\mid n}\phi(n/d)$, which we can write as $n=\sum_{d\mid n}\phi(d)$. This is Theorem 2.17. Then, by Theorem 4.7,

$$\phi(n)=\sum_{d\mid n}\mu(d)\frac{n}{d},\qquad \frac{\phi(n)}{n}=\sum_{d\mid n}\frac{\mu(d)}{d}.$$

Also the function $\mu(d)/d$ is multiplicative, and therefore so is $\phi(n)/n$ by Theorem 4.4. Hence $\phi(n)$ is multiplicative, and we have Theorem 2.15. Finally, using the foregoing equation with n replaced by p^e, we have

$$\phi(p^e)=\sum_{d\mid p^e}\mu(d)\frac{p^e}{d}=\sum_{f=0}^{e}\mu(p^f)\frac{p^e}{p^f}$$
$$=\mu(1)p^e+\mu(p)p^{e-1}=p^e-p^{e-1}=p^e\left(1-\frac{1}{p}\right)$$

if $e \geqq 1$, and hence

$$\phi(p_1^{e_1}p_2^{e_2}\cdots p_r^{e_r}) = p_1^{e_1}p_2^{e_2}\cdots p_r^{e_r}\left(1-\frac{1}{p_1}\right)\left(1-\frac{1}{p_2}\right)\cdots\left(1-\frac{1}{p_r}\right)$$

since $\phi(n)$ is multiplicative. This is Theorem 2.16.

PROBLEMS

1. Find a positive integer n such that $\mu(n) + \mu(n+1) + \mu(n+2) = 3$.
2. Prove that $\mu(n)\mu(n+1)\mu(n+2)\mu(n+3) = 0$ if n is a positive integer.
3. Evaluate $\sum_{j=1}^{\infty} \mu(j!)$.
4. Prove Theorem 4.8 by defining $G(n)$ as $\sum_{d|n} f(d)$, then applying Theorem 4.7 to write $f(n) = \sum_{d|n} \mu(d)G(n/d)$. Thus $\sum_{d|n} \mu(d)G(n/d) = \sum_{d|n} \mu(d)F(n/d)$. Use this to show that $F(1) = G(1)$, $F(2) = G(2)$, $F(3) = G(3)$, and so on.
5. If k denotes the number of distinct prime factors of a positive integer n, prove that $\sum_{d|n} |\mu(d)| = 2^k$.
6. If $F(n) = \sum_{d|n} f(d)$ for every positive integer n, prove that $f(n) = \sum_{d|n} \mu(n/d)F(d)$.
7. Let n have the distinct prime factors $p_1, p_2, \cdots, p_k$. Prove that

$$\sum_{d|n} \mu(d)\tau(d) = (-1)^k.$$

Similarly, evaluate $\sum_{d|n} \mu(d)\sigma(d)$.
8. If n is any even integer, prove that $\sum_{d|n} \mu(d)\phi(d) = 0$.
9. By use of the algebraic identity $(x+1)^2 - x^2 = 2x + 1$, establish that $(n+1)^2 - 1^2 = \sum_{x=1}^{n} \{(x+1)^2 - x^2\} = \sum_{x=1}^{n} (2x+1)$ and so derive the result $\sum_{x=1}^{n} x = n(n+1)/2$.
10. By use of the algebraic identity $(x+1)^3 - x^3 = 3x^2 + 3x + 1$, establish that $(n+1)^3 - 1^3 = \sum_{x=1}^{n} \{(x+1)^3 - x^3\} = \sum_{x=1}^{n} (3x^2 + 3x + 1)$, and so derive the result $\sum_{x=1}^{n} x^2 = n(n+1)(2n+1)/6$. (The results of this and the previous problem can be established by other methods, mathematical induction, for example.)
11. Let $S(n)$ denote the sum of the squares of the positive integers $\leqq n$ and prime to n. Prove that

$$\sum_{j=1}^{n} j^2 = \sum_{d|n} d^2 S\left(\frac{n}{d}\right) = \sum_{d|n} \frac{n^2}{d^2} S(d).$$

Suggestion: separate the integers $\leqq n$ into classes, so that all integers k such that $(k, n) = d$ are in the same class.
12. Combine the results of the two preceding problems to get

$$\sum_{d|n} \frac{S(d)}{d^2} = \frac{1}{6}\left(2n + 3 + \frac{1}{n}\right).$$

Then apply the Moebius inversion formula to get

$$\frac{S(n)}{n^2} = \sum_{d|n} \frac{1}{6}\,\mu(d)\left(\frac{2n}{d} + 3 + \frac{d}{n}\right).$$

13. Prove that $f(n) = n\mu(n)$ is a multiplicative function and that $\sum_{d|n} d\mu(d) = (-1)^k\phi(n)p_1p_2\cdots p_k/n$ where $p_1, p_2, \cdots, p_k$ are the distinct prime factors of n.
14. Combine the results of the two preceding problems to get $S(n) = n^2\phi(n)/3 + (-1)^k\phi(n)p_1p_2\cdots p_k/6$ for $n > 1$, where as before $p_1p_2, \cdots, p_k$ are the prime factors of n. *Suggestion:* use the formula $\sum_{d|n} \mu(d)/d = \phi(n)/n$.
15. Given any positive integer k, prove that there exist infinitely many integers n such that

$$\mu(n+1) = \mu(n+2) = \mu(n+3) = \cdots = \mu(n+k).$$

4.4 The Multiplication of Arithmetic Functions

There are two standard kinds of multiplication of arithmetic functions. The first is the straightforward one, where the product fg of two functions f and g is defined by $(fg)(n) = f(n)g(n)$. If f and g are multiplicative so is fg. We now turn to a more fruitful definition of multiplication.

If f and g are arithmetic functions, define the *Dirichlet product* (or *convolution*) $f * g$ as the arithmetic function whose value at any positive integer n is given by

$$(f * g)(n) = \sum_{d|n} f(d)g(n/d) = \sum_{d_1d_2=n} f(d_1)g(d_2),$$

where the first sum is over all positive integer divisors d of n, and the second sum is over all ordered pairs d_1, d_2 of positive integers with product n. It is clear that the two sums are equal and that multiplication is commutative, $f * g = g * f$, because the set of ordered pairs d_2, d_1 is the same as the set of ordered pairs d_1, d_2.

Furthermore, if h is another arithmetic function, the associative property $(f * g) * h = f * (g * h)$ can be established as follows. For any positive integer n a straightforward application of the definition gives

$$((f * g) * h)(n) = \sum_{d_1d_2d_3=n} f(d_1)g(d_2)h(d_3)$$

where the sum is over all ordered triples d_1, d_2, d_3 of positive integers with product n, and $(f * (g * h))(n)$ is seen to be the same thing.

In addition to the functions μ, τ, ϕ, σ already examined in this chapter we will need the functions I, U, E defined by $I(1) = U(1) = E(1) = 1$ and for all $n > 1$,

$$I(n) = 0, \qquad U(n) = 1, \qquad E(n) = n.$$

It is easy to verify that these are multiplicative functions. The function I is seen to have the property $I * f = f * I = f$ for every arithmetic function f, and so I is called a multiplicative identity for Dirichlet multiplication of arithmetic functions. Furthermore, I is *the* multiplicative identity because if there were another function I_1 such that $I_1 * f = f * I_1 = f$ for every arithmetic function f, we could write $I_1 = I * I_1 = I$.

If two arithmetic functions f and g have the property $f * g = g * f = I$, we say that f and g are multiplicative inverses of each other and we write $f = g^{-1}$ and $g = f^{-1}$.

Theorem 4.9 *An arithmetic function f has a multiplicative inverse if and only if $f(1) \neq 0$. If an inverse exists it is unique.*

Proof. If f has an inverse f^{-1} then $f * f^{-1} = I$ and $f(1)f^{-1}(1) = (f * f^{-1})(1) = I(1) = 1$, so that $f(1) \neq 0$. Conversely if $f(1) \neq 0$ we can calculate $f^{-1}(1)$ from these equations. Next we can calculate $f^{-1}(2)$ from $0 = I(2) = (f * f^{-1})(2) = f(2)f^{-1}(1) + f(1)f^{-1}(2)$. Continuing by mathematical induction we see that if $f^{-1}(j)$ has been evaluated for $j = 1, 2, \cdots, n - 1$ then $f^{-1}(n)$ is determined by

$$0 = I(n) = (f * f^{-1})(n) = \sum_{d \mid n} f(d)f^{-1}(n/d),$$

because this sum contains $f^{-1}(n)$ only in the term $f(1)f^{-1}(n)$. Thus this equation can be solved to give a unique value for $f^{-1}(n)$.

Next we note that Theorem 4.6, if restated in the notation of this section, can be written as follows.

Theorem 4.10 *The functions μ and U are multiplicative inverses, thus $\mu * U = U * \mu = I$, $\mu = U^{-1}$ and $U = \mu^{-1}$.*

Furthermore, Theorem 4.7 and 4.8 can be stated more succinctly in the following way.

Theorem 4.11 *If f and F are any arithmetic functions such that $F = U * f$, then $f = \mu * F$ and conversely.*

It may be noted that we do not need to use Theorems 4.7 and 4.8 to prove this result, because it follows from Theorem 4.10 and the associative property of Dirichlet multiplication. That is, if we multiply $F = U * f$ by μ we get $\mu * F = f$, and the process is reversed by multiplying by U.

Theorem 4.12 *The set of all arithmetic functions f with $f(1) \neq 0$ forms a group under Dirichlet multiplication. Similarly, the set of all multiplicative arithmetic functions is a group.*

Proof. First consider the arithmetic functions f with $f(1) \neq 0$. This set is closed under multiplication because if $f(1) \neq 0$ and $g(1) \neq 0$ then $(f * g)(1) = f(1)g(1) \neq 0$. We have already observed that the associative property holds for all arithmetic functions. The identity element is the function I, and multiplicative inverses have been taken care of in Theorem 4.9.

Turning to the multiplicative functions, we note that the associative property holds as before, and that the function I is multiplicative. So we need only prove first that the product of two multiplicative functions is multiplicative, and second that the inverse of a multiplicative function is multiplicative.

First consider two multiplicative functions f and g. To prove that $f * g$ is multiplicative we parallel the notation and the proof of Theorem 4.4, which is not surprising since Theorem 4.4 is just the special case of what we are now proving, with g replaced by the very special function U. Thus for relatively prime positive integers m and n we get

$$\begin{aligned}(f * g)(mn) &= \sum_{d \mid nm} f(d)g(nm/d) \\ &= \sum_{d_1 \mid n}\sum_{d_2 \mid m} f(d_1 d_2)g(nm/d_1 d_2) \\ &= \sum_{d_1 \mid n}\sum_{d_2 \mid m} f(d_1)f(d_2)g(n/d_1)g(m/d_2) \\ &= \left\{\sum_{d_1 \mid n} f(d_1)g(n/d_1)\right\}\left\{\sum_{d_2 \mid m} f(d_2)g(m/d_2)\right\} \\ &= (f * g)(n) \cdot (f * g)(m).\end{aligned}$$

Finally, to prove that if f is multiplicative so also is f^{-1}, we define a function g as follows. First set $g(p^j) = f^{-1}(p^j)$ for every prime p and every integer $j > 0$. Now we want g to be multiplicative, so for any integer $n = \prod p_i^{\alpha_i}$ we define $g(n) = \prod g(p_i^{\alpha_i})$, where the primes p_i are distinct. Now f and g are both multiplicative, so $f * g$ is multiplicative by the preceding paragraph. Also for any prime power p^k we see that

$$(f * g)(p^k) = \sum_{d_1 d_2 = p^k} f(d_1)g(d_2) = \sum_{d_1 d_2 = p^k} f(d_1)f^{-1}(d_2) = (f * f^{-1})p^k = I(p^k).$$

Thus $f * g$ coincides with I on prime powers, both are multiplicative functions, and so $f * g = I$ by the basic observations made immediately after Definition

4.2. But the inverse of f is unique by Theorem 4.9, and it follows that $f^{-1} = g$, and so f^{-1} is multiplicative since g is.

Theorem 4.13 *The following relations hold among the functions I, U, E, μ, τ, ϕ, σ:*

$$(1)\ \mu = U^{-1} \qquad (2)\ \tau = U * U \qquad (3)\ \phi = \mu * E$$
$$(4)\ \sigma = U * E \qquad (5)\ \sigma = \phi * \tau \qquad (6)\ \sigma * \phi = E * E.$$

Proof. Item (1) was proved in Theorem 4.10, and is repeated for completeness. Since all the functions are multiplicative, we need prove the results for prime powers only. Thus items (2), (3), (4) are established by noting that

$$(U * U)(p^k) = \sum_{d \mid p^k} U(d)U(p^k/d) = \sum_{d \mid p^k} 1 = \tau(p^k),$$

$$(\mu * E)(p^k) = \sum_{d \mid p^k} \mu(d)E(p^k/d) = E(p^k) - E(p^{k-1}) = p^k - p^{k-1} = \phi(p^k),$$

$$(U * E)(p^k) = \sum_{d \mid p^k} E(d)U(p^k/d) = \sum_{d \mid p^k} E(d) = \sum_{d \mid p^k} = \sigma(p^k).$$

Finally, items (5) and (6) follow at once from the others; for example $\phi * \tau = (\mu * E) * (U * U) = U * E$.

The identities in Theorem 4.13 can be elaborated in terms of any positive integer n. Thus item (3) becomes

$$\phi(n) = \sum_{d \mid n} \mu(d)E(n/d) = \sum_{d \mid n} \mu(d) \cdot n/d = n \sum_{d \mid n} \mu(d)/d.$$

Item 3 implies, on multiplication by U, that $E = U * \phi$, and this gives Theorem 2.17 when applied to n. Thus we have given an independent proof of Theorem 2.17.

Items (5) and (6) can be written in the forms

$$\sigma(n) = \sum_{d \mid n} \phi(d)\tau(n/d),$$

and

$$\sum_{d \mid n} \sigma(d)\phi(n/d) = \sum_{d \mid n} E(d)E(n/d) = \sum_{d \mid n} n = n\tau(n).$$

Finally, this approach by algebraic structure is now used to derive a more general inversion formula than Theorems 4.7 and 4.8. Let $\beta(x)$ be a complex-valued function defined for all real $x \geqq 1$. If f is any arithmetic function define the product $f\beta$ by

$$(4.1) \qquad (f\beta)(x) = \sum_{n=1}^{[x]} f(n)\beta(x/n),$$

for all real $x \geqq 1$. Thus $f\beta$, like β itself, is a function whose domain is all real $x \geqq 1$ and whose range is the complex numbers, or a subset thereof. If g is another arithmetic function it turns out that $g(f\beta) = (g * f)\beta$. To prove this we first evaluate $g(f\beta)$ for any real $x \geqq 1$, thus

$$(g(f\beta))(x) = \sum_{j=1}^{[x]} g(j) \cdot (f\beta)(x/j) = \sum_{j=1}^{[x]} \sum_{k=1}^{[x/j]} g(j)f(k)\beta(x/jk).$$

We collect terms with equal values of jk in this double sum. For any fixed value of jk, say $jk = n$ we note that $n = jk \leqq j[x/j] \leqq j(x/j) = x$. Conversely, if n is any positive integer $\leqq x$, so $n \leqq [x]$, and if $n = jk$ is any factoring of n we see that $j \leqq jk \leqq [x]$ and $k = n/j \leqq [x]/j = [x/j]$ by part (f) of Theorem 4.1. Hence we can rewrite the double sum above in the form

$$\sum_{n=1}^{[x]} \sum_{j \mid n} g(j)f(n/j)\beta(x/n) = \sum_{n=1}^{[x]} (g * f)(n) \cdot \beta(x/n) = ((g * f)\beta)(x).$$

Having established that $g(f\beta) = (g * f)\beta$, we take f to be the Moebius function μ, and we write γ for $\mu\beta$, so that $\gamma(x) = (\mu\beta)(x)$ for all real $x \geqq 1$. Then we see that

$$U\gamma = U(\mu\beta) = (U * \mu)\beta = I\beta = \beta,$$

where the last step here follows at once from Definition (4.1). Conversely $U\gamma = \beta$ implies $\mu(U\gamma) = \mu\beta$ or $\gamma = \mu\beta$. Thus we have proved the following result.

Theorem 4.14 *Let $\beta(x)$ and $\gamma(x)$ be complex-valued functions defined for all $x \geqq 1$. If $\beta(x) = \sum_{j=1}^{[x]} \gamma(x/j)$ for all $x \geqq 1$, then*

$$(4.2) \qquad \gamma(x) = \sum_{j=1}^{[x]} \mu(j)\beta(x/j).$$

for all $x \geqq 1$, and conversely.

This result implies, but is not implied by, Theorems 4.7 and 4.8. To see that Theorem 4.14 implies Theorem 4.7, suppose we are given any arithmetic function f. Define $\gamma(x)$ in this way: For any positive integer k take $\gamma(k) = f(k)$; for any noninteger $x \geqq 1$ define $\gamma(x) = 0$. Next $\beta(x)$ is defined in terms of $\gamma(x)$ by the first equation in Theorem 4.14, so that $\beta(x) = 0$ if x is not an integer, and if x is an integer, say $x = n$,

$$\beta(n) = \sum_{j=1}^{[n]} \gamma(n/j) = \sum_{j=1}^{n} \gamma(n/j) = \sum_{j \mid n} f(n/j) = \sum_{j \mid n} f(j).$$

Then (4.2) gives us the conclusion of Theorem 4.7 if we define the arithmetic function F by taking $F(n) = \beta(n)$ for all positive integers n.

To establish that Theorem 4.14 implies Theorem 4.8, the argument is very similar, except that the arithmetic function F is now the starting point, and β is identified with F on the positive integers, with $\beta(x) = 0$ for $x \geqq 1$ but x not an integer. Then (4.2) is used to define $\gamma(x)$, and finally f is defined by $f(n) = \gamma(n)$ for all positive integers n.

PROBLEMS

1. Prove that I, E and U are totally multiplicative but that μ, τ, ϕ and σ are not.
2. If f and g are totally multiplicative, does it follow that $f * g$ is also? Does it follow that f^{-1} is totally multiplicative?
3. Prove that $\mu * \tau = U$ and hence $\sum_{d \mid n} \mu(d)\tau(n/d) = 1$.
4. Prove that $E = \sigma * \mu$ and hence $n = \sum_{d \mid n} \mu(d)\sigma(n/d)$.
5. Prove that $\sigma * U = E * \tau$ and hence $\sum_{d \mid n} \sigma(d) = n \sum_{d \mid n} \tau(d)/d$.
6. Define the arithmetic function Δ by $\Delta(n) = \sum_{j=1}^{n} (j, n)$, where (j, n) denotes the g.c.d. of j and n. Evaluate $\Delta(n)$ by using the fact that the number of integers j among $1, 2, \cdots, n$ such that $(j, n) = d$ is $\phi(n/d)$, and thus establish that $\Delta = E * \phi$. Then use Theorem 4.12 to prove that Δ is a multiplicative function. Finally, prove that the analogous function to Δ with the l.c.m. in place of g.c.d., namely $\sum_{j=1}^{n} [j, n]$, is not multiplicative.
7. Let f, g, h be three arithmetic functions satisfying $f(n) = \sum_{d \mid n} g(d)$, $g(n) = \sum_{d \mid n} f(d)h(n/d)$, $f(1) \neq 0$ for all positive integers n. Prove that h is the Moebius function.
8. Prove that the set of all arithmetic functions f with $f(1) = 1$ is a group.
9. Prove that the set of all arithmetic functions forms a ring, with Dirichlet multiplication, and addition defined by $(f + g)(n) = f(n) + g(n)$ for all positive integers n.
10. Prove that $g(f\beta) = f(g\beta)$, where the product is defined as in (4.1).

4.5 Recurrence Functions

A particular sort of arithmetic function can be defined as follows. If a, b, x_0, x_1 are arbitrary numbers, perhaps even complex, we let $f(0) = x_0$, $f(1) = x_1$, and $f(n + 1) = af(n) + bf(n - 1)$ for $n \geqq 1$. This determines

$f(n)$ uniquely, depending only on a, b, x_0, x_1. For convenience we will write x_n in place of $f(n)$ and we have $x_{n+1} = ax_n + bx_{n-1}$. Such a relation is called a *recurrence* or *recursion* formula.

In order to get a simpler relationship we write this last equation in the form

$$x_{n+1} - kx_n = (a - k)(x_n - kx_{n-1}) + (b + ak - k^2)x_{n-1}.$$

If k_1 and k_2 are the roots of $k^2 - ak - b = 0$, then $k_1 + k_2 = a$ and we have

$$x_{n+1} - k_1x_n = k_2(x_n - k_1x_{n-1}),$$
$$x_{n+1} - k_2x_n = k_1(x_n - k_2x_{n-1}),$$

and hence

$$x_{n+1} - k_2x_n = k_2^n(x_1 - k_1x_0),$$
$$x_{n+1} - k_1x_n = k_1^n(x_1 - k_2x_0).$$

Substracting we find $(k_2 - k_1)x_n = (x_1 - k_1x_0)k_2^n - (x_1 - k_2x_0)k_1^n$. Therefore, if $k_2 \neq k_1$, we have

$$x_n = \frac{(x_1 - k_1x_0)k_2^n - (x_1 - k_2x_0)k_1^n}{k_2 - k_1}, \qquad k_2 \neq k_1. \tag{4.3}$$

Thus we have a formula for finding the value of x_n directly in terms of a, b, x_0, x_1 without having to compute the values of $x_2, x_3, \cdots, x_{n-1}$. However, this formula has no meaning if $k_2 = k_1$. In this case we might try holding k_1 and n fixed and letting k_2 approach k_1, hoping that this will suggest a solution that we can then verify. We regard the above equation as having the form

$$x_n = \frac{g(k_2)}{h(k_2)}$$

with $g(k_1) = h(k_1) = 0$, and we apply L'Hospital's rule in the form

$$\lim_{k_2 \to k_1} \frac{g(k_2)}{h(k_2)} = \lim_{k_2 \to k_1} \frac{g'(k_2)}{h'(k_2)}$$

if the limit on the right exists. This gives us

$$\lim_{k_2 \to k_1} x_n = \lim_{k_2 \to k_1} \{n(x_1 - k_1x_0)k_2^{n-1} + x_0k_1^n\} = nx_1k_1^{n-1} - nx_0k_1^n + x_0k_1^n.$$

Now we can set $y_n = nx_1k_1^{n-1} - nx_0k_1^n + x_0k_1^n$ and check whether or not y_n actually is the solution. Since $k_1 = k_2$ is the only root of $k^2 - ak - b = 0$ we have $a = 2k_1$, $b = -k_1^2$. We have $y_0 = x_0$, $y_1 = x_1$, and, for $n \geqq 1$,

$$\begin{aligned} y_{n+1} &= (n + 1)x_1k_1^n - (n + 1)x_0k_1^{n+1} + x_0k_1^{n+1} \\ &= 2k_1(nx_1k_1^{n-1} - nx_0k_1^n + x_0k_1^n) \\ &\qquad - k_1^2\{(n - 1)x_1k_1^{n-2} - (n - 1)x_0k_1^{n-1} + x_0k_1^{n-1}\} \\ &= ay_n + by_{n-1}. \end{aligned}$$

Now, for $n \geqq 1$, we have $y_{n+1} - x_{n+1} = a(y_n - x_n) + b(y_{n-1} - x_{n-1})$ and $y_0 - x_0 = 0$, $y_1 - x_1 = 0$. This implies $y_{n+1} - x_{n+1} = 0$ for $n + 1 = 2, 3, 4, \cdots$, and we have

$$x_n = y_n = nx_1k_1^{n-1} - (n-1)x_0k_1^n, \qquad k_1 = k_2.$$

We have found formulas for x_n in both cases. If a, b, x_0, x_1 are integers then so are all the x_n, since $x_n = ax_{n-1} + bx_{n-2}$.

The *Fibonacci* numbers $F_0, F_1, \cdots$ are defined by $F_0 = 0$, $F_1 = 1$, $F_{n+1} = F_n + F_{n-1}$. In this case we have $k^2 - k - 1 = 0$, $k_1 = (1 - \sqrt{5})/2$, $k_2 = (1 + \sqrt{5})/2$, and, by (4.1),

$$F_n = \frac{1}{\sqrt{5}}\left\{\left(\frac{1+\sqrt{5}}{2}\right)^n - \left(\frac{1-\sqrt{5}}{2}\right)^n\right\}.$$

As another example we consider the sequence 0, 1, 3, 8, 21, $\cdots$, for which $x_0 = 0$, $x_1 = 1$, $a = 3$, $b = -1$. Then $k_1 = (3 - \sqrt{5})/2$, $k_2 = (3 + \sqrt{5})/2$, and

$$x_n = \frac{1}{\sqrt{5}}\left\{\left(\frac{3+\sqrt{5}}{2}\right)^n - \left(\frac{3-\sqrt{5}}{2}\right)^n\right\}.$$

But $0 < \{(3 - \sqrt{5})/2\}^n \leqq 1$, and x_n is an integer so that in this case we can write the solution as

$$x_n = \left[\frac{1}{\sqrt{5}}\left(\frac{3+\sqrt{5}}{2}\right)^n\right].$$

PROBLEMS

1. Without using the results of this section find a formula for x_n if $x_{n+1} = ax_n$. Also if $x_{n+1} = bx_{n-1}$.

2. Find a formula for x_n if $x_{n+1} = 2x_n - x_{n-1}$, $x_0 = 0$, $x_1 = 1$. Also if $x_0 = 1$, $x_1 = 1$. Then do the same for $x_{n+1} = 2x_n + 3x_{n-1}$.

3. Write the first ten terms of the Fibonacci series. Prove in general that any two consecutive terms are relatively prime.

4. Prove that the Fibonacci numbers satisfy the inequalities

$$\left(\frac{1+\sqrt{5}}{2}\right)^{n-1} < F_{n+1} < \left(\frac{1+\sqrt{5}}{2}\right)^n,$$

if $n > 1$. *Suggestion:* write α for $(1 + \sqrt{5})/2$, and observe that $\alpha^2 = \alpha + 1 > F_2 + F_1 = F_3$, $\alpha^3 = \alpha^2 + \alpha > F_3 + F_2 = F_4$. Then use induction.

5. Prove that for $n \geqq 2$,

$$F_n = \binom{n-1}{0} + \binom{n-2}{1} + \binom{n-3}{2} + \binom{n-4}{3} + \cdots + \binom{n-j}{j-1},$$

where the sum of the binomial coefficients on the right terminates with the largest j such that $2j \leqq n + 1$. *Suggestion:* use the fact that

$$\binom{m}{r} = \binom{m-1}{r-1} + \binom{m-1}{r}.$$

6. Prove that $F_1 + F_2 + F_3 + \cdots + F_n = F_{n+2} - 1$.
7. Prove that $F_{n-1}F_{n+1} - F_n^2 = (-1)^n$.
8. Prove that $F_{m+n} = F_{m-1}F_n + F_mF_{n+1}$ for any positive integers m and n. Then prove that $F_m \mid F_n$ if $m \mid n$. *Suggestion:* let $n = mq$ and use induction on q.
9. Consider the sequence $1, 2, 3, 5, 8, \cdots = F_2, F_3, F_4, F_5, F_6, \cdots$. Prove that every positive integer can be written as a sum of distinct terms from this sequence. *Suggestion:* for k not in the sequence, let n be such that $F_{n-1} < k < F_n$. Show that $0 < k - F_{n-1} < F_{n-2}$. Prove the statement by induction.
10. Let $f(n)$ denote the number of sequences $a_1, a_2, \cdots, a_n$ that can be constructed where each a_j is $+1$, -1, or 0, subject to the restrictions that no two consecutive terms can be $+1$, and no two consecutive terms can be -1. Prove that $f(n)$ is the integer nearest to $\frac{1}{2}(1 + \sqrt{2})^{n+1}$. *Suggestion:* prove that $f(n) = 2f(n-1) + f(n-2)$.
11. (*a*) Let S_n be a set of integers x, $1 \leqq x \leqq n$, such that no member of S_n divides another member of S_n. Show that S_n can have $[(n+1)/2]$ members but no more.
(*b*) Find the maximum size of S_n if the x are also restricted to be odd.
(*c*) Show that if 2^ka, a odd, is in the S_n of (*a*) and if S_n is of maximum size, then $n < 3^{k+1}a$.
(*d*) Show that an S_n of (*a*) can be less than maximum size, and yet be maximal in the sense that no new member can be adjoined to S_n.
12. Let $f(n)$ be the sum of the first n terms of the sequence $0, 1, 1, 2, 2, 3, 3, 4, 4, \cdots$. Construct a table for $f(n)$. Prove that $f(n) = [n^2/4]$. For x, y integers $x > y$, prove that $xy = f(x+y) - f(x-y)$. Thus the process of multiplication can be replaced by an addition, a subtraction, looking up two numbers in the table, and subtracting them.
13. Use the ideas of this section to find a formula for x_n if $x_0 = 1$, $x_1 = 2$, $x_2 = 1$, and $x_{n+1} = x_n + 4x_{n-1} - 4x_{n-2}$. *Suggestion:* consider the expression $x_{n+1} + kx_n + lx_{n-1}$.

NOTES ON CHAPTER 4

The observation following Theorem 4.14 that this result implies the Moebius inversion formula was pointed out by our colleague Robert B. Burckel.

The evaluation of certain determinants whose elements are number theoretic functions is given in Special Topics, page 264 ff.

5

Some Diophantine Equations

5.1 Diophantine Equations

There are many problems and puzzles whose solutions require more than just finding all the solutions of some equation. They are worded in such a way that the desired solutions must satisfy other conditions. For example, we notice that $\frac{26}{65} = \frac{2\not{6}}{\not{6}5} = \frac{2}{5}$ is correct although this cancelling of the 6 violates the rules of algebra. Our problem is to find all the positive fractions that behave in this way. That is, we are to determine x, y, z in such a way that $\frac{10x + y}{10y + z} = \frac{x}{z}$. This reduces to $(y - x)z = 10(y - z)x$, but we are only interested in solutions such that x, y, and z are positive integers less than 10. We will not carry out the solution. It is not hard to see that the solutions are $\frac{19}{95}, \frac{16}{64}, \frac{26}{65}, \frac{49}{98}$, and the fractions of the form $\frac{10x + x}{10x + x}$.

In the foregoing problem the equation $(y - x)z = 10(y - z)x$ is an indeterminate equation. It has many algebraic solutions, and we are required to sort out the solutions in which x, y, and z are positive integers less than 10. Such a problem is called a *Diophantine* problem and we say we are solving a *Diophantine* equation. This particular problem is merely a curiosity, but there are many important Diophantine equations. In general, the added restrictions are that the solutions are to be integers or, sometimes, rational. Frequently the solutions are also to be positive.

There are endless varieties of Diophantine equations, and there is no general method of solution. We will discuss a few of the simplest equations. We will also consider some related problems. For example Theorem 5.6 states, in effect, that the equation $x_1^2 + x_2^2 + x_3^2 + x_4^2 = n$ has at least one solution, with x_1, x_2, x_3, x_4 integers, for each positive integer n. However, we will not attempt to find all the integral solutions.

Pell's equation, $x^2 - dy^2 = N$, will be discussed in Chapter 7.

5.2 The Equation $ax + by = c$

Any linear equation in two variables having integral coefficients can be put in the form $ax + by = c$. The problem is trivial unless neither a nor b is zero, so we can suppose $a \neq 0$, $b \neq 0$. We let g denote (a, b). Then Theorem 1.3 shows that there exist integers x_0 and y_0 such that $ax_0 + by_0 = g$. Numerical values for x_0 and y_0 can be obtained conveniently by applying the Euclidean algorithm, Theorem 1.11, to the integers $|a|$ and $|b|$.

Now if $g \nmid c$, then $ax + by = c$ clearly has no solution in integers. If $g \mid c$, we use our solution x_0, y_0 of $ax + by = g$ to get a solution $x_1 = (c/g)x_0$, $y_1 = (c/g)y_0$ of $ax + by = c$. In order to find all integral solutions we let r, s denote any integral solution. Then we have $ar + bs = c = ax_1 + by_1$, and hence

$$\frac{a}{g}(r - x_1) = -\frac{b}{g}(s - y_1). \tag{5.1}$$

But $(a/g, b/g) = 1$ by Corollary 1.7, and hence $(a/g) \mid (s - y_1)$ and $(b/g) \mid (r - x_1)$ by Theorem 1.9. This implies $s - y_1 = (a/g)u$ and $r - x_1 = (b/g)t$ for some integers u, t, and then (5.1) implies $u = -t$. Therefore every integral solution r, s of $ax + by = c$ can be written in the form $r = x_1 + (b/g)t$, $s = y_1 - (a/g)t$. Since these values clearly satisfy $ax + by = c$, we have solved the Diophantine equation. *Note that the equation has solutions if and only if (a, b) divides c.*

If a, b, c have a common divisor, it can be divided out so we can presume that $(a, b, c) = 1$. With this hypothesis our earlier conclusion can be stated this way: that $ax + by = c$ is solvable if and only if $(a, b) = 1$. Assuming that $(a, b) = 1$, thus $g = 1$ in the previous discussion, we see that every solution in integers of $ax + by = c$ is given by $x = x_1 + bt$, $y = y_1 - at$, where x_1, y_1 is any particular solution and t is an arbitrary integer.

PROBLEMS

1. Prove that all solutions of $3x + 5y = 1$ can be written in the form $x = 2 + 5t$, $y = -1 - 3t$; also in the form $x = 2 - 5t$, $y = -1 + 3t$; also in the form $x = -3 + 5t$, $y = 2 - 3t$. Prove that $x = a + bt$, $y = c + dt$ is a form of the general solution if and only if a, c is a solution and either $b = 5$, $d = -3$, or $b = -5$, $d = 3$.

2. Find all solutions of $10x - 7y = 17$.

3. If $ax + by = c$ is solvable, prove that it has a solution x_0, y_0 with $0 \leqq x_0 < |b|$.

4. Prove that $ax + by = a + c$ is solvable if and only if $ax + by = c$ is solvable.

5. Prove that the conclusion that $ax + by = c$ is solvable if and only if (a, b) divides c can be drawn from the conclusions of Theorem 2.13, with a, m, b of that theorem replaced by a, b, c.

6. Prove that $ax + by = c$ is solvable if and only if $(a, b) = (a, b, c)$.

7. Given that $ax + by = c$ has two solutions, (x_0, y_0) and (x_1, y_1) with $x_1 = 1 + x_0$, and given that $(a, b) = 1$, prove that $b = \pm 1$.

8. Interpreted geometrically, the solutions of $ax + by = c$ in integers are certain points on the straight line represented by the equation in an x, y coordinate system. If $(a, b) = 1$, prove that any segment of the line of length $(a^2 + b^2)^{1/2}$ contains at least one of these points with integral coordinates.

9. Find necessary and sufficient conditions that

$$x + b_1 y + c_1 z = d_1, \qquad x + b_2 y + c_2 z = d_2$$

have at least one simultaneous solution in integers x, y, z, assuming that the coefficients are integers with $b_1 \neq b_2$.

10. Give an independent proof that $ax + by = c$ has at least one solution in integers x, y if $(a, b) \mid c$, by using induction on $\max(a, b)$. *Suggestion:* if $0 < a < b$ then $ax + by = c$ is solvable if and only if $a(x + y) + (b - a)y = c$ is solvable.

5.3 Positive Solutions

Let us suppose that a, b, and c are positive, that $(a, b) \mid c$, and that we are asked to find all solutions r, s of $ax + by = c$ in positive integers. We solve as in Section 5.2, and have only to restrict t in such a way as to make r and s positive. We merely restrict t to the range $-(g/b)x_1 < t < (g/a)y_1$. The smallest allowable value for t is $[-(g/b)x_1 + 1]$ and the largest value is $-[(-g/a)y_1 + 1]$. The number of solutions is then

$$N = -\left[-\frac{g}{a}y_1 + 1\right] - \left[-\frac{g}{b}x_1 + 1\right] + 1$$
$$= -\left(\left[-\frac{g}{a}y_1\right] + \left[-\frac{g}{b}x_1\right] + 1\right)$$

from which we find, by use of Theorem 4.1*d*, that

$$-\left(\left[-\frac{g}{a}y_1 - \frac{g}{b}x_1\right] + 1\right) \leqq N \leqq -\left[-\frac{g}{a}y_1 - \frac{g}{b}x_1\right].$$

Since $-(g/a)y_1 - (g/b)x_1 = -(g/ab)(by_1 + ax_1) = -(gc)/(ab)$ we finally have

$$-\left[-\frac{gc}{ab}\right] - 1 \leqq N \leqq -\left[-\frac{gc}{ab}\right].$$

These inequalities do not constitute a precise formula for N, but they do specify two consecutive integers one of which must be N. Note that there will always be at least one positive solution of $ax + by = c$ if $g \mid c$ and $gc > ab$.

PROBLEMS

1. Find all solutions in positive integers:
(*a*) $5x + 3y = 52$;
(*b*) $15x + 7y = 111$;
(*c*) $40x + 63y = 521$;
(*d*) $123x + 57y = 531$;
(*e*) $12x + 501y = 1$;
(*f*) $12x + 501y = 274$;
(*g*) $97x + 98y = 1000$.
2. Prove that $101x + 37y = 3819$ has a positive solution in integers.
3. Given that $(a, b) = 1$ and that a and b are of opposite sign, prove that $ax + by = c$ has infinitely many positive solutions for any value of c.
4. Let a, b, c be positive integers. Prove that there is no solution of $ax + by = c$ in positive integers if $a + b > c$.
5. The theory in the text states that one of the formulas

$$N = -\left[-\frac{gc}{ab}\right] - 1, \qquad N = -\left[-\frac{gc}{ab}\right]$$

is correct. Prove that neither formula is correct in all cases.
6. Let a, b, c be positive integers such that $(a, b) = 1$. Assuming that c/ab is not an integer, prove that the number N of solutions of $ax + by = c$ in positive integers is $[c/ab]$ or $[c/ab] + 1$. Assuming further that c/a is an integer, prove that $N = [c/ab]$. *Suggestion:* if c/a is an integer, then a specific solution of $ax + by = c$ can be found easily, for example, $x_1 = c/a$, $y_1 = 0$.
7. Let a, b, c be positive integers such that $(a, b) = 1$. Assuming that c/ab is an integer, prove that $N = -1 + c/ab$.
8. Modify the theory of this section so as to treat the number, say N_0, of solutions in non-negative integers of $ax + by = c$, where a, b, c are positive integers such that $(a, b) \mid c$. *Suggestion:* observe that t is restricted to the range $-(g/b)x_1 \leqq t \leqq (g/a)y_1$. The smallest allowable value for t is $-[(g/b)x_1]$ and the largest value is $[(g/a)y_1]$. The final inequalities turn out to be

$$\left[\frac{gc}{ab}\right] \leqq N_0 \leqq \left[\frac{gc}{ab}\right] + 1.$$

9. Let a and b be positive integers satisfying $(a, b) = 1$. Consider the set S of integers $\{ax + by\}$, where x and y range over all non-negative integers. Prove that the set S contains all integers greater than $c = ab - a - b$, but not c itself.

10. In the preceding problem, restrict x and y to be positive integers, yielding the set $S' = \{ax + by\}$. What is the largest integer missing from the set S'?

5.4 Other Linear Equations

Consider the equation

$$a_1x_1 + a_2x_2 + \cdots + a_kx_k = c, \qquad k > 2, \tag{5.2}$$

and let g denote the greatest common divisor $(a_1, a_2, \cdots, a_k)$. If the equation has a solution, then clearly $g \mid c$. Conversely, by Theorem 1.5 there exist $y_1, y_2, \cdots, y_k$ such that $a_1y_1 + a_2y_2 + \cdots + a_ky_k = g$. If $g \mid c$ then $c = gr$ for some integer r and $x_1 = ry_1, x_2 = ry_2, \cdots, x_k = ry_k$ is a solution of (5.2). Thus (5.2) has solutions if and only if $g \mid c$.

To find solutions of (5.2) we reduce it to the case of Section 5.2 with two unknowns. We can suppose that the a_i are not zero, and that $(a_1, a_2, \cdots, a_k) \mid c$. We write

$$x_{k-1} = \alpha u + \beta v, \qquad x_k = \gamma u + \delta v, \tag{5.3}$$

where we shall choose integers $\alpha, \beta, \gamma, \delta$ in such a way that $\alpha\delta - \beta\gamma = 1$. Then we will have $u = \delta x_{k-1} - \beta x_k$ and $v = -\gamma x_{k-1} + \alpha x_k$. Thus, u, v are integers if and only if x_{k-1}, x_k are. If we take

$$\beta = \frac{a_k}{(a_{k-1}, a_k)}, \qquad \delta = \frac{-a_{k-1}}{(a_{k-1}, a_k)},$$

then $(\beta, \delta) = 1$, and we can solve $\alpha\delta - \beta\gamma = 1$ for α, γ by the method of Section 5.2. However, we need only one pair of values for α, γ, not the general solution.

Equation (5.2) now reduces to

$$a_1x_1 + a_2x_2 + \cdots + a_{k-2}x_{k-2} + (a_{k-1}\alpha + a_k\gamma)u = c, \tag{5.4}$$

with one less unknown, and we note that

$$a_{k-1}\alpha + a_k\gamma = -(a_{k-1}, a_k)\alpha\delta + (a_{k-1}, a_k)\beta\gamma = -(a_{k-1}, a_k),$$
$$(a_1, a_2, \cdots, a_{k-2}, (a_{k-1}, a_k)) = (a_1, a_2, \cdots, a_k).$$

Thus the new equation, (5.4), has the same properties as equation (5.2), that the g.c.d. of its coefficients divides c and that no coefficient is zero. If $k > 3$ this reduction process can be applied to equation (5.4) to produce an equation with $k - 2$ variables, and hence repetitions of the process finally lead to an equation with two unknowns.

Furthermore we may note from Section 5.2 that if a linear equation in two unknowns has a solution, its general solution is given in terms of a single parameter t. Similarly the solutions of (5.2), with k unknowns, are expressed in terms of $k - 1$ parameters. This can be proved by induction on k. For if any equation such as (5.4), in $k - 1$ unknowns, has solutions $x_1, x_2, \cdots, x_{k-2}$, u in terms of $k - 2$ parameters, $v_1, v_2, \cdots, v_{k-2}$, then, by (5.3), the solutions of (5.2) are given by $x_1, x_2, \cdots, x_{k-2}, \alpha u + \beta v, \gamma u + \delta v$. These involve the $k - 1$ parameters $v_1, v_2, \cdots, v_{k-2}, v$. It is easy to see that the solutions have the form $x_i = b_i + d_{i,1}v_1 + d_{i,2}v_2 + \cdots + d_{i,k-1}v_{k-1}$ where we have written v_{k-1} for v.

If we are asked to solve a system of s equations in r unknowns,

$$(5.5) \qquad a_{j,1}x_1 + a_{j,2}x_2 + \cdots + a_{j,r}x_r = c_j, \qquad j = 1, 2, \cdots, s,$$

we start by solving the first equation, $j = 1$. If it has a solution, it will be of the form

$$x_i = b_i + d_{i,1}v_1 + d_{i,2}v_2 + \cdots + d_{i,r-1}v_{r-1}, \qquad i = 1, 2, \cdots, r.$$

We substitute these in the remaining equations and solve the second equation, $j = 2$, for $v_1, v_2, \cdots, v_{r-1}$ in terms of $r - 2$ new parameters. Repetition of this process allows us to solve the system (5.5). Of course, if we encounter an equation that has no solutions, then the system (5.5) has no solutions.

PROBLEM

1. Solve the equations:

(*a*) $x + 2y + 3z = 1$ (*b*) $x + 2y + 3z = 10$
(*c*) $5x - 2y - 4z = 1$ (*d*) $5x - 2y - 4z = 10$
(*e*) $3x - 6y + 5z = 11$ (*f*) $6x + 48y - 78z = 5$

5.5 The Equation $x^2 + y^2 = z^2$

We wish to solve the equation $x^2 + y^2 = z^2$ in positive integers. Consider such a solution x, y, z and write g for (x, y). Then $g^2 \mid z^2$ and hence $g \mid z$, and since $(x, y, z) = ((x, y), z)$ holds in general, we see that $(x, y, z) = g$. By symmetry we have $(x, y, z) = (x, y) = (y, z) = (x, z) = g$, and

$$\left(\frac{x}{g}\right)^2 + \left(\frac{y}{g}\right)^2 = \left(\frac{z}{g}\right)^2, \qquad \left(\frac{x}{g}, \frac{y}{g}\right) = \left(\frac{y}{g}, \frac{z}{g}\right) = \left(\frac{x}{g}, \frac{z}{g}\right) = 1.$$

A solution x_1, y_1, z_1 having the property that these three are relatively prime in pairs is called a *primitive* solution. Thus every solution x, y, z can be written in the form gx_1, gy_1, gz_1 where x_1, y_1, z_1 is some primitive solution. Conversely, if x_1, y_1, z_1 is a primitive solution, then gx_1, gy_1, gz_1 is a solution if g is a positive integer. Thus we need look only for primitive solutions, and this we now do.

Now x and y cannot both be even. They cannot both be odd either, for if they were we would have $x^2 \equiv 1 \pmod 4$, $y^2 \equiv 1 \pmod 4$, and therefore $z^2 \equiv 2 \pmod 4$, which is impossible.

Since x and y enter the equations symmetrically we can now restrict our attention to primitive solutions for which y is even, x and z odd. Then we have

$$\frac{z+x}{2}\frac{z-x}{2} = \left(\frac{y}{2}\right)^2. \tag{5.6}$$

Now

$$\left(\frac{z+x}{2}, \frac{z-x}{2}\right) \Bigg| \left(\frac{z+x}{2} + \frac{z-x}{2}\right) = z$$

and

$$\left(\frac{z+x}{2}, \frac{z-x}{2}\right) \Bigg| \left(\frac{z+x}{2} - \frac{z-x}{2}\right) = x$$

and therefore

$$\left(\frac{z+x}{2}, \frac{z-x}{2}\right) = 1.$$

This with equation (5.6) shows that $(z + x)/2 = r^2$ and $(z - x)/2 = s^2$ for some positive integers r, s. We also see that $(r, s) = 1$, $r > s$, $x = r^2 - s^2$, $y = 2rs$, $z = r^2 + s^2$. Also, since z is odd, r and s are of opposite parity, one is even, the other odd.

Conversely, let r and s be any two integers such that $(r, s) = 1$, $r > s > 0$, r and s of opposite parity. Then if $x = r^2 - s^2$, $y = 2rs$, $z = r^2 + s^2$, we have x, y, z positive and

$$x^2 + y^2 = (r^2 - s^2)^2 + (2rs)^2 = (r^2 + s^2)^2 = z^2.$$

It is easy to see that $(x, y) = 1$ and that y is even. Therefore, x, y, z is a primitive solution with y even. Thus we have the following result.

Theorem 5.1 *The positive primitive solutions of $x^2 + y^2 = z^2$ with y even are $x = r^2 - s^2$, $y = 2rs$, $z = r^2 + s^2$, where r and s are arbitrary integers of opposite parity with $r > s > 0$ and $(r, s) = 1$.*

It may be noted that if r and s are any integers whatsoever, then x, y, z defined by $x = r^2 - s^2$, $y = 2rs$, $z = r^2 + s^2$ give a solution of $x^2 + y^2 = z^2$, but not necessarily a primitive solution.

PROBLEMS

1. Find all primitive solutions of $x^2 + y^2 = z^2$ having $0 < z < 30$.
2. Prove that if $x^2 + y^2 = z^2$, then one of x, y is a multiple of 3 and one of x, y, z is a multiple of 5.
3. Any solution of $x^2 + y^2 = z^2$ in positive integers is called a *Pythagorean triple* because there is a right triangle whose sides have corresponding lengths. Find all Pythagorean triples whose terms form (*a*) an arithmetic progression, (*b*) a geometric progression.
4. If n is any integer $\geqq 3$, show that there is a Pythagorean triple with n as one of its members.
5. For which integers n are there solutions to the equation $x^2 - y^2 = n$?
6. Prove that every integer n can be expressed in the form $n = x^2 + y^2 - z^2$.
7. Prove that $x^2 + y^2 = z^4$ has infinitely many solutions with $(x, y, z) = 1$.
8. Show that all solutions of $x^2 + 2y^2 = z^2$ in positive integers with $(x, y, z) = 1$ are given by $x = |u^2 - 2v^2|$, $y = 2uv$, $z = u^2 + 2v^2$, where u and v are arbitrary positive integers such that u is odd and $(u, v) = 1$. *Suggestion:* any solution has y even, because y odd implies $z^2 - x^2 \equiv 2 \pmod 8$, which is impossible. Hence x and z are odd, and the proof of Theorem 5.1 can be used as a model.
9. Prove that no Pythagorean triple of integers belongs to an isosceles right triangle, but that there are infinitely many primitive Pythagorean triples for which the acute angles of the corresponding triangles are, for any given positive ε, within ε of $\pi/4$.

5.6 The Equation $x^4 + y^4 = z^2$

We prove now the impossibility of solving the equation $x^4 + y^4 = z^2$ in positive integers. The argument is indirect, in that a solution in positive integers is assumed, and this will lead to a contradiction. Assuming at least one solution, we presume that x, y, z denotes a positive solution such that no other positive solution has a smaller value of z. The contradiction is obtained by deriving another positive solution in integers with a smaller z value. First we establish that g.c.d. $(x, y, z) = 1$. For if a prime p divides each of x, y, z then $p^4 \mid (x^4 + y^4)$ so that $p^4 \mid z^2$ and $p^2 \mid z$. It would follow that $(x/p)^4 + (y/p)^4 = (z/p^2)^2$, contradicting the minimal character of z.

In particular we see that x, y and z are not all even. If x and y were odd we would have $x^4 \equiv y^4 \equiv 1 \pmod 8$ and so $z^2 \equiv 2 \pmod 8$ which is impossible. If x and y were even then z would be even, and so we conclude that one of x and y is even and the other odd. We presume that x is even and y is odd, and note that z is also odd.

Next we write $y^4 = (z - x^2)(z + x^2)$ and observe that $z - x^2$ and $z + x^2$ are relatively prime. For if p divides both $z - x^2$ and $z + x^2$ it divides their product y^4, their sum $2z$ and their difference $2x^2$. This is impossible because y is odd and g.c.d. $(x, y, z) = 1$. Hence $z - x^2$ and $z + x^2$ are fourth powers of positive integers, u^4 and v^4 say, with $(u, v) = 1$. Thus we have

$$z - x^2 = u^4, \qquad z + x^2 = v^4, \qquad (v^2 - u^2)(v^2 + u^2) = 2x^2.$$

Now u and v are odd because $z - x^2$ and $z + x^2$ are odd, and hence $v^2 + u^2 \equiv 2 \pmod 4$. Also no odd prime p divides both $v^2 - u^2$ and $v^2 + u^2$ since $(u, v) = 1$, and so $v^2 - u^2$ is a perfect square, where $v^2 + u^2$ is twice a square, say $v^2 - u^2 = a^2$, $v^2 + u^2 = 2b^2$ with positive a and b. By Theorem 5.1 we get $v = r^2 + s^2$, $u = r^2 - s^2$ and $a = 2rs$ where r and s are positive integers. Now $v^2 + u^2 = 2b^2$ implies $r^4 + s^4 = b^2$, and since this has the same form as $x^4 + y^4 = z^2$, we have a contradiction if we prove that $z > b$.

Now $u = v = 1$ is impossible because this implies $x^2 = 0$, and so $u^4 + v^4 > u^2 + v^2$ and we have

$$z = \tfrac{1}{2}(u^4 + v^4) > \tfrac{1}{2}(u^2 + v^2) = b^2 \geqq b,$$

which completes the proof of the following proposition.

Theorem 5.2 *The only integral solutions of $x^4 + y^4 = z^2$ are the trivial solutions $x = 0$, y, $z = \pm y^2$ and x, $y = 0$, $z = \pm x^2$.*

The method used in the proof of this theorem is sometimes called "proof by descent" or "Fermat's method of infinite descent." This type of proof, which also occurs at other places in number theory, is based on the principle that every nonempty set of positive integers contains a least element.

The fact that $x^4 + y^4 = z^2$ has no positive solutions implies that $x^4 + y^4 = z^4$ has no positive solutions. This is a particular case of a famous statement of Fermat, in which he asserted in a marginal note that he could prove that for every integer $n > 2$ the equation $x^n + y^n = z^n$ has no solutions in integers other than the trivial ones in which at least one variable is zero. This proposition, although still a conjecture for many values of n, is known as *Fermat's last theorem* or Fermat's big theorem, as contrasted with Fermat's little theorem (Theorem 2.7). In addition to the case $n = 4$, a proof for $n = 3$ is given in §9.10 using simple ideas in algebraic number theory.

PROBLEMS

1. For any positive integer $n \equiv 0 \pmod 4$, prove that $x^n + y^n = z^n$ has no solutions with $xy \neq 0$.

2. Prove that $x^4 + 4y^4 = z^2$ has no solutions with $xy \neq 0$. *Suggestion:* use the method of proof of Theorem 5.2.

3. Prove that $x^4 - y^4 = z^2$ has no solutions with $yz \neq 0$. *Suggestion:* note that the equation implies $z^4 + 4(xy)^4 = (x^4 + y^4)^2$.
4. Consider an integral right triangle, that is, a right triangle the lengths of whose sides form a Pythagorean triple. Prove that the area is not a perfect square.
5. Prove that there are no positive integers a and b such that both $a^2 + b^2$ and $a^2 - b^2$ are perfect squares.

5.7 Sums of Four and Five Squares

Our aim in this section is not to solve a Diophantine equation but only to prove that the equation $x_1^2 + x_2^2 + x_3^2 + x_4^2 = n$ has at least one integral solution whenever n is a positive integer. The solution may include zeros among the x's, as for example with $n = 5$ where x_1, x_2, x_3, x_4 would be $\pm 1, \pm 2, 0, 0$ in some order. So as a simple extension of the four square result we establish that every positive integer (apart from 12 exceptions between 1 and 33) are sums of five positive squares.

Lemma 5.3 *We have*

$$(5.7)\quad (x_1^2 + x_2^2 + x_3^2 + x_4^2)(y_1^2 + y_2^2 + y_3^2 + y_4^2)$$
$$= (x_1y_1 + x_2y_2 + x_3y_3 + x_4y_4)^2 + (x_1y_2 - x_2y_1 + x_3y_4 - x_4y_3)^2$$
$$+ (x_1y_3 - x_3y_1 + x_4y_2 - x_2y_4)^2 + (x_1y_4 - x_4y_1 + x_2y_3 - x_3y_2)^2.$$

Proof. This identity, discovered by Euler, can be verified by just multiplying out both sides.

This identity shows that if X and Y can be expressed as sums of four squares, then so can their product XY. Since $1 = 1^2 + 0^2 + 0^2 + 0^2$ and $2 = 1^2 + 1^2 + 0^2 + 0^2$, we need prove only that every odd prime can be expressed as a sum of four squares. The proof can be broken up into two steps.

Theorem 5.4 *Let p denote any odd prime. There is an integer m such that $1 \leqq m < p$ and $mp = x_1^2 + x_2^2 + x_3^2 + x_4^2$ for some integers x_1, x_2, x_3, x_4.*

Proof. Consider the sets S_1 consisting of $0^2, 1^2, 2^2, \cdots, \{(p-1)/2\}^2$ and S_2 consisting of $-0^2 - 1, -1^2 - 1, -2^2 - 1, \cdots, -\{(p-1)/2\}^2 - 1$. Since $x^2 \equiv y^2 \pmod{p}$ implies $p \mid (x - y)$ or $p \mid (x + y)$, we see that no two numbers of S_1 are congruent modulo p. Also no two numbers of S_2 are congruent modulo p. Now S_1 and S_2 together consist of $p + 1$ integers. Since there are only p distinct residue classes modulo p we see that some number of S_1, say x^2, is congruent modulo p to some number, $-y^2 - 1$, of S_2. Then $x^2 + y^2 + 1 \equiv 0 \pmod{p}$, $0 \leqq x \leqq (p-1)/2$, $0 \leqq y \leqq (p-1)/2$, and we have

$$x^2 + y^2 + 1 = mp, \qquad 1 \leqq m = \frac{1}{p}(x^2 + y^2 + 1)$$
$$\leqq \frac{1}{p}\left(2\left(\frac{p-1}{2}\right)^2 + 1\right) < \frac{1}{p}\left(\frac{p^2}{2} + 1\right) < p.$$

Theorem 5.5 *If m is the least integer satisfying Theorem 5.4, then $m = 1$.*

Proof. There certainly is at least one such m. If m is even, then so is $mp = x_1^2 + x_2^2 + x_3^2 + x_4^2$, and hence either none, two, or four of the x_i are even. If exactly two of the x_i are even, we can number the x_i in such a way that x_1 and x_2 are the even ones. Then in all cases $x_1 \pm x_2$ and $x_3 \pm x_4$ are even, and

$$\left(\frac{x_1 + x_2}{2}\right)^2 + \left(\frac{x_1 - x_2}{2}\right)^2 + \left(\frac{x_3 + x_4}{2}\right)^2 + \left(\frac{x_3 - x_4}{2}\right)^2 = \frac{m}{2}p.$$

Therefore m is not least if it is even.

Now consider the possibility $m > 1$. Since m is odd we have $3 \leqq m < p$. For $1 \leqq i \leqq 4$, we define numbers y_i by

$$(5.8) \qquad y_i \equiv x_i \pmod{m}, \qquad -\frac{m-1}{2} \leqq y_i \leqq \frac{m-1}{2}.$$

Then $y_1^2 + y_2^2 + y_3^2 + y_4^2 \equiv x_1^2 + x_2^2 + x_3^2 + x_4^2 \equiv 0 \pmod{m}$ since $x_1^2 + x_2^2 + x_3^2 + x_4^2 = mp$, and we can write

$$(5.9) \quad y_1^2 + y_2^2 + y_3^2 + y_4^2 = mn, \qquad 0 \leqq n \leqq \frac{1}{m} 4\left(\frac{m-1}{2}\right)^2 < m.$$

If n were zero, we would have $y_1 = y_2 = y_3 = y_4 = 0$ by (5.9), and then $x_1 \equiv x_2 \equiv x_3 \equiv x_4 \equiv 0 \pmod{m}$ by (5.8). This would imply $mp = x_1^2 + x_2^2 + x_3^2 + x_4^2 \equiv 0 \pmod{m^2}$, hence $p \equiv 0 \pmod{m}$, which is impossible since $3 \leqq m < p$. Therefore we have $n > 0$.

Using Lemma 5.3 we see that

$$(5.10) \qquad m^2np = (x_1^2 + x_2^2 + x_3^2 + x_4^2)(y_1^2 + y_2^2 + y_3^2 + y_4^2)$$
$$= A_1^2 + A_2^2 + A_3^2 + A_4^2,$$

where the A_i denote the expressions whose squares appear on the right side of (5.7). Using (5.8) we easily find that $A_i \equiv 0 \pmod{m}$ for $i = 1, 2, 3, 4$. Dividing (5.10) by m^2 we get

$$np = \left(\frac{A_1}{m}\right)^2 + \left(\frac{A_2}{m}\right)^2 + \left(\frac{A_3}{m}\right)^2 + \left(\frac{A_4}{m}\right)^2$$

with $0 < n < m$. This shows that m is not least if $m > 1$. Hence $m = 1$ and the theorem is proved.

Collecting our results and noting that $7 = 2^2 + 1^2 + 1^2 + 1^2$ requires four squares, we obtain the first sentence of the following theorem.

Theorem 5.6 *Every positive integer is a sum of four squares, and fewer than four squares will not suffice in general. Every sufficiently large positive integer is a sum of five positive squares of integers; this result is false if "five" is replaced by "four."*

It is possible to be quite specific about what is meant by "sufficiently large" here; in fact, every positive integer except 1, 2, 3, 4, 6, 7, 9, 10, 12, 15, 18, 33 is a sum of five positive squares. These exceptional cases can be verified with ease, and on the other hand it is readily checked that all the rest of the positive integers up to 169 *can* be expressed in five square form. We now prove that this holds for every integer $\geqq 170$.

For any integer $n \geqq 170$, we write $n - 169$ as a sum of four squares by the first part of Theorem 5.6,

$$n - 169 = x^2 + y^2 + z^2 + w^2.$$

We may presume that $x \geqq y \geqq z \geqq w \geqq 0$. If these integers are all positive we write $n = 13^2 + x^2 + y^2 + z^2 + w^2$. Second, if x, y, z are positive but $w = 0$, we write $n = 12^2 + 5^2 + x^2 + y^2 + z^2$. Third, if x, y are positive but $z = w = 0$, we note that $n = 12^2 + 4^2 + 3^2 + x^2 + y^2$. Finally if x is the only positive one, we get $n = 10^2 + 8^2 + 2^2 + 1^2 + x^2$. In every case n has been expressed as a sum of five positive squares.

To complete the proof of Theorem 5.6 we must show that it is false that every sufficiently large positive integer is a sum of four positive squares; that is, we must show that there are infinitely many exceptions. To do this, we first prove that *for any positive integer k, the integer $2k$ is a sum of four positive squares if and only if $8k$ is such a sum.* Assuming that $2k = x^2 + y^2 + z^2 + w^2$ we get $8k = (2x)^2 + (2y)^2 + (2z)^2 + (2w)^2$. Conversely, assume that $8r = u^2 + v^2 + s^2 + t^2$. Then u, v, s, t are all even because otherwise it would be false that $u^2 + v^2 + s^2 + t^2 \equiv 0 \pmod 8$, in view of the illustration

$$u = 2m + 1, \qquad u^2 = (2m + 1)^2 \equiv 1 \pmod 8.$$

It follows that $2r = (u/2)^2 + (v/2)^2 + (s/2)^2 + (t/2)^2$.

Next we observe that 8 is not a sum of four positive squares of integers, and taking $2r = 8$, $8r = 32$ we conclude that 32 is not a sum of four positive squares. This leads to an infinite chain of integers 8, 32, 128, 512, $\cdots$, 2^{2c+1}, $\cdots$ which are not sums of four positive squares.

PROBLEMS

1. Prove that no integer of the form $8k + 7$ can be expressed as a sum of three squares.

2. Prove that no integer of the form $4^m(8k + 7)$ is expressible as a sum of three squares. (*Remark:* these are the only integers not expressible as sums of three squares, but the proof of this result is not given in this book.)
3. Verify that every positive integer $\leqq 169$, apart from the 12 exceptions listed following Theorem 5.6, is expressible as a sum of five positive squares of integers.
4. Prove that every positive integer, apart from a few exceptions, is expressible as a sum of six positive integral squares. What are the exceptions?

5.8 Waring's Problem

Waring considered the question as to whether the results of Section 5.7 can be generalized to higher powers. He conjectured that 9 cubes, 19 fourth powers, etc. would suffice. Hilbert proved that for each positive integer k there is a number g such that g kth powers will suffice. If $g(k)$ denotes the least number of kth powers that will suffice, then $g(2) = 4$, as we have just seen. It is known that $g(3) = 9$, and the value of $g(k)$ is also known for $k \geqq 6$ apart from a possible exceptional circumstance relating to the fractional part of $(3/2)^k$. For the cases $k = 4$ and 5 it is known that $19 \leqq g(4) \leqq 35$ and $37 \leqq g(5) \leqq 54$. However, except for $g(2)$ and $g(3)$, the known proofs of these results involve much more complicated methods. The proofs lean heavily on the theory of functions of a complex variable, and they belong to the part of number theory called the analytic theory of numbers. There is a good account of the history of the problem in the notes at the end of Chapter 21 of Hardy and Wright's *An Introduction to the Theory of Numbers.*

PROBLEM

1. Prove that $g(k) \geqq 2^k + [3^k/2^k] - 1$. *Suggestion:* express the number $n = 2^k[3^k/2^k] - 1$ as a sum of kth powers.

5.9 Sum of Fourth Powers

There is one small result connected with Waring's problem that we can obtain simply. We will prove $g(4) \leqq 50$. This is far from the known result, $g(4) \leqq 35$, but it is of some interest because the proof is elementary, and because it proves the existence of $g(4)$.

Summing the identity $(x_i + x_j)^4 + (x_i - x_j)^4 = 2x_i^4 + 12x_i^2x_j^2 + 2x_j^4$ we find

$$\sum_{1 \leqq i < j \leqq 4} ((x_i + x_j)^4 + (x_i - x_j)^4)$$

$$= 2\sum_{i=1}^{4}(4 - i)x_i^4 + 6\sum_{i=1}^{4}\sum_{\substack{j=1\\ j\neq i}}^{4} x_i^2x_j^2 + 2\sum_{j=1}^{4}(j-1)x_j^4$$

$$= 2\sum_{i=1}^{4}(4 - i + i - 1)x_i^4 + 6\sum_{i=1}^{4}\sum_{\substack{j=1\\ j\neq i}}^{4} x_i^2x_j^2$$

$$= 6\sum_{i=1}^{4} x_i^4 + 6\sum_{i=1}^{4}\sum_{\substack{j=1\\ j\neq i}}^{4} x_i^2x_j^2$$

$$= 6\sum_{i=1}^{4}\sum_{j=1}^{4} x_i^2x_j^2 = 6\left(\sum_{i=1}^{4} x_i^2\right)^2.$$

The left side is a sum of 12 fourth powers. This with Theorem 5.6 shows that every number of the form $6m^2$ is a sum of 12 fourth powers. But Theorem 5.6 also shows that every positive integer of the form $6l$ can be expressed as a sum of four numbers of the form $6m^2$, and hence as a sum of 48 fourth powers.

Furthermore the integers $j = 0, 1, 2, 81, 16, 17$ form a complete residue system modulo 6, and each one is the sum of at most 2 fourth powers. Then every integer $n > 81$ can be written in the form $6l + j$ with l positive, and hence can be expressed as a sum of at most 50 fourth powers. For $0 < n \leqq 50$ we have $n = \sum_{i=1}^{n} 1^4$, and for $50 < n \leqq 81$ we have $n = 2^4 + 2^4 + 2^4 + \sum_{i=1}^{n-48} 1^4$. In all cases n is expressed as a sum of at most 50 fourth powers.

5.10 Sum of Two Squares

Not every positive integer is a sum of two squares. We will determine just which integers are sums of two squares and will find the number of solutions of $x^2 + y^2 = n$.

A solution x, y of $x^2 + y^2 = n$ will be called *primitive* if $(x, y) = 1$. We restrict n to be positive, and we let

$N(n)$ = number of solutions of $x^2 + y^2 = n$,
$P(n)$ = number of non-negative, primitive solutions of $x^2 + y^2 = n$,
$Q(n)$ = number of primitive solutions of $x^2 + y^2 = n$.

In counting solutions we consider x_1, y_1 and x_2, y_2 as distinct if $x_1 \neq x_2$ or $y_1 \neq y_2$. It may be noted that, for $n > 1$, $P(n)$ is actually the number of positive primitive solutions of $x^2 + y^2 = n$, since neither x nor y can be zero.

Theorem 5.7 *We have* $N(1) = Q(1) = 4$, $P(1) = 2$, *and, for* $n > 1$, $Q(n) = 4P(n)$ *and*

$$N(n) = \sum_{d^2 \mid n} Q\left(\frac{n}{d^2}\right).$$

Proof. Since $1 = (\pm 1)^2 + 0^2 = 0^2 + (\pm 1)^2$ and there are no other solutions, we have $N(1) = Q(1) = 4$ and $P(1) = 2$.

If $n > 1$ and x, y is a non-negative primitive solution, then $x \geqq 1$, $y \geqq 1$, and $\pm x$, $\pm y$ is a primitive solution for all choices of the signs. From this it follows that $Q(n) = 4P(n)$.

If x, y is any solution of $x^2 + y^2 = n$, and if $g = (x, y)$, then $g^2 \mid n$, $(x/g, y/g) = 1$, and $(x/g)^2 + (y/g)^2 = n/g^2$. From this it is easy to see that

$$N(n) = \sum_{d^2 \mid n} Q\left(\frac{n}{d^2}\right).$$

Theorem 5.8 *Suppose* $n > 1$. *Each non-negative primitive solution of* $x^2 + y^2 = n$ *determines a unique* s *modulo* n *such that* $sy \equiv x \pmod{n}$. *Furthermore,* $s^2 \equiv -1 \pmod{n}$, *and different non-negative primitive solutions determine different* s *modulo* n.

Proof. If x, y is a non-negative primitive solution, then $(y, n) = 1$ and hence $sy \equiv x \pmod{n}$ determines a unique s modulo n. Furthermore if y' is a solution of $yy' \equiv 1 \pmod{n}$, then $s \equiv xy' \pmod{n}$, and we have $s^2 \equiv x^2y'^2 \equiv -y^2y'^2 \equiv -1 \pmod{n}$.

We must still show that different solutions determine different s. Suppose that x, y and u, v are both non-negative primitive solutions and that $sy \equiv x \pmod{n}$ and $sv \equiv u \pmod{n}$. Then we have $xv \equiv syv \equiv yu \pmod{n}$. But since $n > 1$, any non-negative primitive solution is a positive solution and so we have $1 \leqq x < \sqrt{n}$, $1 \leqq v < \sqrt{n}$. Hence $1 \leqq xv < n$, and similarly $1 \leqq yu < n$. Therefore $xv = yu$, and hence $x = u$, $y = v$, since $(x, y) = (u, v) = 1$ and all the numbers are positive.

Theorem 5.9 *Suppose* $n > 1$, $s^2 \equiv -1 \pmod{n}$. *There is a non-negative primitive solution* x, y *of* $x^2 + y^2 = n$ *such that* $sy \equiv x \pmod{n}$.

Proof. Consider the set of integers $u - sv$ where u and v run through all integral values such that $0 \leqq u \leqq \sqrt{n}$, $0 \leqq v \leqq \sqrt{n}$. There are $(1 + [\sqrt{n}])^2 > n$ different pairs u, v. Therefore, there are two pairs u_1, v_1 and u_2, v_2 such that $u_1 - sv_1 \equiv u_2 - sv_2 \pmod{n}$. Let us set $v_2 - v_1 = v_0$, $u_2 - u_1 = u_0$. We then have $sv_0 \equiv u_0 \pmod{n}$ and $|u_0| \leqq \sqrt{n}$, $|v_0| \leqq \sqrt{n}$. Furthermore, since u_1, v_1 and u_2, v_2 are different pairs, u_0 and v_0 cannot both be zero. Also, we can show that at least one of $|u_0|$ and $|v_0|$ is less than $\sqrt{n}$. This is obvious if n is not a square. If n is a square and $|u_0| = |v_0| = \sqrt{n}$, we have $s\sqrt{n} \equiv \pm\sqrt{n} \pmod{n}$, and hence $s \equiv \pm 1 \pmod{\sqrt{n}}$, $s^2 \equiv 1 \pmod{\sqrt{n}}$. But $s^2 \equiv -1 \pmod{n}$, so we have $s^2 \equiv -1 \pmod{\sqrt{n}}$, and hence $1 \equiv -1 \pmod{\sqrt{n}}$. Therefore $\sqrt{n} = 2$ and $n = 4$, but this cannot occur since there is no integer s such that $s^2 \equiv -1 \pmod{4}$.

The bounds for u_0 and v_0 imply the inequality $1 \leqq u_0^2 + v_0^2 < 2n$. The congruence $sv_0 \equiv u_0 \pmod{n}$ implies $u_0^2 + v_0^2 \equiv s^2v_0^2 + v_0^2 \equiv v_0^2(s^2 + 1) \equiv 0 \pmod{n}$. Together these imply $u_0^2 + v_0^2 = n$.

Now let $g = (u_0, v_0)$. Then $g^2 \mid n$ and $s(v_0/g) \equiv (u_0/g) \pmod{n/g}$, and hence

$$\frac{n}{g^2} \equiv \frac{u_0^2 + v_0^2}{g^2} \equiv \left(s\frac{v_0}{g}\right)^2 + \left(\frac{v_0}{g}\right)^2 \equiv -\left(\frac{v_0}{g}\right)^2 + \left(\frac{v_0}{g}\right)^2 \equiv 0 \left(\bmod \frac{n}{g}\right).$$

This is possible only if $g = 1$, and we have $(u_0, v_0) = 1$.

Finally, if u_0 and v_0 have the same sign we let $x = |u_0|$, $y = |v_0|$. If u_0 and v_0 have opposite signs we let $x = |v_0|$, $y = |u_0|$. In both cases we see that x, y is a non-negative primitive solution. In the first case we have $sy \equiv s(\pm v_0) \equiv \pm u_0 \equiv x \pmod{n}$. In the second case we have $sy \equiv s(\pm u_0) \equiv \pm s(sv_0) \equiv \mp v_0 \equiv x \pmod{n}$.

The last two theorems show that there is a one to one correspondence between the non-negative primitive solutions of $x^2 + y^2 = n$ and the solutions of the congruence $s^2 \equiv -1 \pmod{n}$. Combining this with Theorem 5.7 we have the following.

Theorem 5.10 *Let $R(n)$ denote the number of roots of $s^2 \equiv -1 \pmod{n}$. Then $P(n) = R(n)$ for $n > 1$, $P(1) = 2$, $R(1) = 1$, $Q(1) = 4$, $Q(n) = 4R(n)$ for $n \geqq 1$, and $N(n) = 4\sum_{d^2 \mid n} R(n/d^2)$.*

Theorem 5.11 *The functions $R(n)$ and $N(n)/4$ are multiplicative functions.*

Proof. The fact that $R(n)$ is multiplicative follows directly from Theorem 2.18. To show that $N(n)/4$ is multiplicative, we consider any two positive relatively prime integers n_1 and n_2. Then

$$\begin{aligned}
\frac{1}{4}N(n_1n_2) &= \sum_{d^2 \mid n_1n_2} R\left(\frac{n_1n_2}{d^2}\right) = \sum_{d_1^2 \mid n_1} \sum_{d_2^2 \mid n_2} R\left(\frac{n_1}{d_1^2}\frac{n_2}{d_2^2}\right) \\
&= \sum_{d_1^2 \mid n_1} \sum_{d_2^2 \mid n_2} R\left(\frac{n_1}{d_1^2}\right) R\left(\frac{n_2}{d_2^2}\right) = \sum_{d_1^2 \mid n_1} R\left(\frac{n_1}{d_1^2}\right) \sum_{d_2^2 \mid n_2} R\left(\frac{n_2}{d_2^2}\right) \\
&= \frac{1}{4}N(n_1)\frac{1}{4}N(n_2).
\end{aligned}$$

Theorem 5.12 *Let $h(1) = 1$, $h(2^e) = 0$, $h(p^e) = (-1)^{\{(p-1)/2\}e}$, p an odd prime, $e \geqq 1$. Let $h(n)$ for composite n be determined in such a way that $h(n)$ is a multiplicative function. Then $N(n) = 4\sum_{d\mid n} h(d)$.*

Proof. It is interesting to note that $h(n)$ as defined above is actually totally multiplicative.

By Theorem 4.4, $\sum_{d\mid n} h(d)$ is multiplicative. Since $N(n)/4$ is also multiplicative, we have only to verify that $N(n) = 4\sum_{d\mid n} h(d)$ for n a prime power. The congruence $s^2 \equiv -1 \pmod 2$ has the one solution $s \equiv 1 \pmod 2$. For $e > 1$ the congruence $s^2 \equiv -1 \pmod{2^e}$ has no solutions since $s^2 \equiv -1 \pmod 4$ has none. Therefore by Theorem 5.10,

$$N(2^e) = 4\sum_{f=0}^{[e/2]} R(2^{e-2f}) = 4$$

since nonzero terms will occur only for $e - 2f = 0$ or 1. Correspondingly we have

$$4\sum_{d\mid 2^e} h(d) = 4\sum_{f=0}^{e} h(2^f) = 4h(1) = 4.$$

We now consider an odd prime p. From Theorem 2.11 we see that $s^2 \equiv -1 \pmod p$ has two solutions if $p \equiv 1 \pmod 4$, and no solutions if $p \equiv 3 \pmod 4$. We apply Section 2.6 to the polynomial $f(x) = x^2 + 1$, with $f'(x) = 2x$. Since $(2s, p) = 1$ we find that $s^2 \equiv -1 \pmod{p^e}$ has the same number of solutions for all $e \geqq 1$. That is

$$R(p^e) = R(p) = \begin{cases} 2 \text{ if } p \equiv 1 \pmod 4 \\ 0 \text{ if } p \equiv 3 \pmod 4, \end{cases} \qquad e \geqq 1.$$

Then, by Theorem 5.10, if e is even we have

$$(5.11) \qquad N(p^e) = 4\sum_{f=0}^{e/2} R(p^{e-2f}) = 4\frac{e}{2}R(p) + 4R(1)$$

$$= \begin{cases} 4e + 4 \text{ if } p \equiv 1 \pmod 4 \\ 4 \qquad\quad \text{if } p \equiv 3 \pmod 4, \end{cases}$$

and if e is odd we have

$$(5.12) \qquad N(p^e) = 4\sum_{f=0}^{(e-1)/2} R(p^{e-2f}) = 4\frac{e+1}{2}R(p)$$

$$= \begin{cases} 4e + 4 \text{ if } p \equiv 1 \pmod 4 \\ 0 \qquad\quad \text{if } p \equiv 3 \pmod 4. \end{cases}$$

Corresponding to this we also have

$$(5.13)\qquad 4\sum_{d\mid p^e} h(d) = 4\sum_{f=0}^{e} h(p^f) = 4h(1) + 4\sum_{f=1}^{e}(-1)^{[(p-1)/2]f}$$

$$= \begin{cases} 4 + 4e & \text{if } p \equiv 1 \pmod 4 \\ 4 & \text{if } p \equiv 3 \pmod 4,\ e \text{ even} \\ 0 & \text{if } p \equiv 3 \pmod 4,\ e \text{ odd.} \end{cases}$$

A comparison of (5.11) and (5.12) with (5.13) completes the proof of the theorem.

Corollary 5.13 *$N(n)$ is four times the excess of the number of divisors of n of the form $4j + 1$ over those of the form $4j + 3$.*

Proof. If $d = 1$, then $h(d) = 1$. If d is even, then $h(d) = 0$. If d is odd and d is the product of the primes $p_1, p_2, \cdots, p_k$, not necessarily distinct, then $h(d) = 1$ or -1 according as an even or an odd number of the p_i are of the form $4j + 3$. But also, in this case, $d \equiv p_1p_2\cdots p_k \equiv 1$ or $3 \pmod 4$, according as an even or an odd number of the p_i are of the form $4j + 3$. Since $N(n) = 4\sum_{d\mid n} h(d)$, the corollary follows.

Corollary 5.14 *The equation $x^2 + y^2 = n$ is solvable if and only if the canonical factoring of n into prime powers contains no factor p^e with p of the form $4j + 3$ and e odd.*

Proof. This follows at once from (5.11) and (5.12) and Theorem 5.11.

PROBLEMS

1. Evaluate $N(n)$, $P(n)$, and $Q(n)$ for $n = 100$, 101, and 102.
2. Prove that if n is square-free, $N(n) = Q(n)$.
3. Prove that the number of representations of an integer $m > 1$ as a sum of two squares of positive relatively prime integers equals the number of solutions of the congruence $x^2 \equiv -1 \pmod m$.
4. Corollary 5.13 implies that every positive integer has at least as many divisors of the form $4j + 1$ as of the form $4j + 3$. Prove this fact directly.
5. For a given positive integer K, prove that there is an integer n such that
(*a*) $N(n) = K$ if and only if $K \equiv 0 \pmod 4$;
(*b*) $P(n) = K$ if and only if K has the form 2^m with $m \geqq 0$;
(*c*) $Q(n) = K$ if and only if K has the form 2^m with $m \geqq 2$.
6. Prove that if an integer n is divisible by a prime of the form $4k + 3$, then $Q(n) = 0$.
7. Suppose that q is any positive divisor of n, and that n is expressible as $n = a^2 + b^2$ with $(a, b) = 1$. Prove that there exist integers c and d such that $c^2 + d^2 = q$ with $(c, d) = 1$.

5.11 The Equation $4x^2 + y^2 = n$

The ideas and methods of Section 5.10 can be greatly extended. The quadratic form $x^2 + y^2$ is only a special case of an extended theory. However, it is probably the most interesting case, and we will do no more than consider a few direct consequences and applications.

In this section we will restrict our attention to positive $n \equiv 1 \pmod 4$. If $x^2 + y^2 = n$, then one of x and y is odd, the other is even. If x is even, we let $u = x/2$, $v = y$; if y is even, we let $u = y/2$, $v = x$. In both cases we have $4u^2 + v^2 = n$. Since $x^2 + y^2 = y^2 + x^2$ and $x \neq y$, we see that exactly two solutions of $x^2 + y^2 = n$ correspond to one solution of $4u^2 + v^2 = n$. Also $(x, y) = (2u, v) = (u, v)$ since v is odd. If we define $N'(n)$, $P'(n)$, $Q'(n)$ for the equation $4x^2 + y^2 = n$, just as we defined $N(n)$, $P(n)$, $Q(n)$ for $x^2 + y^2 = n$, we have

$$N'(n) = \frac{N(n)}{2}, \qquad P'(n) = \frac{P(n)}{2}, \qquad Q'(n) = \frac{Q(n)}{2}.$$

Theorem 5.15 *Let n be an integer, $n > 1$, $n \equiv 1 \pmod 4$. If n is a prime, then $4x^2 + y^2 = n$ has exactly one non-negative solution, and it is a primitive solution. If n is not a prime, then $4x^2 + y^2 = n$ has either no primitive solutions, more than one non-negative primitive solution, or it has one non-negative primitive solution and at least one non-negative nonprimitive solution.*

Proof. If n is a prime, we use Theorems 5.12 and 5.10 to obtain

$$N'(n) = \frac{N(n)}{2} = 2(h(n) + h(1)) = 4, \qquad P'(n) = \frac{P(n)}{2} = \frac{R(n)}{2} = 1.$$

Therefore $4x^2 + y^2 = n$ has just one non-negative primitive solution, say u, v. Then, changing signs of u and v, we find three other solutions. Since $N'(n) = 4$ we see that $4x^2 + y^2 = n$ has just one non-negative solution, and it is primitive.

If n is composite, and if some prime $p \equiv 3 \pmod 4$ divides n, then, by Theorem 5.10, $Q'(n) = Q(n)/2 = 2R(n) = 0$. Thus $4x^2 + y^2 = n$ has no primitive solutions in this case.

If $n = p_1^{e_1}p_2^{e_2} \cdots p_r^{e_r}$, $p_i \equiv 1 \pmod 4$, $e_i > 0$, $r > 1$, then $P'(n) = P(n)/2 = R(n)/2 = 2^{r-1} > 1$, and $4x^2 + y^2 = n$ has more than one non-negative primitive solution.

If $n = p^e$, $e > 1$, $p \equiv 1 \pmod 4$, then

$$N'(n) = \frac{N(n)}{2} = 2\sum_{f=0}^{e} h(p^f) = 2(e + 1) \geqq 6$$

and $P'(n) = P(n)/2 = R(p^e)/2 = 1$. Therefore $4x^2 + y^2 = n$ has just one non-negative primitive solution, and it has more than four solutions. It must have some nonprimitive solution, u, v. Then $|u|$, $|v|$ is a non-negative non-primitive solution. This completes the proof.

The problem of deciding whether or not some number is a prime has always been of interest to many. If the number is large, it can present quite a problem, and it has led to the development of various methods and machines. No matter how far these go, there is always someone trying to push a little further. Theorem 5.15 is a criterion for primality. Given $n \equiv 1 \pmod 4$, we look for non-negative solutions of $4x^2 + y^2 = n$. If we find a nonprimitive solution or two non-negative solutions, we can stop looking further; n is composite. If we find no solutions, n is again composite. If we find exactly one non-negative solution, and it is primitive, then n is a prime.

If we find a nonprimitive solution x, y of $4x^2 + y^2 = n$ we not only know that n is composite, but also we know a factor (x, y) of n. If we find two non-negative primitive solutions u, v and μ, ν we can also find a factor of n. Since $2u$, v and 2μ, ν are different non-negative primitive solutions of $x^2 + y^2 = n$, they determine different s and t such that $sv \equiv 2u \pmod n$, $t\nu \equiv 2\mu \pmod n$ by Theorem 5.8. Then $s^2 \equiv t^2 \equiv -1 \pmod n$, $s \not\equiv t \pmod n$, and

$$2(s - t)(u\nu + v\mu) \equiv (s - t)(sv\nu + vt\nu) \equiv (s - t)(s + t)v\nu$$
$$\equiv (s^2 - t^2)v\nu \equiv 0 \pmod n,$$

which implies that $u\nu + v\mu$ and n have a common factor >1. If g denotes $(u\nu + v\mu, n)$, then since $\sqrt{n}/2$ is not an integer,

$$1 < g \leqq u\nu + v\mu \leqq \left[\frac{\sqrt{n}}{2}\right][\sqrt{n}] + \left[\frac{\sqrt{n}}{2}\right][\sqrt{n}] < n,$$

and g is a proper factor of n.

When looking for non-negative solutions of $4x^2 + y^2 = n$, we can restrict x to $0 \leqq x \leqq \sqrt{n}/2$ and we need only check whether $n - 4x^2$ is a perfect square. But we don't even need to try all these values of x. Since y is odd, we have $y^2 \equiv 1 \pmod 8$. Therefore x must satisfy $n - 4x^2 \equiv 1 \pmod 8$, which is equivalent to $x^2 \equiv (n - 1)/4 \pmod 2$. Thus x is even if $n \equiv 1 \pmod 8$, x is odd if $n \equiv 5 \pmod 8$. In either case we have halved the number of x that must be tried.

Going further, if c is odd and positive then, $n - 4x^2 \equiv y^2 \pmod c$, and we can exclude all x for which $n - 4x^2$ is a quadratic nonresidue modulo c. For example 2 and 3 are nonresidues modulo 5, and we can exclude x for which $n - 4x^2 \equiv 2$ or $3 \pmod 5$, that is $x^2 \equiv 2 - n$ or $3 - n \pmod 5$. If $n \equiv 3 \pmod 5$ this excludes x such that $x^2 \equiv -1$ or $0 \pmod 5$, that is $x \equiv 0, 2, 3 \pmod 5$.

Let us consider a simple example. We take $n = 4993$. Then we can restrict x to $0 \leqq x < 71/2 < 36$. Since $n \equiv 1 \pmod 8$, we take x even. Since $n \equiv 3 \pmod 5$ we can exclude all $x \equiv 0, 2, 3 \pmod 5$. Writing down the even numbers $0, 2, 4, \cdots, 34$ and scratching out the $x \equiv 0, 2, 3 \pmod 5$, we find that we need test only $x = 4, 6, 14, 16, 24, 26, 34$. This list can be cut down further if we use other values of c. For example, $c = 7$ eliminates 6 and 34, and $c = 11$ eliminates 14. This is hardly worthwhile for an n of this size. In any event we have only a few values to check. It will be found that $x = 16$ gives $n - 4x^2 = 3969 = 63^2$ and that there are no other solutions. Therefore 4993 is a prime.

For larger values of n one would probably use more values of c. If one is going to use this method very much, it is advisable to make short tables showing which x are excluded by various values of c. If the method is carried out systematically, and if a table of squares is available, it is a useful test for primality of n that are not too large. This test was based on the equation $4x^2 + y^2 = n$, and it is useful only for $n \equiv 1 \pmod 4$. There are other tests, based on other equations, valid for other n.

5.12 The Equation $ax^2 + by^2 + cz^2 = 0$

Although the theorem we give concerning this equation goes back to Legendre, our proof is a recent one, adapted from a paper of Mordell (see the Notes at the end of this chapter).

Theorem 5.16 *Let a, b, c be nonzero integers such that the product abc is square-free. Necessary and sufficient conditions that $ax^2 + by^2 + cz^2 = 0$ have a solution in integers x, y, z, not all zero, are that a, b, c do not have the same sign, and that $-bc$, $-ac$, $-ab$ are quadratic residues modulo a, b, c, respectively.*

Before giving the proof of this result we establish two lemmas.

Lemma 5.17 *Let λ, μ, ν be positive real numbers with product $\lambda\mu\nu = m$ an integer. Then any congruence $\alpha x + \beta y + \gamma z \equiv 0 \pmod m$ has a solution x, y, z, not all zero, such that $|x| \leqq \lambda$, $|y| \leqq \mu$, $|z| \leqq \nu$.*

Proof. Let x range over the values $0, 1, \cdots, [\lambda]$, y over $0, 1, \cdots, [\mu]$, and z over $0, 1, \cdots, [\nu]$. This gives us $(1 + [\lambda])(1 + [\mu])(1 + [\nu])$ different triples x, y, z. Now $(1 + [\lambda])(1 + [\mu])(1 + [\nu]) > \lambda\mu\nu = m$ by Theorem 4.1*a* and hence there must be some two triples x_1, y_1, z_1 and x_2, y_2, z_2 such that $\alpha x_1 + \beta y_1 + \gamma z_1 \equiv \alpha x_2 + \beta y_2 + \gamma z_2 \pmod m$. Then we have $\alpha(x_1 - x_2) + \beta(y_1 - y_2) + \gamma(z_1 - z_2) \equiv 0 \pmod m$, $|x_1 - x_2| \leqq [\lambda] \leqq \lambda$, $|y_1 - y_2| \leqq \mu$, $|z_1 - z_2| \leqq \nu$.

Lemma 5.18 *Suppose that $ax^2 + by^2 + cz^2$ factors into linear factors modulo m and also modulo n; that is*

$$ax^2 + by^2 + cz^2 \equiv (\alpha_1 x + \beta_1 y + \gamma_1 z)(\alpha_2 x + \beta_2 y + \gamma_2 z) \pmod{m},$$
$$ax^2 + by^2 + cz^2 \equiv (\alpha_3 x + \beta_3 y + \gamma_3 z)(\alpha_4 x + \beta_4 y + \gamma_4 z) \pmod{n}.$$

If $(m, n) = 1$ then $ax^2 + by^2 + cz^2$ factors into linear factors modulo mn.

Proof. Using Theorem 2.14, we can choose $\alpha, \beta, \gamma, \alpha', \beta', \gamma'$ to satisfy

$$\alpha \equiv \alpha_1, \beta \equiv \beta_1, \gamma \equiv \gamma_1, \alpha' \equiv \alpha_2, \beta' \equiv \beta_2, \gamma' \equiv \gamma_2 \pmod{m},$$
$$\alpha \equiv \alpha_3, \beta \equiv \beta_3, \gamma \equiv \gamma_3, \alpha' \equiv \alpha_4, \beta' \equiv \beta_4, \gamma' \equiv \gamma_4 \pmod{n}.$$

Then the congruence

$$ax^2 + by^2 + cz^2 \equiv (\alpha x + \beta y + \gamma z)(\alpha' x + \beta' y + \gamma' z)$$

holds modulo m and modulo n, and hence it holds modulo mn.

Proof of Theorem 5.16. If $ax^2 + by^2 + cz^2 = 0$ has a solution x_0, y_0, z_0 not all zero, then a, b, c are not of the same sign. Dividing x_0, y_0, z_0 by (x_0, y_0, z_0) we have a solution x_1, y_1, z_1 with $(x_1, y_1, z_1) = 1$.

Next we prove that $(c, x_1) = 1$. If this were not so there would be a prime p dividing both c and x_1. Then $p \nmid b$ since $p \mid c$ and abc is square-free. Therefore $p \mid by_1^2$ and $p \nmid b$, hence $p \mid y_1^2$, $p \mid y_1$, and then $p^2 \mid (ax_1^2 + by_1^2)$ so that $p^2 \mid cz_1^2$. But c is square-free so $p \mid z_1$. We have concluded that p is a factor of x_1, y_1, and z_1 contrary to $(x_1, y_1, z_1) = 1$. Consequently, we have $(c, x_1) = 1$.

Let u be chosen to satisfy $ux_1 \equiv 1 \pmod{c}$. Then the equation $ax_1^2 + by_1^2 + cz_1^2 = 0$ implies $ax_1^2 + by_1^2 \equiv 0 \pmod{c}$, and multiplying this by $u^2 b$ we get $u^2 b^2 y_1^2 = -ab \pmod{c}$. Thus we have established that $-ab$ is a quadratic residue modulo c. A similar proof shows that $-bc$ and $-ac$ are quadratic residues modulo a and b respectively.

Conversely, let us assume that $-bc, -ac, -ab$ are quadratic residues modulo a, b, c respectively. Note that this property does not change if a, b, c are replaced by their negatives. Since a, b, c are not of the same sign, we can change the signs of all of them, if necessary, in order to have one positive and two of them negative. Then, perhaps with a change of notation, we can arrange it so that a is positive and b and c are negative.

Define r as a solution of $r^2 \equiv -ab \pmod{c}$, and a_1 as a solution of $aa_1 \equiv 1 \pmod{c}$. These solutions exist because of our assumptions on a, b, c. Then we can write

$$ax^2 + by^2 \equiv aa_1(ax^2 + by^2) \equiv a_1(a^2x^2 + aby^2) \equiv a_1(a^2x^2 - r^2y^2)$$
$$\equiv a_1(ax - ry)(ax + ry) \equiv (x - a_1 ry)(ax + ry) \pmod{c},$$
$$ax^2 + by^2 + cz^2 \equiv (x - a_1 ry)(ax + ry) \pmod{c}.$$

Thus $ax^2 + by^2 + cz^2$ is the product of two linear factors modulo c, and similarly modulo a and modulo b. Applying Lemma 5.18 twice, we conclude that $ax^2 + by^2 + cz^2$ can be written as the product of two linear factors modulo abc. That is, there exist numbers $\alpha, \beta, \gamma, \alpha', \beta', \gamma'$ such that

$$(5.14) \quad ax^2 + by^2 + cz^2 \equiv (\alpha x + \beta y + \gamma z)(\alpha' x + \beta' y + \gamma' z) \pmod{abc}.$$

Now we apply Lemma 5.17 to the congruence

$$(5.15) \qquad \alpha x + \beta y + \gamma z \equiv 0 \pmod{abc},$$

using $\lambda = \sqrt{bc}$, $\mu = \sqrt{|ac|}$, $\nu = \sqrt{|ab|}$. Thus we get a solution x_1, y_1, z_1 of the congruence (5.15) with $|x_1| \leqq \sqrt{bc}$, $|y_1| \leqq \sqrt{|ac|}$, $|z_1| \leqq \sqrt{|ab|}$. But abc is square-free, so $\sqrt{bc}$ is an integer only if it is 1, and similarly for $\sqrt{|ac|}$ and $\sqrt{|ab|}$. Therefore we have

$|x_1| \leqq \sqrt{bc}$, $x_1^2 \leqq bc$ with equality possible only if $b = c = -1$,

$|y_1| \leqq \sqrt{|ac|}$, $y_1^2 \leqq -ac$ with equality possible only if $a = 1$, $c = -1$,

$|z_1| \leqq \sqrt{|ab|}$, $z_1^2 \leqq -ab$ with equality possible only if $a = 1$, $b = -1$.

Hence, since a is positive and b and c are negative, we have, unless $b = c = -1$,

$$ax_1^2 + by_1^2 + cz_1^2 \leqq ax_1^2 < abc,$$

and

$$ax_1^2 + by_1^2 + cz_1^2 \geqq by_1^2 + cz_1^2 > b(-ac) + c(-ab) = -2abc.$$

Leaving aside the special case when $b = c = -1$, we have

$$-2abc < ax_1^2 + by_1^2 + cz_1^2 < abc.$$

Now x_1, y_1, z_1 is a solution of (5.15) and so also, because of (5.14), a solution of

$$ax^2 + by^2 + cz^2 \equiv 0 \pmod{abc}.$$

Thus the above inequalities imply that

$$ax_1^2 + by_1^2 + cz_1^2 = 0 \quad \text{or} \quad ax_1^2 + by_1^2 + cz_1^2 = -abc.$$

In the first case we have our solution of $ax^2 + by^2 + cz^2 = 0$. In the second case we readily verify that x_2, y_2, z_2, defined by $x_2 = -by_1 + x_1z_1$, $y_2 = ax_1 + y_1z_1$, $z_2 = z_1^2 + ab$, form a solution. In case $x_2 = y_2 = z_2 = 0$ then $z_1^2 + ab = 0$, $z_1^2 = -ab$ and $z_1 = \pm 1$ because ab, like abc, is square-free. Then $a = 1$, $b = 1$, and $x = 1$, $y = -1$, $z = 0$ is a solution.

Finally we must dispose of the special case $b = c = -1$. The conditions on a, b, c now imply that -1 is a quadratic residue modulo a, in other words, that $R(a)$ of Theorem 5.10 is positive. By Theorem 5.10 this implies that

$Q(a)$ is positive and hence that the equation $y^2 + z^2 = a$ has a solution y_1, z_1. Then $x = 1$, $y = y_1$, $z = z_1$ is a solution of $ax^2 + by^2 + cz^2 = 0$ since $b = c = -1$.

PROBLEM

1. Show that in the proof of Theorem 5.16 we have established more than the theorem stated, that the following stronger result is implied. Let a, b, c be non-zero integers not of the same sign such that the product abc is square-free. Then the following three conditions are equivalent.
(*a*) $ax^2 + by^2 + cz^2 = 0$ has a solution x, y, z not all zero;
(*b*) $ax^2 + by^2 + cz^2$ factors into linear factors modulo abc;
(*c*) $-bc$, $-ac$, $-ab$ are quadratic residues modulo a, b, c respectively.

5.13 Binary Quadratic Forms

A *form* is a homogeneous polynomial, that is, a polynomial in several variables, all of whose terms are of the same degree. A *quadratic* form f has terms of degree two, and is therefore an expression of the type

$$f(x_1, x_2, \cdots, x_n) = \sum_{i,j=1}^{n} a_{ij} x_i x_j. \tag{5.16}$$

We restrict our attention to quadratic forms with integral coefficients a_{ij}. If $f(x_1, x_2, \cdots, x_n)$ assumes only positive values whenever $x_1, x_2, \cdots, x_n$ are replaced by any set of integers other than $0, 0, \cdots, 0$, then f is said to be a *positive* form. Similarly f is called a *negative* form if its value is negative when $x_1, x_2, \cdots, x_n$ are replaced by integers not all zero. A *definite* form is one which is either positive or negative. For example $x_1^2 + x_2^2$, or $x^2 + y^2$ in other notation, is a positive form, and $-x^2 - 3y^2$ is a negative form; both are definite forms. The form $x^2 - y^2$ is an indefinite form. Sometimes positive forms are called positive definite, and negative forms negative definite. Clearly if f is positive then $-f$ is negative, and conversely. Hence there is no need to study both positive and negative forms because the properties of one kind follow from the properties of the other.

A quadratic form (5.16) is said to represent an integer m if there are integers $b_1, b_2, \cdots, b_n$ such that $f(b_1, b_2, \cdots, b_n) = m$. For example, $x^2 + y^2$ represents 5 but not 6. Every quadratic form represents 0 because $f(0, 0, \cdots, 0) = 0$. The form f is called a *zero* form if $f(b_1, b_2, \cdots, b_n) = 0$ for some integers $b_1, b_2, \cdots, b_n$, not all zero. By definition a definite form is not a zero form. Theorem 5.16 gave necessary and sufficient conditions that $ax^2 + by^2 + cz^2$ be a zero form for a wide class of integers a, b, c.

A positive form is said to be *universal* if it represents all positive integers. Thus, according to Theorem 5.6, $x_1^2 + x_2^2 + x_3^2 + x_4^2$ is a universal form. Although we will not prove it, it is a fact that no form of the type $ax^2 + by^2 + cz^2$ can be universal. The arithmetic theory of quadratic forms includes such problems as to determine which forms are universal, to determine or characterize the class of integers represented by a quadratic form that is not universal, and to determine how many ways an integer can be represented by a quadratic form. For example, Corollary 5.14 determined the class of integers represented by the form $x^2 + y^2$, and Corollary 5.13 determined the number of representations.

A form involving two variables is called a *binary* form. The rest of this chapter deals with binary quadratic forms, that is forms of the type

$$f(x, y) = ax^2 + bxy + cy^2. \tag{5.17}$$

We do not intend to do more than just give an introduction to a part of the theory dealing with binary quadratic forms. The use of matrices would simplify a few of our proofs. However, for the little that we shall do, the simplification is not enough to offset the work that would be required in introducing matrices.

Theorem 5.19 *The quadratic form $f(x, y) = ax^2 + bxy + cy^2$ is positive if and only if its discriminant $b^2 - 4ac$ is negative, $a > 0$, and $c > 0$.*

Proof. Since $f(1, 0) = a$ and $f(0, 1) = c$, we see that f is not positive if $a \leqq 0$ or if $c \leqq 0$. We can now suppose $a > 0$, $c > 0$, and can write

$$f(x, y) = \frac{1}{4a}((2ax + by)^2 + (4ac - b^2)y^2). \tag{5.18}$$

This shows that $f(-b, 2a) = (4ac - b^2)a$ and hence, since $a > 0$, that f is not positive if $4ac - b^2 \leqq 0$. On the other hand, if $b^2 - 4ac < 0$, it shows that $f(x, y)$ is never negative no matter what integers may be substituted for x and y. Moreover $f(x, y) = 0$ then holds if and only if $2ax + by = 0$ and $(4ac - b^2)y^2 = 0$. These equations imply that $y = 0$ and $x = 0$ if $a > 0$, $b^2 - 4ac < 0$.

The quadratic form $x^2 - dy^2$ with $d > 0$ has discriminant $4d > 0$, and it is clearly indefinite. Pell's equation $x^2 - dy^2 = n$, $d > 0$, d not a perfect square, will be discussed in Section 7.8. For fixed d and n it turns out that the equation has either no solution or an infinite number of solutions. On the other hand it is easy to see that $x^2 - y^2 = n$ has no solutions if $n \equiv 2 \pmod 4$, and only a finite number of solutions, $x = (t + n/t)/2$, $y = (t - n/t)/2$, $t \mid n$, $t \equiv n/t \pmod 2$, if $n \not\equiv 2 \pmod 4$. The first situation does not arise in the case of definite forms, as shown in the following theorem.

Theorem 5.20 *Let f be a positive quadratic form. Then the number of representations of an integer m by f is finite, possibly zero.*

Proof. We will show that there is only a finite number of pairs of integers for which $f(x, y) \leqq m$. We can suppose $m > 0$. Using (5.18) we see that $f(x, y) \leqq m$ implies $(4ac - b^2)y^2 \leqq 4am$, and hence

$$-2\left(\frac{am}{4ac - b^2}\right)^{\frac{1}{2}} \leqq y \leqq 2\left(\frac{am}{4ac - b^2}\right)^{\frac{1}{2}}.$$

This restricts y to a finite number of values. For each such y we then have $(2ax + by)^2 \leqq 4am - (4ac - b^2)y^2$, which then restricts $2ax + by$ and hence x to a finite number of values.

Corollary 5.21 *Let f be a positive quadratic form. Then the smallest positive integer represented by f can be found in a finite number of steps.*

Proof. Since a is represented by f we just take $m = a - 1$ in the proof of Theorem 5.20, find all y and x that are allowed, and then determine the least value of $f(x, y)$.

For example, if $f(x, y) = 5x^2 + 14xy + 11y^2$, then $b^2 - 4ac = -24 < 0$ and y is limited to $-\sqrt{30}/3 \leqq y \leqq \sqrt{30}/3$. Hence $y = -1, 0, 1$. For $y = -1$ we have $(10x - 14)^2 \leqq 56$ from which we find $x = 1$ or 2, $f(x, -1) = 2$ or 3. For $y = 0$ we find $x = 0$, $f(0, 0) = 0$. For $y = 1$ we find $x = -2$ or -1, $f(x, 1) = 3$ or 2. Therefore the least positive integer represented by f is 2.

If in the quadratic form $f(x, y) = 5x^2 + 14xy + 11y^2$ we replace x by $-X + 2Y$ and y by $X - Y$, we get the form $F(X, Y) = 2X^2 + 3Y^2$. Now the transformation

$$x = -X + 2Y, \qquad y = X - Y \tag{5.19}$$

can be solved for X and Y to give the inverse transformation

$$X = x + 2y, \qquad Y = x + y.$$

The transformation (5.19) has the special property that its inverse also has integral coefficients; both (5.19) and its inverse are integral transformations. If X and Y are replaced by integers, then (5.19) gives a pair of integers for x and y to make the value of $f(x, y)$ the same as that of $F(X, Y)$. The inverse transformation works the other way, giving integers X and Y corresponding to integers x and y.

It follows that any integer represented by f can also be represented by F, and conversely. Furthermore, the number of representations is the same in both cases. The form F is considerably simpler than f. By trial, it is easy to verify that all the solutions of $F(X, Y) = 14$ are $(1, 2)$, $(1, -2)$, $(-1, 2)$, $(-1, -2)$. Then (5.19) gives us the solutions of $f(x, y) = 14$, namely $(3, -1)$, $(-5, 3)$, $(5, -3)$, $(-3, 1)$.

The quadratic form F is said to be equivalent to f. Anything we can say about representations by F can be carried over to f by virtue of (5.19) and its inverse. If we study F there is no need to study f or any other equivalent form. This example suggests the desirability of studying transformations and equivalence of forms more closely.

PROBLEMS

1. Determine the class of integers represented by each of the following forms:
(*a*) $2x^2 + 2y^2$;
(*b*) $2x^2 - 2y^2$;
(*c*) $x^2 - xy$;
(*d*) $2x^2 + 2y^2 + 2z^2 + 2t^2$.

2. Prove that $x^2 - 2xy + y^2$ is a zero form. Determine the class of integers represented by this form.

3. If C is any class of integers, finite or infinite, let mC denote the class obtained by multiplying each integer of C by the integer m. Prove that if C is the class of integers represented by any form f, then mC is the class represented by mf.

4. Prove that $ax^2 + bxy + cy^2$ is a positive form if and only if $a > 0$ and $b^2 - 4ac < 0$. (Observe that this problem shows that the condition $c > 0$ in Theorem 5.19 is superfluous.)

5. Prove that $ax^2 + bxy + cy^2$ is a positive form if and only if $c > 0$ and $b^2 - 4ac < 0$.

6. Prove that the form $ax^2 + bxy + cy^2$ is negative if and only if $a < 0$ and $b^2 - 4ac < 0$.

7. Prove that the form $ax^2 + bxy + cy^2$ is definite if and only if $b^2 - 4ac < 0$.

8. Prove that $x^2 + 7xy + y^2$ is neither a definite form nor a zero form.

9. Find all solutions of the Diophantine equations
(*a*) $5x^2 + 14xy + 11y^2 = 35$;
(*b*) $5x^2 + 14xy + 11y^2 = 46$.

10. Find the least positive integer represented by the positive form $7x^2 + 25xy + 23y^2$.

11. If $f(x, y)$ is a positive binary quadratic form, prove that $f(\alpha, \beta)$ is positive for every pair of real numbers α, β except $0, 0$.

12. Let $f(x, y) = ax^2 + bxy + cy^2$ be a quadratic form with $a > 0$ and $b^2 - 4ac = 0$. Given also that $(a, b, c) = 1$, prove that the integers represented by f are precisely the numbers m^2, with $m = 0, 1, 2, \cdots$. Hence, removing the restriction that $(a, b, c) = 1$, prove that the integers represented by f are all numbers of the form dm^2, where $d = (a, b, c)$.

5.14 Equivalence of Quadratic Forms

Theorem 5.22 *The inverse of an integral transformation*

$$x = \alpha X + \beta Y, \qquad y = \gamma X + \delta Y \tag{5.20}$$

is also an integral transformation if and only if $\Delta = \pm 1$ where Δ is the determinant of the transformation,

$$\Delta = \begin{vmatrix} \alpha\beta \\ \gamma\delta \end{vmatrix} = \alpha\delta - \beta\gamma.$$

Proof. The inverse exists if and only if (5.20) can be solved uniquely for X and Y, hence if and only if $\Delta \neq 0$. Then if the inverse does exist we can solve to get

$$X = \frac{\delta}{\Delta}\, x - \frac{\beta}{\Delta}\, y, \qquad Y = -\frac{\gamma}{\Delta}\, x + \frac{\alpha}{\Delta}\, y. \tag{5.21}$$

If $\Delta = \pm 1$ then (5.21) is an integral transformation.

Now suppose that (5.21) is an integral transformation. Let p be any prime, and let p^k be the highest power of p that divides Δ. If p does not divide Δ, then $k = 0$. Since the coefficients in (5.21) are integers, we conclude that

$$p^k \mid \delta, \qquad p^k \mid \beta, \qquad p^k \mid \gamma, \qquad p^k \mid \alpha.$$

It follows that

$$p^{2k} \mid \alpha\delta, \qquad p^{2k} \mid \beta\gamma, \qquad p^{2k} \mid \Delta.$$

But p^k is the highest power of p that divides Δ, so we have $2k \leqq k$, which implies $k = 0$. Thus no prime divides Δ, and hence $\Delta = \pm 1$.

This theorem shows that if F is a quadratic form obtained from a form f by means of (5.20) with $\Delta = \pm 1$, then, just as in Section 5.13, the problem of representing integers by f can be reduced to that of representing them by F.

Definition 5.1 *A quadratic form $f(x, y) = ax^2 + bxy + cy^2$ is equivalent to a form $g(x, y) = Ax^2 + Bxy + Cy^2$ if there is an integral transformation* (5.20), *with determinant $\Delta = \pm 1$, that carries $f(x, y)$ into $g(X, Y)$. In case f is equivalent to g we write $f \sim g$.*

Applying (5.20) to f and multiplying out we find

$$A = a\alpha^2 + b\alpha\gamma + c\gamma^2, \qquad B = 2a\alpha\beta + b(\alpha\delta + \beta\gamma) + 2c\gamma\delta, \tag{5.22}$$
$$C = a\beta^2 + b\beta\delta + c\delta^2,$$

if (5.20) carries $f(x, y)$ into $g(X, Y)$.

The next theorem shows that this equivalence is a true equivalence relation: it is reflexive, symmetric, and transitive.

Theorem 5.23 *Let f, g, and h be binary quadratic forms. Then*

(*a*) $f \sim f$,

(*b*) *if* $f \sim g$ *then* $g \sim f$,

(*c*) *if* $f \sim g$ *and* $g \sim h$ *then* $f \sim h$.

Proof. (*a*) The identity transformation $x = X$, $y = Y$ has determinant 1, and it carries $f(x, y)$ into $f(X, Y)$.

(*b*) We can suppose that (5.20) is the transformation that carries $f(x, y)$ into $g(X, Y)$. Then, by (5.21) the inverse transformation that carries $g(X, Y)$ back into $f(x, y)$ has determinant

$$\frac{\delta}{\Delta}\frac{\alpha}{\Delta} - \frac{\beta}{\Delta}\frac{\gamma}{\Delta} = \frac{1}{\Delta} = \pm 1.$$

(*c*) We suppose that (5.20) with $\Delta = \pm 1$ carries $f(x, y)$ into $g(X, Y)$ and that the transformation $X = \alpha_1 u + \beta_1 v$, $Y = \gamma_1 u + \delta_1 v$, with $\Delta_1 = \alpha_1\delta_1 - \beta_1\gamma_1 = \pm 1$ carries $g(X, Y)$ into $h(u, v)$. If we eliminate X and Y, we find

$$x = (\alpha\alpha_1 + \beta\gamma_1)u + (\alpha\beta_1 + \beta\delta_1)v,$$
$$y = (\gamma\alpha_1 + \delta\gamma_1)u + (\gamma\beta_1 + \delta\delta_1)v,$$

and this transformation carries $f(x, y)$ into $h(u, v)$. An easy computation shows that the determinant of this transformation is equal to $(\alpha\delta - \beta\gamma)(\alpha_1\delta_1 - \beta_1\gamma_1) = \pm 1$. Therefore $f(x, y)$ is equivalent to $h(x, y)$.

Theorem 5.24 *If* $f \sim g$, *then the discriminants of f and g are equal.*

Proof. We suppose that (5.20) carries $f(x, y)$ into $g(X, Y)$ as in Definition 5.1. It is possible to compute the discriminant $B^2 - 4AC$ of g, using (5.22), to obtain $B^2 - 4AC = (\alpha\delta - \beta\gamma)^2(b^2 - 4ac)$, but it is simpler to use the ordinary rule for multiplying determinants,

$$\begin{vmatrix} \alpha & \gamma \\ \beta & \delta \end{vmatrix}\begin{vmatrix} 2a & b \\ b & 2c \end{vmatrix}\begin{vmatrix} \alpha & \beta \\ \gamma & \delta \end{vmatrix} = \begin{vmatrix} 2a\alpha + b\gamma & b\alpha + 2c\gamma \\ 2a\beta + b\delta & b\beta + 2c\delta \end{vmatrix}\begin{vmatrix} \alpha & \beta \\ \gamma & \delta \end{vmatrix}$$
$$= \begin{vmatrix} 2A & B \\ B & 2C \end{vmatrix}.$$

Evaluating the determinants we find $(\alpha\delta - \beta\gamma)^2(4ac - b^2) = 4AC - B^2$. Since $(\alpha\delta - \beta\gamma)^2 = \Delta^2 = 1$ we have $b^2 - 4ac = B^2 - 4AC$.

In light of Theorem 5.23 we can separate all binary quadratic forms into equivalence classes, putting two forms f and g into the same class if $f \sim g$. Part (*b*) of Theorem 5.23 assures that f and g represent the same set of integers if f and g are in the same class, and that there is a one to one correspondence between the representations by f and those by g. In other words it suffices to consider just one representative form from each equivalence class. Clearly, if one form in a class is positive then so are all the others in that class. Also two forms in the same class have the same discriminants.

The next thing that it is desirable to have is some way of picking one unique representative form from each equivalence class. In the case of positive binary quadratic forms, this can be done in a fairly simple way. The whole theory connected with positive forms is quite elegant. For this reason we will restrict our attention to positive forms from now on.

Definition 5.2 *A positive binary quadratic form $ax^2 + bxy + cy^2$ is a reduced form if $0 \leqq b \leqq a \leqq c$.*

For example, $x^2 + y^2$ is a reduced form.

Theorem 5.25 *To each positive binary quadratic form there corresponds an equivalent reduced form.*

Proof. Consider any positive binary quadratic form $g(x, y) = Ax^2 + Bxy + Cy^2$. We shall show that there is an equivalent form $ax^2 + bxy + cy^2$ with $0 \leqq |b| \leqq a \leqq c$. This will suffice because in case b is negative, the transformation $x = -X$, $y = Y$ will carry $ax^2 + bxy + cy^2$ into a reduced form.

Let a denote the least positive integer represented by g. In Corollary 5.21 we saw how a can be found. There are integers α and γ such that $A\alpha^2 + B\alpha\gamma + C\gamma^2 = a$. Note that $(\alpha, \gamma) = 1$; for if $(\alpha, \gamma) = d$, then $g(\alpha/d, \gamma/d) = a/d^2$, which implies $d = 1$ since a was minimal. Then by Theorem 1.3 there are integers β and δ such that $\alpha\delta - \beta\gamma = 1$, and we use these to construct the transformation $x = \alpha u + \beta v$, $y = \gamma u + \delta v$ of determinant 1. This transformation carries $g(x, y)$ into

$$\begin{aligned} h(u, v) &= A(\alpha u + \beta v)^2 + B(\alpha u + \beta v)(\gamma u + \delta v) + C(\gamma u + \delta v)^2 \\ &= au^2 + kuv + mv^2, \end{aligned}$$

say, where a is the integer defined above.

Now, for any integer j the transformation $u = x - jy$, $v = y$ has determinant 1, and it carries $h(u, v)$ into

$$f(x, y) = ax^2 + (k - 2aj)xy + (aj^2 - kj + m)y^2.$$

If we take j to be the nearest integer to $k/2a$, we have

$$-\frac{1}{2} \leqq \frac{k}{2a} - j \leqq \frac{1}{2}, \qquad -a \leqq k - 2aj \leqq a, \qquad |k - 2aj| \leqq a$$

and we can write

$$f(x, y) = ax^2 + bxy + cy^2, \qquad |b| \leqq a.$$

Now $g \sim f$, g is positive, and $f(0, 1) = c$. Therefore g represents c, and c is positive, hence $c \geqq a$.

Theorem 5.26 *If two reduced forms are equivalent they are identical.*

Proof. Let $ax^2 + bxy + cy^2$ and $Ax^2 + Bxy + Cy^2$ be equivalent reduced forms. We can suppose that $a \geqq A$ and that (5.20) is the transformation carrying one into the other. Then the coefficients satisfy (5.22).

First we prove that $a = A$. We have $0 \leqq b \leqq a \leqq c$, and using the simple inequality $\alpha^2 + \gamma^2 \geqq 2\,|\alpha\gamma|$ we find

$$(5.23)\qquad \begin{aligned} A &= a\alpha^2 + b\alpha\gamma + c\gamma^2 \geqq a\alpha^2 + c\gamma^2 - b\,|\alpha\gamma| \\ &\geqq a\alpha^2 + a\gamma^2 - b\,|\alpha\gamma| \geqq 2a\,|\alpha\gamma| - b\,|\alpha\gamma| \geqq a\,|\alpha\gamma|. \end{aligned}$$

Since $a \geqq A > 0$ it follows that $|\alpha\gamma| \leqq 1$. If $|\alpha\gamma| = 0$ we have

$$(5.24)\qquad A = a\alpha^2 + c\gamma^2 \geqq a\alpha^2 + a\gamma^2 \geqq a$$

because α and γ are not both zero. If $|\alpha\gamma| = 1$, then (5.23) reduces immediately to the same result, $A \geqq a$. Thus we have $A \geqq a$ in both cases, and this with $a \geqq A$ implies that $a = A$.

Now that we have $a = A > 0$ we need only prove $b = B$ or $c = C$, since one follows from the other because $b^2 - 4ac = B^2 - 4AC$ by Theorem 5.24. Thus the case $c = C$ requires no further proof. Since $a = A$ we can interchange the forms if necessary and can assume $c > C$. Then we have $c > a$ since $C \geqq A = a$. This rules out the possibility $|\alpha\gamma| = 1$, for if $|\alpha\gamma| = 1$ then $c\gamma^2 > a\gamma^2$, and (5.23) would then imply $A > a\,|\alpha\gamma| = a$. Now that we have $|\alpha\gamma| = 0$, we can prove that $\gamma = 0$, for otherwise we would again have $c\gamma^2 > a\gamma^2$ and (5.24) would imply $A > a$.

We have narrowed the possibilities down to $a = A$, $a < c$, $\gamma = 0$. Since the determinant $\Delta = \alpha\delta - \beta\gamma$ is ± 1, we also have $\alpha\delta = \pm 1$. Then by (5.22) we have $B = 2a\alpha\beta \pm b$. There are two cases.

First, suppose that $B = 2a\alpha\beta + b$. Then $0 \leqq b \leqq a$ and $0 \leqq B \leqq A = a$ imply that $-a \leqq B - b \leqq a$. But $B - b = 2a\alpha\beta$ is a multiple of $2a$ so we have $B - b = 0$, $b = B$.

Second, suppose that $B = 2a\alpha\beta - b$. This time we find $0 \leqq B + b \leqq 2a$ and $B + b$ is a multiple of $2a$. We have either $B + b = 0$ or $B + b = 2a$. Since $0 \leqq b \leqq a$ and $0 \leqq B \leqq a$ we have $b = B = 0$ if $B + b = 0$, and $b = B = a$ if $B + b = 2a$. Again we have $b = B$.

This completes the proof since we have $a = A$, $b = B$, $c = C$ in every case.

The last two theorems show that Definition 5.2 does just what we wanted. It supplies us with one and just one representative for each equivalence class of positive binary quadratic forms. It should be noted that Definition 5.1 is not the only way in which the positive binary quadratic forms can be put into equivalence classes. In fact, some authors demand that the transformation have determinant 1 in their definition of equivalence. They then use a slightly different definition of a reduced form in order that Theorems 5.25 and 5.26 still hold under their definitions of equivalence and of reduced forms.

Theorem 5.27 *There are only finitely many reduced forms having a given discriminant.*

Proof. Let $-d$ be any negative integer. If $ax^2 + bxy + cy^2$ is a reduced form with discriminant $-d$, then $b^2 - 4ac = -d$ and $0 \leqq b \leqq a \leqq c$. Thus we have $d = 4ac - b^2 \geqq 4ac - ac \geqq 3a^2$ as well as $0 \leqq b \leqq a$, so that there are at most $\sqrt{d/3}$ possible values for $a > 0$, and at most $a + 1$ values for b corresponding to each a. Finally there is at most one c, for each a, b, such that $4ac - b^2 = d$.

Let us find all reduced forms with $d \leqq 16$. We have $1 \leqq a \leqq \sqrt{16/3}$, $a = 1$ or 2. Corresponding to $a = 1$, $b = 0$ we have $d = 4c$, and we can take $c = 1, 2, 3$, or 4. Corresponding to $a = 1$, $b = 1$ we have $d = 4c - 1$, and again we can take $c = 1, 2, 3$, or 4. Similarly we find $a = 2$, $b = 0$, $d = 8c$, $c = 2$, and $a = 2$, $b = 1$, $d = 8c - 1$, $c = 2$, and $a = 2$, $b = 2$, $d = 8c - 4$, $c = 2$. Listing the reduced forms according to the values of d we get the following table.

$$d = 3,\ x^2 + xy + y^2$$
$$d = 4,\ x^2 + y^2$$
$$d = 7,\ x^2 + xy + 2y^2$$
$$d = 8,\ x^2 + 2y^2$$
$$d = 11,\ x^2 + xy + 3y^2$$
$$d = 12,\ x^2 + 3y^2,\ 2x^2 + 2xy + 2y^2$$
$$d = 15,\ x^2 + xy + 4y^2,\ 2x^2 + xy + 2y^2$$
$$d = 16,\ x^2 + 4y^2,\ 2x^2 + 2y^2.$$

For $d = 3, 4, 7, 8, 11$, there is just one reduced form of discriminant $-d$. For these values of d, any positive form with discriminant $-d$ will represent the same set of integers as the corresponding reduced form represents. For example, Corollary 5.14 determined the set of integers represented by $x^2 + y^2$, and this is the set represented by any positive form of discriminant -4.

The case $d = 12$ is a little different since there are two reduced forms of discriminant -12. But in the first form $x^2 + 3y^2$ the coefficients are relatively prime whereas in the second form $2x^2 + 2xy + 2y^2$ the coefficients are all divisible by 2. From (5.22) it is obvious that any form equivalent to

$2x^2 + 2xy + 2y^2$ will also have all its coefficients divisible by 2. Also by (5.22) any positive form with discriminant -12, having all its coefficients divisible by 2, will be equivalent to a reduced form having all its coefficients divisible by 2, hence it is equivalent to $2x^2 + 2xy + 2y^2$. Therefore we can make the statement: Suppose f is a positive form of discriminant -12. If the coefficients of f are relatively prime, then f represents the same class as does $x^2 + 3y^2$. If the coefficients are not relatively prime, then f represents the same class as does $2x^2 + 2xy + 2y^2$. The case $d = 16$ can be treated in a similar way.

The case $d = 15$ is still different, and harder. Here we have two reduced forms, and in both of them the coefficients are relatively prime. There are ways of distinguishing between forms equivalent to one or the other reduced form without having to actually produce the reduced form by going through the steps of the proof of Theorem 5.25. However, we shall not go into this.

This is as far as we will pursue the topic of quadratic forms. There is more that can be done. For one thing, we have not discussed the question of how to determine the class of integers represented by a reduced form. The methods used to obtain Corollary 5.14 can be extended to give some information. A further study of the transformation introduced in the present section is also very useful.

PROBLEMS

1. Prove that the following forms are equivalent:

$$ax^2 + bxy + cy^2, \qquad cx^2 + bxy + ay^2,$$
$$ax^2 - bxy + cy^2, \qquad cx^2 - bxy + ay^2.$$

2. Find the reduced form equivalent to

(*a*) $3x^2 + 7xy + 5y^2$;

(*b*) $2x^2 - 5xy + 4y^2$;

(*c*) $2x^2 + xy + 6y^2$;

(*d*) $3x^2 + xy + y^2$.

3. Prove that there are no binary quadratic forms with discriminant congruent to 2 or 3 modulo 4.

4. Find the reduced form equivalent to $7x^2 + 25xy + 23y^2$.

5. Prove that to any given positive binary quadratic form there corresponds infinitely many equivalent forms.

6. Prove that there is only one reduced form of discriminant -43. Hence prove that any two positive binary quadratic forms of discriminant -43 are equivalent.

7. Prove that any two positive binary quadratic forms of discriminant -67 are equivalent.

8. Denote the positive form $ax^2 + bxy + cy^2$ by $[a, b, c]$. Prove that, as a variation of the method of Theorem 5.25, the reduced form equivalent to $[a, b, c]$ can be obtained by a finite sequence of operations of three types:
(1) in case $a > c$, the replacement of $[a, b, c]$ by $[c, b, a]$;
(2) in case $|b| > a$, the replacement of $[a, b, c]$ by $[a, b - 2aj, c_1]$, where j is chosen so that $|b - 2aj| \leqq a$, and c_1 is determined by the equality of the discriminants, $b^2 - 4ac = (b - 2aj)^2 - 4ac_1$;
(3) in case $b < 0$, the replacement of $[a, b, c]$ by $[a, -b, c]$.

NOTES ON CHAPTER 5

The methods of this chapter are not suitable for dealing with Pell's equation, discussed in Section **7.8**, and the impossibility of $x^3 + y^3 = z^3$ in positive integers, given in Section **9.10**.

The Diophantine equations $x^{-n} + y^{-n} = z^{-n}$, where n is a positive integer, are shown to be very closely related to $x^n + y^n = z^n$ in the Special Topics, page 261,

For further reading on the subject of this chapter, see the book "Diophantine Equations" by L. J. Mordell listed in the General References on page 269, and the following:

J. Hunter, *Number Theory*, Edinburgh, Oliver and Boyd, 1964.

L. J. Mordell, "On the equation $ax^2 + by^2 - cz^2 = 0$," *Monatsh. Math.*, Bd. **55,** 323–327 (1951).

Th. Skolem, "A simple proof of the solvability of the Diophantine equation $ax^2 + by^2 + cz^2 = 0$," *Norsk Vid. Selsk. Forh.*, Trondheim **24,** 102–107 (1952).

H. S. Vandiver, "Fermat's last theorem, its history, and the nature of the known results concerning it," *Amer. Math. Monthly*, **53,** 555–578 (1946).

6

Farey Fractions and Irrational Numbers

A rational number is one which is expressible as the quotient of two integers. Real numbers which are not rational are said to be irrational. In this chapter the Farey fractions are presented; they give a useful classification of the rational numbers. Some results on irrational numbers are given in Section 6.3, and this material can be read independently of the other two sections. The discussion of irrational numbers is limited to number theoretic considerations, with no attention given to questions that belong more properly to analysis or the foundations of mathematics.

6.1 Farey Sequences

Let us construct a table in the following way. In the first row we write 0/1 and 1/1. For $n = 2, 3, \cdots$ we use the rule: Form the nth row by copying the $(n - 1)$st in order, but insert the fraction $(a + a')/(b + b')$ between the consecutive fractions a/b and a'/b' of the $(n - 1)$st row if $b + b' \leqq n$. Thus, since $1 + 1 \leqq 2$ we insert $(0 + 1)/(1 + 1)$ between 0/1 and 1/1 and obtain 0/1, 1/2, 1/1, for the second row. The third row is 0/1, 1/3, 1/2, 2/3, 1/1. To obtain the fourth row we insert $(0 + 1)/(1 + 3)$ and $(2 + 1)/(3 + 1)$ but not $(1 + 1)/(3 + 2)$ and $(1 + 2)/(2 + 3)$. The first five rows of the table are:

$$
\begin{array}{ccccccccccc}
\frac{0}{1} & & & & & & & & & & \frac{1}{1} \\
\frac{0}{1} & & & & & \frac{1}{2} & & & & & \frac{1}{1} \\
\frac{0}{1} & & & \frac{1}{3} & & \frac{1}{2} & & \frac{2}{3} & & & \frac{1}{1} \\
\frac{0}{1} & & \frac{1}{4} & \frac{1}{3} & & \frac{1}{2} & & \frac{2}{3} & \frac{3}{4} & & \frac{1}{1} \\
\frac{0}{1} & \frac{1}{5} & \frac{1}{4} & \frac{1}{3} & \frac{2}{5} & \frac{1}{2} & \frac{3}{5} & \frac{2}{3} & \frac{3}{4} & \frac{4}{5} & \frac{1}{1}
\end{array}
$$

Up to this row, at least, the table has a number of interesting properties. All the fractions that appear are in reduced form; all reduced fractions a/b such that $0 \leqq a/b \leqq 1$ and $b \leqq n$ appear in the nth row; if a/b and a'/b' are consecutive fractions in the nth row, then $a'b - ab' = 1$ and $b + b' > n$. We will prove all these properties for the entire table.

Theorem 6.1 *If a/b and a'/b' are consecutive fractions in the* n*th row, say with a/b to the left of a'/b', then $a'b - ab' = 1$.*

Proof. It is true for $n = 1$. Suppose it is true for the $(n - 1)$st row. Any consecutive fractions in the nth row will be either a/b, a'/b' or a/b, $(a + a')/(b + b')$, or $(a + a')/(b + b')$, a'/b' where a/b and a'/b' are consecutive fractions in the $(n - 1)$st row. But then we have $a'b - ab' = 1$, $(a + a')b - a(b + b') = a'b - ab' = 1$, $a'(b + b') - (a + a')b' = a'b - ab' = 1$, and the theorem is proved by mathematical induction.

Corollary 6.2 *Every a/b in the table is in reduced form, that is, $(a, b) = 1$.*

Corollary 6.3 *The fractions in each row are listed in order of their size.*

Theorem 6.4 *If a/b and a'/b' are consecutive fractions in any row, then among all rational fractions with values between these two, $(a + a')/(b + b')$ is the unique fraction with smallest denominator.*

Proof. In the first place, the fraction $(a + a')/(b + b')$ will be the first fraction to be inserted between a/b and a'/b' as we continue to further rows of the table. It will first appear in the $(b + b')$th row. Therefore we have

$$\frac{a}{b} < \frac{a + a'}{b + b'} < \frac{a'}{b'}$$

by Corollary 6.3.

Now consider any fraction x/y between a/b and a'/b' so that $a/b < x/y < a'/b'$. Then

$$(6.1)\qquad \frac{a'}{b'} - \frac{a}{b} = \left(\frac{a'}{b'} - \frac{x}{y}\right) + \left(\frac{x}{y} - \frac{a}{b}\right)$$
$$= \frac{a'y - b'x}{b'y} + \frac{bx - ay}{by} \geqq \frac{1}{b'y} + \frac{1}{by} = \frac{b + b'}{bb'y},$$

and therefore

$$\frac{b + b'}{bb'y} \leqq \frac{a'b - ab'}{bb'} = \frac{1}{bb'},$$

which implies $y \geqq b + b'$. If $y > b + b'$ then x/y does not have least denominator among fractions between a/b and a'/b'. If $y = b + b'$, then the inequality in (6.1) must become equality and we have $a'y - b'x = 1$ and $bx - ay = 1$. Solving, we find $x = a + a'$, $y = b + b'$, and hence $(a + a')/(b + b')$ is the unique rational fraction lying between a/b and a'/b' with denominator $b + b'$.

Theorem 6.5 *If $0 \leqq x \leqq y$, $(x, y) = 1$, then the fraction x/y appears in the yth and all later rows.*

Proof. This is obvious if $y = 1$. Suppose it is true for $y = y_0 - 1$, with $y_0 > 1$. Then if $y = y_0$, the fraction x/y cannot be in the $(y - 1)$st row by definition and so it must lie in value between two consecutive fractions a/b and a'/b' of the $(y - 1)$st row. Thus $a/b < x/y < a'/b'$. Since

$$\frac{a}{b} < \frac{a + a'}{b + b'} < \frac{a'}{b'}$$

and a/b, a'/b' are consecutive, the fraction $(a + a')/(b + b')$ is not in the $(y - 1)$st row and hence $b + b' > y - 1$ by our induction hypothesis. But $y \geqq b + b'$ by Theorem 6.4, and so we have $y = b + b'$. Then the uniqueness part of Theorem 6.4 shows that $x = a + a'$. Therefore $x/y = (a + a')/(b + b')$ enters in the yth row, and it is then in all later rows.

Corollary 6.6 *The nth row consists of all reduced rational fractions a/b such that $0 \leqq a/b \leqq 1$ and $0 < b \leqq n$. The fractions are listed in order of their size.*

Definition 6.1 *The sequence of all reduced fractions with denominators not exceeding n, listed in order of their size, is called the Farey sequence of order n.*

The nth row of our table gives that part of the Farey sequence of order n that lies between 0 and 1, and so the entire Farey sequence of order n can be obtained from the nth row by adding and subtracting integers. For example, the Farey sequence of order 2 is

$$\dots, \frac{-3}{1}, \frac{-5}{2}, \frac{-2}{1}, \frac{-3}{2}, \frac{-1}{1}, \frac{-1}{2}, \frac{0}{1}, \frac{1}{2}, \frac{1}{1}, \frac{3}{2}, \frac{2}{1}, \frac{5}{2}, \frac{3}{1}, \dots.$$

This definition of the Farey sequences seems to be the most convenient. However, some authors prefer to restrict the fractions to the interval from 0 to 1; they define the Farey sequences to be just the rows of our table.

Any reduced fraction with positive denominator $\leqq n$ is a member of the Farey sequence of order n and can be called a Farey fraction of order n. Note that consecutive fractions a/b and a'/b' in the Farey sequence of order n satisfy the equality of Theorem 6.1 and also the inequality $b + b' > n$.

PROBLEMS

1. Let a/b and a'/b' be the fractions immediately to the left and the right of the fraction 1/2 in the Farey sequence of order n. Prove that $b = b' = 1 + 2[(n-1)/2]$, that is, b is the greatest odd integer $\leqq n$. Also prove that $a + a' = b$.

2. Prove that the number of Farey fractions a/b of order n satisfying the inequalities $0 \leqq a/b \leqq 1$ is $1 + \sum_{j=1}^{n} \phi(j)$, and that their sum is exactly half of this value.

3. Let a/b, a'/b', a''/b'' be any three consecutive fractions in the Farey sequence of order n. Prove that $a'/b' = (a + a'')/(b + b'')$.

4. Let a/b and a'/b' run through all pairs of adjacent fractions in the Farey sequence of order $n > 1$. Prove that

$$\min\left(\frac{a'}{b'} - \frac{a}{b}\right) = \frac{1}{n(n-1)} \quad \text{and} \quad \max\left(\frac{a'}{b'} - \frac{a}{b}\right) = \frac{1}{n}.$$

5. Consider two rational numbers a/b and c/d such that $ad - bc = 1$, $b > 0$, $d > 0$. Define n as max (b, d), and prove that a/b and c/d are adjacent fractions in the Farey sequence of order n.

6. Prove that the two fractions described in the preceding problem are not necessarily adjacent in the Farey sequence of order $n + 1$.

7. Consider the fractions from 0/1 to 1/1 inclusive in the Farey sequence of order n. Reading from left to right, let the denominators of these fractions be $a_1, a_2, \cdots, a_k$ so that $a_1 = 1$ and $a_k = 1$. Prove that $\sum_{j=1}^{n-1} (a_j a_{j+1})^{-1} = 1$.

6.2 Rational Approximations

Theorem 6.7 *If a/b and c/d are Farey fractions of order n such that no other Farey fraction of order n lies between them, then*

$$\left|\frac{a}{b} - \frac{a+c}{b+d}\right| = \frac{1}{b(b+d)} \leqq \frac{1}{b(n+1)},$$

and

$$\left|\frac{c}{d} - \frac{a+c}{b+d}\right| = \frac{1}{d(b+d)} \leqq \frac{1}{d(n+1)}.$$

Proof. We have

$$\left|\frac{a}{b} - \frac{a+c}{b+d}\right| = \frac{|ad-bc|}{b(b+d)} = \frac{1}{b(b+d)} \leqq \frac{1}{b(n+1)}$$

by Theorem 6.1 and the fact that $b + d \geqq n + 1$. The second formula is proved in a similar way.

Theorem 6.8 *If n is a positive integer and x is real, there is a rational number a/b such that $0 < b \leqq n$ and*

$$\left|x - \frac{a}{b}\right| \leqq \frac{1}{b(n+1)}.$$

Proof. Consider the set of all Farey fractions of order n and all of the fractions $(a + c)/(b + d)$ as described in Theorem 6.7. For some Farey fractions a/b and c/d, the number x will lie between or on, and so by interchanging a/b and c/d if necessary, we can say that x lies in the closed interval between a/b and $(a + c)/(b + d)$. Then, by Theorem 6.7,

$$\left|x - \frac{a}{b}\right| \leqq \left|\frac{a}{b} - \frac{a+c}{b+d}\right| \leqq \frac{1}{b(n+1)}.$$

Theorem 6.9 *If ξ is real and irrational, there are infinitely many distinct rational numbers a/b such that*

$$\left|\xi - \frac{a}{b}\right| < \frac{1}{b^2}.$$

Proof. For each $n = 1, 2, \cdots$ we can find an a_n and a b_n by Theorem 6.8 such that $0 < b_n \leqq n$ and

$$\left|\xi - \frac{a_n}{b_n}\right| \leqq \frac{1}{b_n(n+1)} < \frac{1}{b_n^2}.$$

Many of the a_n/b_n may be equal to each other, but there will be infinitely many distinct ones. For if there were not infinitely many distinct ones, there would be only a finite number of distinct values taken by $|\xi - a_n/b_n|$, $n = 1, 2, 3, \cdots$. Then there would be a least one among these values, and it would be the value of $|\xi - a_n/b_n|$ for some n, say $n = k$. We would have $|\xi - a_n/b_n| \geqq |\xi - a_k/b_k|$ for all $n = 1, 2, 3, \cdots$. But $|\xi - a_k/b_k| > 0$ since ξ is irrational, and we can find an n sufficiently large that

$$\frac{1}{n+1} < \left|\xi - \frac{a_k}{b_k}\right|.$$

This leads to a contradiction since we would now have

$$\left|\xi - \frac{a_k}{b_k}\right| \leqq \left|\xi - \frac{a_n}{b_n}\right| \leqq \frac{1}{b_n(n+1)} \leqq \frac{1}{n+1} < \left|\xi - \frac{a_k}{b_k}\right|.$$

The condition that ξ be irrational is necessary in the theorem. For if x is any rational number, we can write $x = r/s$, $s > 0$. Then if a/b is any fraction such that $a/b \neq r/s$, $b > s$, we have

$$\left|\frac{r}{s} - \frac{a}{b}\right| = \frac{|rb - as|}{sb} \geqq \frac{1}{sb} > \frac{1}{b^2}.$$

Hence all fractions a/b, $b > 0$, satisfying $|x - a/b| < 1/b^2$ have denominators $b \leqq s$, and there can only be a finite number of such fractions.

The result of Theorem 6.9 can be improved, as Theorem 6.11 will show. Different proofs of Theorems 6.11 and 6.12 are given in Section 7.6.

Lemma 6.10 *If x and y are positive integers then not both of the two inequalities*

$$\frac{1}{xy} \geqq \frac{1}{\sqrt{5}}\left(\frac{1}{x^2} + \frac{1}{y^2}\right) \quad \textit{and} \quad \frac{1}{x(x+y)} \geqq \frac{1}{\sqrt{5}}\left(\frac{1}{x^2} + \frac{1}{(x+y)^2}\right)$$

can hold.

Proof. The two inequalities can be written as

$$\sqrt{5}\,xy \geqq y^2 + x^2, \qquad \sqrt{5}\,x(x+y) \geqq (x+y)^2 + x^2.$$

Adding these inequalities, we get $\sqrt{5}\,(x^2 + 2xy) \geqq 3x^2 + 2xy + 2y^2$, hence $2y^2 - 2(\sqrt{5} - 1)xy + (3 - \sqrt{5})x^2 \leqq 0$. Multiplying this by 2 we put it in the form $4y^2 - 4(\sqrt{5} - 1)xy + (5 - 2\sqrt{5} + 1)x^2 \leqq 0$, $(2y - (\sqrt{5} - 1)x)^2 \leqq 0$. This is impossible for positive integers x and y because $\sqrt{5}$ is irrational.

Theorem 6.11 (*Hurwitz*) *Given any irrational number ξ, there exist infinitely many differemt rational numbers h/k such that*

$$\left|\xi - \frac{h}{k}\right| < \frac{1}{\sqrt{5}\,k^2}. \tag{6.2}$$

Proof. Let n be a positive integer. There exist two consecutive fractions a/b and c/d in the Farey sequence of order n, such that $a/b < \xi < c/d$. Either $\xi < (a + c)/(b + d)$ or $\xi > (a + c)/(b + d)$.

Case I. $\xi < (a + c)/(b + d)$. Suppose that

$$\xi - \frac{a}{b} \geqq \frac{1}{b^2\sqrt{5}}, \qquad \frac{a+c}{b+d} - \xi \geqq \frac{1}{(b+d)^2\sqrt{5}}, \qquad \frac{c}{d} - \xi \geqq \frac{1}{d^2\sqrt{5}}.$$

Adding inequalities we obtain

$$\frac{c}{d} - \frac{a}{b} \geqq \frac{1}{d^2\sqrt{5}} + \frac{1}{b^2\sqrt{5}}, \qquad \frac{a+c}{b+d} - \frac{a}{b} \geqq \frac{1}{(b+d)^2\sqrt{5}} + \frac{1}{b^2\sqrt{5}},$$

hence

$$\frac{1}{bd} = \frac{cb - ad}{bd} = \frac{c}{d} - \frac{a}{b} \geqq \frac{1}{\sqrt{5}}\left(\frac{1}{b^2} + \frac{1}{d^2}\right)$$

and

$$\frac{1}{b(b+d)} = \frac{(a+c)b - (b+d)a}{b(b+d)} \geqq \frac{1}{\sqrt{5}}\left(\frac{1}{b^2} + \frac{1}{(b+d)^2}\right).$$

These two inequalities contradict Lemma 6.10. Therefore at least one of a/b, c/d, $(a+c)/(b+d)$ will serve as h/k in this case.

Case II. $\xi > (a+c)/(b+d)$. Suppose that

$$\xi - \frac{a}{b} \geqq \frac{1}{b^2\sqrt{5}}, \qquad \xi - \frac{a+c}{b+d} \geqq \frac{1}{(b+d)^2\sqrt{5}}, \qquad \frac{c}{d} - \xi \geqq \frac{1}{d^2\sqrt{5}}.$$

Adding as before, we obtain

$$\frac{c}{d} - \frac{a}{b} \geqq \frac{1}{d^2\sqrt{5}} + \frac{1}{b^2\sqrt{5}}, \qquad \frac{c}{d} - \frac{a+c}{b+d} \geqq \frac{1}{d^2\sqrt{5}} + \frac{1}{(b+d)^2\sqrt{5}},$$

hence

$$\frac{1}{bd} \geqq \frac{1}{\sqrt{5}}\left(\frac{1}{d^2} + \frac{1}{b^2}\right), \qquad \frac{1}{d(b+d)} \geqq \frac{1}{\sqrt{5}}\left(\frac{1}{(b+d)^2} + \frac{1}{d^2}\right),$$

which also contradicts Lemma 6.10. Again at least one of a/b, c/d, $(a+c)/(b+d)$ will serve as h/k.

We have shown the existence of some h/k that satisfies (6.2). This h/k depends on our choice of n. In fact h/k is either a/b, c/d, or $(a+c)/(b+d)$, where a/b and c/d are consecutive fractions in the Farey sequence of order n, and $a/b < \xi < c/d$. Using Theorem 6.7 we see that

$$\left|\xi - \frac{h}{k}\right| < \left|\frac{c}{d} - \frac{a}{b}\right| = \left|\frac{c}{d} - \frac{a+c}{b+d}\right| + \left|\frac{a+c}{b+d} - \frac{a}{b}\right|$$

$$\leqq \frac{1}{d(n+1)} + \frac{1}{b(n+1)} \leqq \frac{2}{n+1}.$$

We want to establish that there are infinitely many h/k that satisfy (6.2). Suppose that we have any h_1/k_1 that satisfies (6.2). Then $\left|\xi - \frac{h_1}{k_1}\right|$ is positive, and we can choose $n > 2\Big/\left|\xi - \frac{h_1}{k_1}\right|$. The Farey sequence of order n then yields an h/k that satisfies (6.2) and such that

$$\left|\xi - \frac{h}{k}\right| \leqq \frac{2}{n+1} < \left|\xi - \frac{h_1}{k_1}\right|.$$

This shows that there exist infinitely many rational numbers h/k that satisfy (6.2) since, given any rational number, we can find another that is closer to ξ.

Theorem 6.12 *The constant $\sqrt{5}$ in Theorem 6.11 is the best possible. In other words Theorem 6.11 does not hold if $\sqrt{5}$ is replaced by any larger value.*

Proof. We need only exhibit one ξ for which $\sqrt{5}$ cannot be replaced by a larger value. Let us take $\xi = \frac{1}{2}(1 + \sqrt{5})$. Then

$$(x - \xi)\left(x - \frac{1 - \sqrt{5}}{2}\right) = x^2 - x - 1.$$

For integers h, k with $k > 0$, we then have

$$(6.3)\quad \left|\frac{h}{k} - \xi\right|\left|\frac{h}{k} - \xi + \sqrt{5}\right| = \left|\left(\frac{h}{k} - \xi\right)\left(\frac{h}{k} - \frac{1 - \sqrt{5}}{2}\right)\right| = \left|\frac{h^2}{k^2} - \frac{h}{k} - 1\right| = \frac{1}{k^2}|h^2 - hk - k^2|.$$

The expression on the left in (6.3) is not zero because both ξ and $\sqrt{5} - \xi$ are irrational. The expression $|h^2 - hk - k^2|$ is a non-negative integer. Therefore $|h^2 - hk - k^2| \geqq 1$ and we have

$$(6.4)\quad \left|\frac{h}{k} - \xi\right|\left|\frac{h}{k} - \xi + \sqrt{5}\right| \geqq \frac{1}{k^2}.$$

Now suppose we have an infinite sequence of rational numbers h_j/k_j, $k_j > 0$, and a positive real number m such that

$$(6.5)\quad \left|\frac{h_j}{k_j} - \xi\right| < \frac{1}{mk_j^2}.$$

Then $k_j\xi - \dfrac{1}{mk_j} < h_j < k_j\xi + \dfrac{1}{mk_j}$, and this implies that there are only a finite number of h_j corresponding to each value of k_j. Therefore we have $k_j \to \infty$ as $j \to \infty$. Also, by (6.4), (6.5) and the triangle inequality we have

$$\frac{1}{k_j^2} \leqq \left|\frac{h_j}{k_j} - \xi\right|\left|\frac{h_j}{k_j} - \xi + \sqrt{5}\right| < \frac{1}{mk_j^2}\left(\frac{1}{mk_j^2} + \sqrt{5}\right),$$

hence

$$m < \frac{1}{mk_j^2} + \sqrt{5},$$

and therefore

$$m \leqq \lim_{j\to\infty}\left(\frac{1}{mk_j^2} + \sqrt{5}\right) = \sqrt{5}.$$

PROBLEMS

1. Prove that for every real number x there are infinitely many pairs of integers a, b, with b positive such that $|bx - a| < (\sqrt{5}\,b)^{-1}$.

2. Take ξ as on p. 141. Let $\lambda > 0$ and $\alpha > 2$ be real numbers. Prove that there are only finitely many rationals h/k satisfying

$$\left|\xi - \frac{h}{k}\right| < \frac{1}{\lambda k^\alpha}.$$

3. Suppose $h = a$, $k = b$ is a solution of the inequality (6.2) for some irrational ξ. Prove that only a finite number of pairs h, k in the set $\{h = ma, k = mb;\ m = 1, 2, 3, \cdots\}$ satisfy (6.2).

4. Let $\alpha > 1$ be a real number. Suppose that for some real number β there are infinitely many rational numbers h/k such that $|\beta - h/k| < k^{-\alpha}$. Prove that β is irrational.

5. Prove that the following are irrational: $\sum_{j=1}^{\infty} 2^{-3^j}$, $\sum_{j=1}^{\infty} 2^{-j!}$.

6.3 Irrational Numbers

A rational number, as is well-known, is one which can be expressed as the quotient of two integers, a/b with $b \neq 0$. A real number which is not rational is said to be irrational.

That $\sqrt{2}$ is irrational can be concluded at once from the unique factorization theorem. For if $\sqrt{2}$ could be represented in the form a/b, it would follow that $a^2 = 2b^2$. But this is impossible with integers a and b because the highest power of 2 that divides a^2 is an even power, whereas the highest power of 2 that divides $2b^2$ is an odd power, by the unique factorization theorem. A more general argument for deducing irrationality is formulated next.

Theorem 6.12 *If a polynomial equation with integral coefficients*

$$c_nx^n + c_{n-1}x^{n-1} + \cdots c_2x^2 + c_1x + c_0 = 0, \quad c_n \neq 0, \tag{6.6}$$

has a nonzero rational solution a/b where the integers a and b are relatively prime, then $a \mid c_0$ and $b \mid c_n$.

Proof. Replacing x by a/b in (6.6) and multiplying by b^{n-1}, we note that $c_n a^n/b$ is an integer, and hence $b \mid c_n$ since $(a, b) = 1$. On the other hand, replacing x by a/b in (6.6) and multiplying by b^n/a, we observe that $c_0 b^n/a$ is an integer, so $a \mid c_0$.

Corollary 6.13 *If a polynomial equation (6.6) with $c_n = \pm 1$ has a nonzero rational solution, that solution is an integer which divides c_0.*

Corollary 6.14 *For any integers c and $n > 0$, the only rational solutions, if any, of $x^n = c$ are integers. Thus $x^n = c$ has rational solutions if and only if c is the* nth *power of an integer.*

It follows at once that such numbers as $\sqrt{2}$, $\sqrt{3}$, $\sqrt{5}$ are irrational because there are no integral solutions of $x^2 = 2$, $x^2 = 3$, and $x^3 = 5$.

Another application of Theorem 6.12 can be made to certain values of the trigonometric functions, as follows.

Theorem 6.15 *Let θ be a rational multiple of π, thus $\theta = r\pi$ where r is rational. Then* $\cos\theta$, $\sin\theta$, $\tan\theta$ *are irrational numbers apart from the cases where* $\tan\theta$ *is undefined, and the exceptions*

$$\cos\theta = 0, \pm 1/2, \pm 1; \qquad \sin\theta = 0, \pm 1/2, \pm 1; \qquad \tan\theta = 0, \pm 1.$$

Proof. Let n be any positive integer. First we prove by mathematical induction that there is a polynomial $f_n(x)$ of degree n with integral coefficients and leading coefficient 1 such that $2\cos n\theta = f_n(2\cos\theta)$ holds for all real numbers θ. We note that $f_1(x) = x$, and $f_2(x) = x^2 - 2$ because of the well-known identity $2\cos 2\theta = (2\cos\theta)^2 - 2$. The identity

$$2\cos(n+1)\theta = (2\cos\theta)(2\cos n\theta) - 2\cos(n-1)\theta$$

is easily established by elementary trigonometry, and this reveals that $f_{n+1}(x) = xf_n(x) - f_{n-1}(x)$ which completes the proof by induction.

Next, let the positive integer n be chosen so that nr is also an integer. With $\theta = r\pi$ it follows that

$$f_n(2\cos\theta) = 2\cos n\theta = 2\cos nr\pi = \pm 2,$$

where the plus sign holds if nr is even, the minus sign if odd. Thus $2\cos\theta$ is a solution of $f_n(x) = \pm 2$. Setting aside the cases where $\cos\theta = 0$, we apply Corollary 6.13 to conclude that $2\cos\theta$, if rational, is a nonzero integer. But $-1 \leqq \cos\theta \leqq 1$ so the only possible values of $2\cos\theta$, apart from 0, are ± 1 and ± 2. So Theorem 6.15 has been established in the case of $\cos\theta$.

As to $\sin\theta$, if θ is a rational multiple of π so is $\pi/2 - \theta$, and from the identity $\sin\theta = \cos(\pi/2 - \theta)$ we arrive at the conclusion stated in the theorem.

Finally, the identity $\cos 2\theta = (1 - \tan^2 \theta)/(1 + \tan^2 \theta)$ reveals that if $\tan \theta$ is rational so is $\cos 2\theta$. In view of what was just proved about the cosine function, we need look only at the possibilities $\cos 2\theta = 0, \pm 1/2, \pm 1$. When $\cos 2\theta = 0$ it is readily calculated that $\tan \theta = \pm 1$; when $\cos 2\theta = +1$, $\tan \theta = 0$; when $\cos 2\theta = -1$, $\tan \theta$ is undefined; when $\cos 2\theta = \pm 1/2$, $\tan \theta$ is one of the irrational values $\pm\sqrt{3}$, $\pm 1/\sqrt{3}$. This completes the proof of Theorem 6.15.

The logarithm of any positive rational number to a positive rational base is easily classified as rational or irrational. Consider, for example, $\log_6 9$. If this were a rational number a/b, where a and b are positive integers, this would imply that $9 = 6^{a/b}$ or $9^b = 6^a$. The unique factorization theorem can be applied to separate the primes 2 and 3 to give $9^b = 3^a$ and $1 = 2^a$. These equations imply that $a = b = 0$, and so we conclude that $\log_6 9$ is irrational.

The basic mathematical constants π and e are irrational. A proof of this for e is sufficiently simple that we leave it to the reader in Problems 7 and 8 below. For π the matter is not quite so easy, so we precede the proof with a lemma.

Lemma 6.16 *If n is any positive integer, and $g(x)$ any polynomial with integral coefficients, then $x^n g(x)$ and all its derivatives, evaluated at $x = 0$, are integers divisible by $n!$.*

Proof. Any term in $g(x)$ is of the form cx^j where c and j are integers with $c \neq 0$ and $j \geqq 0$. The corresponding term in $x^n g(x)$ is cx^{j+n}; if we prove the lemma for this single term, the entire lemma will follow because the derivative of a finite sum is the sum of the derivatives.

At $x = 0$, it is readily seen that cx^{j+n} and all its derivatives are zero, with one exception, namely the $(j + n)$th derivative. The $(j + n)$th derivative is $c\{(j + n)!\}$, and since $j \geqq 0$, this is divisible by $n!$.

Theorem 6.17 *π is irrational.*

Proof. Suppose that $\pi = a/b$, where a and b are positive integers. Define the polynomial

$$f(x) = x^n(a - bx)^n/n! = b^n x^n(\pi - x)^n/n!, \tag{6.7}$$

where the second form of $f(x)$ stems from the first by simple algebra. The integer n will be specified later. We apply Lemma 6.16 with $g(x)$ in the form $(a - bx)^n$ to conclude that $x^n(a - bx)^n$ and all its derivatives, evaluated at $x = 0$, are integers divisible by $n!$. Dividing by $n!$, we see that $f(x)$ and all its derivatives, evaluated at $x = 0$, are integers. Denoting the j-th derivative of $f(x)$ by $f^{(j)}(x)$, and writing $f(x) = f^{(0)}(x)$, we can state that $f^{(j)}(0)$ is an integer for every $j = 0, 1, 2, 3, \cdots$.

By the second part of (6.7) we find that $f(\pi - x) = f(x)$, and taking derivatives we get $-f'(\pi - x) = f'(x), f^{(2)}(\pi - z) = f^{(2)}(x)$, and in general $(-1)^j f^{(j)}(\pi - x) = f^{(j)}(x)$. Letting $x = 0$ we obtain the result that $f^{(j)}(\pi)$ is an integer for every $j = 0, 1, 2, 3, \cdots$.

Next the polynomial $F(x)$ is defined by

$$F(x) = f(x) - f^{(2)}(x) + f^{(4)}(x) - f^{(6)}(x) + \cdots + (-1)^n f^{(2n)}(x).$$

Now if this equation is differentiated twice the result is

$$F^{(2)}(x) = f^{(2)}(x) - f^{(4)}(x) + f^{(6)}(x) - f^{(8)}(x) + \cdots + (-1)^{n-1} f^{(2n)}(x) + 0,$$

because $f^{(2n+2)}(x) = 0$ since $f(x)$ is a polynomial of degree $2n$. Adding these equations we get $F(x) + F^{(2)}(x) = f(x)$. Also, by the preceding paragraphs we observe that $F(0)$ and $F(\pi)$ are integers, because they are sums and differences of integers.

Now by elementary calculus it is seen that

$$\frac{d}{dx}\{F'(x) \sin x - F(x) \cos x\} = F''(x) \sin x + F(x) \sin x = f(x) \sin x.$$

Thus we are able to integrate $f(x) \sin x$, to get

$$\int_0^\pi f(x) \sin x \, dx = [F'(x) \sin x - F(x) \cos x]_0^\pi = F(\pi) + F(0). \tag{6.8}$$

A contradiction arises from this equation, because whereas $F(\pi) + F(0)$ is an integer, we demonstrate that the integer n can be chosen sufficiently large in the definition of $f(x)$ in (6.7) that the integral in (6.8) lies strictly between 0 and 1.

From (6.7) we see that from $x = 0$ to $x = \pi$,

$$f(x) < \frac{\pi^n a^n}{n!} \quad \text{and} \quad f(x) \sin x < \frac{\pi^n a^n}{n!}.$$

Also $f(x) \sin x > 0$ in the open interval $0 < x < \pi$, and hence

$$0 < \int_0^\pi f(x) \sin x \, dx < \frac{\pi^n a^n}{n!} \cdot \pi,$$

because the interval of integration is of length π. From elementary calculus it is well-known that for any constant such as πa, the limit of $(\pi a)^n/n!$ is zero as n tends to infinity. Hence we can choose n sufficiently large that the integral in (6.8) lies strictly between 0 and 1, and we have obtained the contradiction stated above. It follows that π is irrational.

PROBLEMS

1. Prove that the irrational numbers are not closed under addition, subtraction, multiplication, or division.

2. Prove that the sum, difference, product, and quotient of two numbers, one irrational and the other a nonzero rational, are irrational.

3. Prove that $\sqrt{2} + \sqrt{3}$ is a root of $x^4 - 10x^2 + 1 = 0$, and hence establish that it is irrational.

4. (*a*) For any positive integer h, note that h^2 ends in an even number of zeros whereas $10h^2$ ends in an odd number of zeros in the ordinary base ten notation. Use this to prove that $\sqrt{10}$ is irrational, by assuming $\sqrt{10} = h/k$ so that $h^2 = 10k^2$. (*b*) Extend this argument to $\sqrt[3]{10}$. (*c*) Extend the argument to prove that $\sqrt{n}$ is irrational, where n is a positive integer not a perfect square, by taking n as the base of the number system instead of ten.

5. (i) Verify the details of the following sketch of an argument that $\sqrt{77}$ is irrational. Suppose that $\sqrt{77}$ is rational, and among its rational representations let a/b be that one having the smallest positive integer denominator b, where a is also an integer. Prove that another rational representation of $\sqrt{77}$ is $(77b - 8a)/(a - 8b)$. Prove that $a - 8b$ is a smaller positive integer than b, which is a contradiction. (ii) Generalize this argument to prove that $\sqrt{n}$ is irrational if n is a positive integer not a perfect square, by assuming $n = a/b$ and then getting another rational representation of n with denominator $a - kb$ where $k = [\sqrt{n}]$, the greatest integer less than $\sqrt{n}$. (An interesting aspect of this problem is that it establishes irrationality by use of the idea that every nonempty set of positive integers has a least member, not by use of the unique factorization theorem.)

6. Let a/b be a positive rational number with $a > 0$, $b > 0$, g.c.d. $(a, b) = 1$. Generalize Corollary 6.14 by proving that for any integer $n > 1$ the equation $x^n = a/b$ has a rational solution if and only if both a and b are nth powers of integers. *Suggestion:* If $(a/b)^{1/n}$ is rational so is $b(a/b)^{1/n}$, which is a root of the equation $x^n = ab^{n-1}$.

7. Prove that a number α is rational if and only if there exists a positive integer k such that $[k\alpha] = k\alpha$. Prove that a number α is rational if and only if there exists a positive integer k such that $[(k!)\alpha] = (k!)\alpha$.

8. Recalling that the mathematical constant e has value $\sum_{j=0}^{\infty} 1/j!$, prove that

$$[(k!)e] = k! \sum_{j=0}^{k} 1/j! < (k!)e.$$

Hence prove that e is irrational.

9. Prove that cos 1 is irrational, where "1" is in radian measure. *Suggestion:* use the infinite series for $\cos x$, and adapt the ideas of the two preceding problems.

6.4 Coverings of the Real Line

The rational numbers are dense on the real line, that is, between any two of them there is another. This is obvious, because if r and s are rational numbers, $(r + s)/2$ lies between them on the real line. It might appear therefore that if every rational number is covered by an interval on the real line, the entire line would be covered. Whether this is so or not depends on the lengths of the intervals, as we shall see.

Given two real numbers α and β with $\alpha < \beta$, the open interval from α to β is the set of real numbers x satisfying $\alpha < x < \beta$; this interval is denoted by (α, β). The closed interval is the set of real numbers x satisfying $\alpha \leqq x \leqq \beta$, denoted by $[\alpha, \beta]$.

Now Theorem 6.11 can be interpreted as stating that if every rational number h/k is covered by an open interval

$$\left(\frac{h}{k} - \frac{1}{\sqrt{5}\,k^2}, \frac{h}{k} + \frac{1}{\sqrt{5}\,k^2}\right)$$

then every irrational number ξ is covered, in fact covered by infinitely many of the intervals. Thus the entire real line is covered.

However, if every rational number h/k is covered by the slightly shorter open interval

$$\left(\frac{h}{k} - \frac{1}{4k^2}, \frac{h}{k} + \frac{1}{4k^2}\right) \tag{6.9}$$

then *not every irrational number is covered.* Specifically we prove that $\sqrt{2}/2$ is not covered. Without loss of generality we may presume that h/k is in lowest terms, i.e., that g.c.d. $(h, k) = 1$, because the interval (6.9) covers any interval with the rational number h/k written in the form $(ch)/(ck)$ where c is a positive integer. Also $k > 0$ may be presumed. First if $k = 1$, the intervals (6.9) extend for a distance $1/4$ on each side of every integer h, so they do not cover $\sqrt{2}/2$ because it is approximately .707. Henceforth we take $k \geqq 2$. Second if $h/k > 1$ or $h/k < 0$ then the interval (6.9) does not overlap the interval $(0, 1)$ because if $h/k > 1$ then

$$\frac{h}{k} - \frac{1}{4k^2} > \frac{h}{k} - \frac{1}{k} = \frac{h-1}{k} \geqq 1,$$

and if $h/k < 0$ then

$$\frac{h}{k} + \frac{1}{4k^2} < \frac{h}{k} + \frac{1}{k} = \frac{h+1}{k} \leqq 0.$$

Third if h/k lies in the interval (0, 1), then $0 < h/k < 1$. Now suppose that for some h/k the interval (6.6) covers $\sqrt{2}/2$, that is,

$$\frac{h}{k} - \frac{1}{4k^2} < \frac{\sqrt{2}}{2} < \frac{h}{k} + \frac{1}{4k^2}.$$

Squaring all three expressions we see that

$$\frac{h^2}{k^2} - \frac{h}{2k^3} + \frac{1}{16k^4} < \frac{1}{2} < \frac{h^2}{k^2} + \frac{h}{2k^3} + \frac{1}{16k^4}.$$

Multiplying by $2k^2$ and then subtracting $2h^2$ from each expression we get

$$(6.10) \qquad -\frac{h}{k} + \frac{1}{8k^2} < k^2 - 2h^2 < \frac{h}{k} + \frac{1}{8k^2}.$$

However we note that

$$-\frac{h}{k^2} + \frac{1}{8k^2} > -\frac{h}{k} > -1 \quad \text{and} \quad \frac{h}{k} + \frac{1}{8k^2} < \frac{h}{k} + \frac{1}{k} \leqq 1.$$

But $k^2 - 2h^2$ in (6.10) is an integer, and a view of these last inequalities, we must have $k^2 - 2h^2 = 0$, which is impossible because $\sqrt{2}$ is irrational.

PROBLEMS

1. Prove that if every rational number h/k is covered by an open interval of the type (6.9) but with the constant 4 replaced by 3, then $\sqrt{2}/2$ is covered.
2. Prove that the irrational numbers are dense on the real line, i.e., between every two irrational numbers there is another irrational.

NOTES ON CHAPTER 6

§**6.3.** The proof of Theorem 6.15 follows that of E. A. Maier, "On the irrationality of certain trigonometric numbers," *Amer. Math. Monthly* **72,** 1012 (1965). Further results on the topic of this section can be found in Ivan Niven, *Irrational Numbers*, Carus Monograph 11, New York, John Wiley and Sons, 1956.

§**6.3.** The relationships between rational and irrational numbers and their decimal expansions are given on page 259 ff in Special Topics. The fact that every positive rational number is expressible as a sum of distinct "unit fractions," is established also in Special Topics, page 260 ff.

§**6.4.** The sharpest possible statement that can be made for coverings of the type considered is this: if every rational number h/k on the real line is covered by the closed interval $[h/k - 1/\beta k^2, h/k + 1/\beta k^2]$ with $\beta = (\sqrt{5} + 3)/2$, then the entire real line is covered by these intervals. For any larger value of β, the entire real line is not covered. This is part of what is established by A. V. Prasad in "Note on a theorem of Hurwitz," *Journal London Math. Soc.*, **23**, 169–171 (1948).

7

Simple Continued Fractions

7.1 The Euclidean Algorithm

Given any rational fraction u_0/u_1, in lowest terms so that $(u_0, u_1) = 1$ and $u_1 > 0$, we apply the Euclidean algorithm as formulated in Theorem 1.11 to get

$$
\begin{aligned}
(7.1) \qquad u_0 &= u_1 a_0 + u_2, & 0 < u_2 < u_1 \\
u_1 &= u_2 a_1 + u_3, & 0 < u_3 < u_2 \\
u_2 &= u_3 a_2 + u_4, & 0 < u_4 < u_3 \\
&\cdots \\
u_{j-1} &= u_j a_{j-1} + u_{j+1}, & 0 < u_{j+1} < u_j \\
u_j &= u_{j+1} a_j.
\end{aligned}
$$

The notation has been altered from that of Theorem 1.11 by the replacement of b, c by u_0, u_1, of $r_1, r_2, \cdots, r_j$ by $u_2, u_3, \cdots, u_{j+1}$, and of $q_1, q_2, \cdots, q_{j+1}$ by $a_0, a_1, \cdots, a_j$. The form (7.1) is a little more suitable for our present purposes. If we write ξ_i in place of u_i/u_{i+1} for all values of i in the range $0 \leqq i \leqq j$, then equations (7.1) become

$$(7.2) \qquad \xi_i = a_i + \frac{1}{\xi_{i+1}}, \qquad 0 \leqq i \leqq j - 1; \qquad \xi_j = a_j.$$

If we take the first two of these equations, those for which $i = 0$ and $i = 1$, and eliminate ξ_1, we get

$$\xi_0 = a_0 + \cfrac{1}{a_1 + \cfrac{1}{\xi_2}}.$$

In this result we replace ξ_2 by its value from (7.2), and then we continue with the replacement of $\xi_3, \xi_4, \cdots$, to get

$$(7.3) \qquad \frac{u_0}{u_1} = \xi_0 = a_0 + \cfrac{1}{a_1 + \cfrac{}{\ddots + \cfrac{1}{a_{j-1} + \cfrac{1}{a_j}}}}.$$

This is a *continued fraction expansion* of ξ_0, or of u_0/u_1. The integers a_i are called the *partial quotients* since they are the quotients in the repeated application of the division algorithm in equations (7.1). We presumed that the rational fraction u_0/u_1 had positive denominator u_1, but we cannot make a similar assumption about u_0. Hence a_0 may be positive, negative, or zero. However, since $0 < u_2 < u_1$, we note that the quotient a_1 is positive, and similarly the subsequent quotients $a_2, a_3, \cdots, a_j$ are positive integers. In case $j \geqq 1$, that is if the set (7.1) contains more than one equation, then $a_j = u_j/u_{j+1}$ and $0 < u_{j+1} < u_j$ imply that $a_j > 1$.

We shall use the notation $\langle a_0, a_1, \cdots, a_j\rangle$ to designate the continued fraction in (7.3). In general, if $x_0, x_1, \cdots, x_j$ are any real numbers, all positive except perhaps x_0, we shall write

$$\langle x_0, x_1, \cdots, x_j\rangle = x_0 + \cfrac{1}{x_1 + \cfrac{}{\ddots + \cfrac{1}{x_{j-1} + \cfrac{1}{x_j}}}}.$$

Such a finite continued fraction is said to be *simple* if all the x_i are integers. The following obvious formulas are often useful:

$$\langle x_0, x_1, \cdots, x_j\rangle = x_0 + \frac{1}{\langle x_1, \cdots, x_j\rangle} = \left\langle x_0, x_1, \cdots, x_{j-2}, x_{j-1} + \frac{1}{x_j}\right\rangle.$$

The symbol $[x_0, x_1, \cdots, x_j]$ is often used to represent a continued fraction. We use the notation $\langle x_0, x_1, \cdots, x_j\rangle$ to avoid confusion with the least common multiple and the greatest integer.

PROBLEMS

1. Expand the rational fractions 17/3, 3/17, and 8/1 into finite simple continued fractions.
2. Prove that the set (7.1) consists of exactly one equation if and only if $u_1 = 1$. Under what circumstances is $a_0 = 0$?
3. Convert into rational numbers: $\langle 2, 1, 4\rangle$; $\langle -3, 2, 12\rangle$; $\langle 0, 1, 1, 100\rangle$.
4. Given positive integers b, c, d with $c > d$, prove that $\langle a, c\rangle < \langle a, d\rangle$ but $\langle a, b, c\rangle > \langle a, b, d\rangle$ for any integer a.
5. Let $a_1, a_2, \cdots, a_n$ and c be positive real numbers. Prove that

$$\langle a_0, a_1, \cdots, a_n\rangle > \langle a_0, a_1, \cdots, a_n + c\rangle$$

holds if n is odd, but is false if n is even.

7.2 Uniqueness

In the last section we saw that such a fraction as 51/22 can be expanded into a simple continued fraction, $51/22 = \langle 2, 3, 7\rangle$. It can be verified that 51/22 can also be expressed as $\langle 2, 3, 6, 1\rangle$, but it turns out that these are the only two representations of 51/22. In general, we note that the simple continued fraction expansion (7.3) has an alternate form,

$$(7.4) \qquad \frac{u_0}{u_1} = \langle a_0, a_1, \cdots, a_{j-1}, a_j\rangle = \langle a_0, a_1, \cdots, a_{j-1}, a_j - 1, 1\rangle.$$

The following result establishes that these are the only two simple continued fraction expansions of a fixed rational number.

Theorem 7.1 *If $\langle a_0, a_1, \cdots, a_j\rangle = \langle b_0, b_1, \cdots, b_n\rangle$ where these finite continued fractions are simple, and if $a_j > 1$ and $b_n > 1$, then $j = n$ and $a_i = b_i$ for $i = 0, 1, \cdots, n$.*

Proof. We write y_i for the continued fraction $\langle b_i, b_{i+1}, \cdots, b_n\rangle$ and observe that

$$(7.5) \quad y_i = \langle b_i, b_{i+1}, \cdots, b_n\rangle = b_i + \frac{1}{\langle b_{i+1}, b_{i+2}, \cdots, b_n\rangle} = b_i + \frac{1}{y_{i+1}}.$$

Thus we have $y_i > b_i$ and $y_i > 1$ for $i = 1, 2, \cdots, n - 1$, and $y_n = b_n > 1$. Consequently $b_i = [y_i]$ for all values of i in the range $0 \leqq i \leqq n$. The hypothesis that the continued fractions are equal can be written in the form $y_0 = \xi_0$, where we are using the notation of equation (7.3). Now the definition of ξ_i as u_i/u_{i+1} implies that $\xi_{i+1} > 1$ for all values of $i \geqq 0$, and so $a_i = [\xi_i]$

for $0 \leqq i \leqq j$ by equations (7.2). It follows from $y_0 = \xi_0$ that, taking integral parts, $b_0 = [y_0] = [\xi_0] = a_0$. By equations (7.2) and (7.5) we get

$$\frac{1}{\xi_1} = \xi_0 - a_0 = y_0 - b_0 = \frac{1}{y_1}, \qquad \xi_1 = y_1, \qquad a_1 = [\xi_1] = [y_1] = b_1.$$

This gives us the start of a proof by mathematical induction. We now establish that $\xi_i = y_i$ and $a_i = b_i$ imply that $\xi_{i+1} = y_{i+1}$ and $a_{i+1} = b_{i+1}$. To see this, we again use equations (7.2) and (7.5) to write

$$\frac{1}{\xi_{i+1}} = \xi_i - a_i = y_i - b_i = \frac{1}{y_{i+1}},$$

$$\xi_{i+1} = y_{i+1}, \quad a_{i+1} = [\xi_{i+1}] = [y_{i+1}] = b_{i+1}.$$

It must also follow that the continued fractions have the same length, that is, that $j = n$. For suppose that, say, $j < n$. From the preceding argument we have $\xi_j = y_j$, $a_j = b_j$. But $\xi_j = a_j$ by (7.2) and $y_j > b_j$ by (7.5), and so we have a contradiction. If we had assumed $j > n$, a symmetrical contradiction would have arisen, and thus j must equal n, and the theorem is proved.

Theorem 7.2 *Any finite simple continued fraction represents a rational number. Conversely any rational number can be expressed as a finite simple continued fraction, and in exactly two ways.*

Proof. The first assertion can be established by mathematical induction on the number of terms in the continued fraction, by use of the formula

$$\langle a_0, a_1, \cdots, a_j \rangle = a_0 + \frac{1}{\langle a_1, a_2, \cdots, a_j \rangle}.$$

The second assertion follows from the development of u_0/u_1 into a finite simple continued fraction in Section 7.1, together with equation (7.4) and Theorem 7.1.

PROBLEM

1. Let $a_0, a_1, \cdots, a_n$ and $b_0, b_1, \cdots, b_{n+1}$ be positive integers. What are the conditions for

$$\langle a_0, a_1, \cdots, a_n \rangle < \langle b_0, b_1, \cdots, b_{n+1} \rangle?$$

7.3 Infinite Continued Fractions

Let $a_0, a_1, a_2, \cdots$ be an infinite sequence of integers, all positive except perhaps a_0. We define two sequences of integers $\{h_n\}$ and $\{k_n\}$ inductively as follows:

$$h_{-2} = 0, \quad h_{-1} = 1, \quad h_i = a_i h_{i-1} + h_{i-2} \qquad \text{for } i \geqq 0,$$
$$k_{-2} = 1, \quad k_{-1} = 0, \quad k_i = a_i k_{i-1} + k_{i-2} \qquad \text{for } i \geqq 0. \tag{7.6}$$

We note that $k_0 = 1$, $k_1 = a_1 k_0 \geqq k_0$, $k_2 > k_1$, $k_3 > k_2$, etc., so that $1 = k_0 \leqq k_1 < k_2 < k_3 < \cdots < k_n < \cdots$.

Theorem 7.3 *For any positive real number* x,

$$\langle a_0, a_1, \cdots, a_{n-1}, x\rangle = \frac{xh_{n-1} + h_{n-2}}{xk_{n-1} + k_{n-2}}.$$

Proof. If $n = 0$, the result is to be interpreted as

$$x = \frac{xh_{-1} + h_{-2}}{xk_{-1} + k_{-2}},$$

which is true by equations (7.6). If $n = 1$, the result is

$$\langle a_0, x\rangle = \frac{xh_0 + h_{-1}}{xk_0 + k_{-1}},$$

which can be verified from (7.6) and the fact that $\langle a_0, x\rangle$ stands for $a_0 + 1/x$. We establish the theorem in general by induction. Assuming that the result holds for $\langle a_0, a_1, \cdots, a_{n-1}, x\rangle$, we see that

$$\begin{aligned}\langle a_0, a_1, \cdots, a_n, x\rangle &= \left\langle a_0, a_1, \cdots, a_{n-1}, a_n + \frac{1}{x}\right\rangle \\ &= \frac{(a_n + 1/x)h_{n-1} + h_{n-2}}{(a_n + 1/x)k_{n-1} + k_{n-2}} \\ &= \frac{x(a_n h_{n-1} + h_{n-2}) + h_{n-1}}{x(a_n k_{n-1} + k_{n-2}) + k_{n-1}} = \frac{xh_n + h_{n-1}}{xk_n + k_{n-1}}.\end{aligned}$$

Theorem 7.4 *If we define* $r_n = \langle a_0, a_1, \cdots, a_n\rangle$ *for all integers* $n \geqq 0$, *then* $r_n = h_n/k_n$.

Proof. We apply Theorem 7.3 with x replaced by a_n and then use equations (7.6) thus:

$$r_n = \langle a_0, a_1, \cdots, a_n\rangle = \frac{a_n h_{n-1} + h_{n-2}}{a_n k_{n-1} + k_{n-2}} = \frac{h_n}{k_n}.$$

Theorem 7.5 *The equations*

$$h_i k_{i-1} - h_{i-1} k_i = (-1)^{i-1} \quad \textit{and} \quad r_i - r_{i-1} = \frac{(-1)^{i-1}}{k_i k_{i-1}}$$

hold for $i \geqq 1$. The identities

$$h_i k_{i-2} - h_{i-2} k_i = (-1)^i a_i \quad \text{and} \quad r_i - r_{i-2} = \frac{(-1)^i a_i}{k_i k_{i-2}}$$

hold for $i \geqq 1$. The fraction h_i/k_i is reduced, that is $(h_i, k_i) = 1$.

Proof. The equations (7.6) imply that $h_{-1}k_{-2} - h_{-2}k_{-1} = 1$. Continuing the proof by induction, we assume that $h_{i-1}k_{i-2} - h_{i-2}k_{i-1} = (-1)^{i-2}$. Again we use equations (7.6) to get $h_i k_{i-1} - h_{i-1}k_i = (a_i h_{i-1} + h_{i-2})k_{i-1} - h_{i-1}(a_i k_{i-1} + k_{i-2}) = -(h_{i-1}k_{i-2} - h_{i-2}k_{i-1}) = (-1)^{i-1}$. This proves the first result stated in the theorem. We divide by $k_{i-1}k_i$ to get the second result, the formula for $r_i - r_{i-1}$. Furthermore, the fraction h_i/k_i is in lowest terms since any factor of h_i and k_i is also a factor of $(-1)^{i-1}$.

The other formulas can be derived in much the same way from (7.6), although we do not need induction in this case. First we observe that $h_0 k_{-2} - h_{-2}k_0 = a_0$, and that in general $h_i k_{i-2} - h_{i-2}k_i = (a_i h_{i-1} + h_{i-2})k_{i-2} - h_{i-2}(a_i k_{i-1} + k_{i-2}) = a_i(h_{i-1}k_{i-2} - h_{i-2}k_{i-1}) = (-1)^i a_i$. The final identity can be obtained by dividing by $k_{i-2}k_i$.

Theorem 7.6 *The values r_n defined in Theorem 7.4 satisfy the infinite chain of inequalities $r_0 < r_2 < r_4 < r_6 < \cdots < r_7 < r_5 < r_3 < r_1$. Stated in words, the r_n with even subscripts form an increasing sequence, those with odd subscripts form a decreasing sequence, and every r_{2n} is less than every r_{2j-1}. Furthermore, $\lim_{n\to\infty} r_n$ exists, and for every $j \geqq 0$, $r_{2j} < \lim_{n\to\infty} r_n < r_{2j+1}$.*

Proof. The identities of Theorem 7.5 for $r_i - r_{i-1}$ and $r_i - r_{i-2}$ imply that $r_{2j} < r_{2j+2}$, $r_{2j-1} > r_{2j+1}$, and $r_{2j} < r_{2j-1}$ because the k_i are positive for $i \geqq 0$ and the a_i are positive for $i \geqq 1$. Thus we have $r_0 < r_2 < r_4 < \cdots$ and $r_1 > r_3 > r_5 > \cdots$. To prove that $r_{2n} < r_{2j-1}$, we put the previous results together in the form

$$r_{2n} < r_{2n+2j} < r_{2n+2j-1} \leqq r_{2j-1}.$$

The sequence $r_0, r_2, r_4, \cdots$ is monotonically increasing and is bounded above by r_1, and so has a limit. Analogously, the sequence $r_1, r_3, r_5, \cdots$ is monotonically decreasing and is bounded below by r_0, and so has a limit. These two limits are equal because, by Theorem 7.5, the difference $r_i - r_{i-1}$ tends to zero as i tends to infinity, since the integers k_i are increasing with i. Another way of looking at this is to observe that (r_0, r_1), (r_2, r_3), (r_4, r_5), $\cdots$ is a chain of nested intervals defining a real number, namely $\lim_{n\to\infty} r_n$.

These theorems suggest the following definition.

Definition 7.1 *An infinite sequence $a_0, a_1, a_2, \cdots$ of integers, all positive except perhaps for a_0, determines an infinite simple continued fraction $\langle a_0, a_1, a_2, \cdots\rangle$. The value of $\langle a_0, a_1, a_2, \cdots\rangle$ is defined to be $\lim_{n\to\infty} \langle a_0, a_1, a_2, \cdots, a_n\rangle$.*

This limit, being the same as $\lim_{n\to\infty} r_n$, exists by Theorem 7.6. Another way of writing this limit is $\lim_{n\to\infty} h_n/k_n$. The rational number $\langle a_0, a_1, \cdots, a_n\rangle = h_n/k_n = r_n$ is called the nth *convergent* to the infinite continued fraction. We say that the infinite continued fraction converges to the value $\lim_{n\to\infty} r_n$. In the case of a finite simple continued fraction $\langle a_0, a_1, \cdots, a_n\rangle$ we similarly call the number $\langle a_0, a_1, \cdots, a_m\rangle$ the mth *convergent* to $\langle a_0, a_1, \cdots, a_n\rangle$.

Theorem 7.7 *The value of any infinite simple continued fraction* $\langle a_0, a_1, a_2, \cdots\rangle$ *is irrational.*

Proof. Writing θ for $\langle a_0, a_1, a_2, \cdots\rangle$ we observe by Theorem 7.6 that θ lies between r_n and r_{n+1}, so that $0 < |\theta - r_n| < |r_{n+1} - r_n|$. Multiplying by k_n, and making use of the result from Theorem 7.5 that $|r_{n+1} - r_n| = (k_n k_{n+1})^{-1}$, we have

$$0 < |k_n\theta - h_n| < \frac{1}{k_{n+1}}.$$

Now suppose that θ were rational, say $\theta = a/b$ with integers a and b, $b > 0$. Then the above inequality would become, upon multiplication by b,

$$0 < |k_n a - h_n b| < \frac{b}{k_{n+1}}.$$

The integers k_n increase with n, so we could choose n sufficiently large so that $b < k_{n+1}$. Then the integer $|k_n a - h_n b|$ would lie between 0 and 1, which is impossible.

Suppose we have two different infinite simple continued fractions, $\langle a_0, a_1, a_2, \cdots\rangle$ and $\langle b_0, b_1, b_2, \cdots\rangle$. Can these converge to the same value? The answer is no, and we establish this in the next two results.

Lemma 7.8 *Let* $\theta = \langle a_0, a_1, a_2, \cdots\rangle$ *be a simple continued fraction. Then* $a_0 = [\theta]$. *Furthermore if* θ_1 *denotes* $\langle a_1, a_2, a_3, \cdots\rangle$ *then* $\theta = a_0 + 1/\theta_1$.

Proof. By Theorem 7.6 we see that $r_0 < \theta < r_1$, that is $a_0 < \theta < a_0 + 1/a_1$. Now $a_1 \geqq 1$, so we have $a_0 < \theta < a_0 + 1$, and hence $a_0 = [\theta]$. Also

$$\theta = \lim_{n\to\infty} \langle a_0, a_1, \cdots, a_n\rangle = \lim_{n\to\infty} \left(a_0 + \frac{1}{\langle a_1, \cdots, a_n\rangle}\right)$$

$$= a_0 + \frac{1}{\lim_{n\to\infty} \langle a_1, \cdots, a_n\rangle} = a_0 + \frac{1}{\theta_1}.$$

Theorem 7.9 *Two distinct infinite simple continued fractions converge to different values.*

Proof. Let us suppose that $\langle a_0, a_1, a_2, \cdots \rangle = \langle b_0, b_1, b_2, \cdots \rangle = \theta$. Then by Lemma 7.8, $[\theta] = a_0 = b_0$ and

$$\theta = a_0 + \frac{1}{\langle a_1, a_2, \cdots \rangle} = b_0 + \frac{1}{\langle b_1, b_2, \cdots \rangle}.$$

Hence $\langle a_1, a_2, \cdots \rangle = \langle b_1, b_2, \cdots \rangle$. Repetition of the argument gives $a_1 = b_1$, and so by mathematical induction $a_n = b_n$ for all n.

PROBLEMS

1. Evaluate the infinite continued fraction $\langle 1, 1, 1, 1, \cdots \rangle$. *Suggestion:* by Lemma 7.8, we see that $\theta = 1 + 1/\theta$ in this case. This gives a quadratic equation, only one of whose roots is positive.
2. Evaluate the infinite continued fractions $\langle 2, 1, 1, 1, 1, \cdots \rangle$ and $\langle 2, 3, 1, 1, 1, 1, \cdots \rangle$. *Suggestion:* use the result of the preceding problem along with Lemma 7.8.
3. Evaluate the infinite continued fractions:

(*a*) $\langle 2, 2, 2, 2, \cdots \rangle$; (*b*) $\langle 1, 2, 1, 2, 1, 2, \cdots \rangle$;
(*c*) $\langle 2, 1, 2, 1, 2, 1, \cdots \rangle$; (*d*) $\langle 1, 3, 1, 2, 1, 2, 1, 2, \cdots \rangle$.

4. For $n \geqq 1$, prove that $k_n/k_{n-1} = \langle a_n, a_{n-1}, \cdots, a_2, a_1 \rangle$. Find and prove a similar continued fraction expansion for h_n/h_{n-1}, assuming $a_0 \geqq 0$.

7.4 Irrational Numbers

We have shown that any infinite simple continued fraction represents an irrational number. Conversely, if we begin with an irrational number ξ, or ξ_0, we can expand it into an infinite simple continued fraction. To do this we define $a_0 = [\xi_0]$, $\xi_1 = 1/(\xi_0 - a_0)$, and next $a_1 = [\xi_1]$, $\xi_2 = 1/(\xi_1 - a_1)$, and so by an inductive definition

$$(7.7) \qquad a_i = [\xi_i], \qquad \xi_{i+1} = \frac{1}{\xi_i - a_i}.$$

The a_i are integers by definition, and the ξ_i are all irrational since the irrationality of ξ_1 is implied by that of ξ_0, that of ξ_2 by that of ξ_1, and so on. Furthermore, $a_i \geqq 1$ for $i \geqq 1$ because $a_{i-1} = [\xi_{i-1}]$ and the fact that ξ_{i-1} is irrational imply that

$$a_{i-1} < \xi_{i-1} < 1 + a_{i-1}, \qquad 0 < \xi_{i-1} - a_{i-1} < 1,$$

$$\xi_i = \frac{1}{\xi_{i-1} - a_{i-1}} > 1, \qquad a_i = [\xi_i] \geqq 1.$$

Next we use repeated application of (7.7) in the form $\xi_i = a_i + 1/\xi_{i+1}$ to get the chain

$$\xi = \xi_0 = a_0 + \frac{1}{\xi_1} = \langle a_0, \xi_1 \rangle$$

$$= \left\langle a_0, a_1 + \frac{1}{\xi_2} \right\rangle = \langle a_0, a_1, \xi_2 \rangle$$

$$\cdots\cdots\cdots\cdots\cdots\cdots\cdots\cdots$$

$$= \left\langle a_0, a_1, \cdots, a_{m-2}, a_{m-1} + \frac{1}{\xi_m} \right\rangle = \langle a_0, a_1, \cdots, a_{m-1}, \xi_m \rangle.$$

This suggests, but does not establish, that ξ is the value of the infinite continued fraction $\langle a_0, a_1, a_2, \cdots \rangle$ determined by the integers a_i.

To prove this we use Theorem 7.3 to write

$$\xi = \langle a_0, a_1, \cdots, a_{n-1}, \xi_n \rangle = \frac{\xi_n h_{n-1} + h_{n-2}}{\xi_n k_{n-1} + k_{n-2}} \tag{7.8}$$

with the h_i and k_i defined as in (7.6). By Theorem 7.5 we get

$$\xi - r_{n-1} = \xi - \frac{h_{n-1}}{k_{n-1}} = \frac{\xi_n h_{n-1} + h_{n-2}}{\xi_n k_{n-1} + k_{n-2}} - \frac{h_{n-1}}{k_{n-1}} \tag{7.9}$$

$$= \frac{-(h_{n-1}k_{n-2} - h_{n-2}k_{n-1})}{k_{n-1}(\xi_n k_{n-1} + k_{n-2})} = \frac{(-1)^{n-1}}{k_{n-1}(\xi_n k_{n-1} + k_{n-2})}.$$

This fraction tends to zero as n tends to infinity because the integers k_n are increasing with n, and ξ_n is positive. Hence $\xi - r_{n-1}$ tends to zero as n tends to infinity and then, by Definition 7.1,

$$\xi = \lim_{n\to\infty} r_n = \lim_{n\to\infty} \langle a_0, a_1, \cdots, a_n \rangle = \langle a_0, a_1, a_2, \cdots \rangle.$$

We summarize the results of the last two sections in the following theorem.

Theorem 7.10 *Any irrational number ξ is uniquely expressible, by the procedure that gave equations (7.7), as an infinite simple continued fraction $\langle a_0, a_1, a_2, \cdots \rangle$. Conversely any such continued fraction determined by integers a_i which are positive for all $i > 0$ represents an irrational number, ξ. The finite simple continued fraction $\langle a_0, a_1, \cdots, a_n \rangle$ has the rational value $h_n/k_n = r_n$, and is called the nth convergent to ξ. Equations (7.6) relate the h_i and k_i to the a_i. For $n = 0, 2, 4, \cdots$ these convergents form a monotonically increasing sequence with ξ as a limit. Similarly, for $n = 1, 3, 5, \cdots$ the convergents form a monotonically decreasing sequence tending to ξ. The denominators k_n of the convergents are an increasing sequence of positive integers for $n > 0$.*

Finally, with ξ_i *defined by* (7.7), *we have* $\langle a_0, a_1, \cdots\rangle = \langle a_0, a_1, \cdots, a_{n-1}, \xi_n\rangle$ *and* $\xi_n = \langle a_n, a_{n+1}, a_{n+2}, \cdots\rangle$.

Proof. Only the last equation is new, and it becomes obvious if we apply to ξ_n the process described at the opening of this section.

PROBLEMS

1. Expand each of the following as infinite simple continued fractions: $\sqrt{2}$, $\sqrt{2} - 1$, $\sqrt{2}/2$, $\sqrt{3}$, $1/\sqrt{3}$.

2. Given that two irrational numbers have identical convergents h_0/k_0, $h_1/k_1, \cdots$, up to h_n/k_n, prove that their continued fraction expansions are identical up to a_n.

3. Let α, β, γ be irrational numbers satisfying $\alpha < \beta < \gamma$. If α and γ have identical convergents $h_0/k_0, h_1/k_1, \cdots$, up to h_n/k_n, prove that β also has these same convergents up to h_n/k_n.

4. Let ξ be an irrational number with continued fraction expansion $\langle a_0, a_1, a_2, a_3, \cdots\rangle$. Let $b_1, b_2, b_3, \cdots$ be any finite or infinite sequence of positive integers. Prove that

$$\lim_{n\to\infty} \langle a_0, a_1, a_2, \cdots, a_n, b_1, b_2, b_3, \cdots\rangle = \xi.$$

5. In the notation used in the text, prove that

$$\xi_n = \langle a_n, a_{n+1}, a_{n+2}, \cdots\rangle.$$

6. Prove that for $n \geqq 1$,

$$\xi - \frac{h_n}{k_n} = (-1)^n k_n^{-2}\{\xi_{n+1} + \langle 0, a_n, a_{n-1}, \cdots, a_2, a_1\rangle\}^{-1}.$$

7. Prove that

$$k_n\,|k_{n-1}\xi - h_{n-1}| + k_{n-1}\,|k_n\xi - h_n| = 1.$$

7.5 Approximations to Irrational Numbers

Continuing to use the notation of the preceding sections, we now show that the convergents h_n/k_n form a sequence of "best" rational approximations to the irrational number ξ.

Theorem 7.11 *We have for any* $n \geqq 0$,

$$\left|\xi - \frac{h_n}{k_n}\right| < \frac{1}{k_n k_{n+1}} \quad \text{and} \quad |\xi k_n - h_n| < \frac{1}{k_{n+1}}.$$

Proof. The second inequality follows from the first by multiplication by k_n. By (7.9) and (7.7) we see that

$$\left|\xi - \frac{h_n}{k_n}\right| = \frac{1}{k_n(\xi_{n+1}k_n + k_{n-1})} < \frac{1}{k_n(a_{n+1}k_n + k_{n-1})}.$$

Using (7.6), we replace $a_{n+1}k_n + k_{n-1}$ by k_{n+1} to obtain the first inequality.

Theorem 7.12 *The convergents h_n/k_n are successively closer to ξ, that is*

$$\left|\xi - \frac{h_n}{k_n}\right| < \left|\xi - \frac{h_{n-1}}{k_{n-1}}\right|.$$

In fact the stronger inequality $|\xi k_n - h_n| < |\xi k_{n-1} - h_{n-1}|$ holds.

Proof. To see that the second inequality is stronger in that it implies the first, we use $k_{n-1} \leqq k_n$ to write

$$\left|\xi - \frac{h_n}{k_n}\right| = \frac{1}{k_n}|\xi k_n - h_n| < \frac{1}{k_n}|\xi k_{n-1} - h_{n-1}|$$

$$\leqq \frac{1}{k_{n-1}}|\xi k_{n-1} - h_{n-1}| = \left|\xi - \frac{h_{n-1}}{k_{n-1}}\right|.$$

Now to prove the stronger inequality we observe that $a_n + 1 > \xi_n$ by (7.7), and so by (7.6),

$$\xi_n k_{n-1} + k_{n-2} < (a_n + 1)k_{n-1} + k_{n-2}$$

$$= k_n + k_{n-1} \leqq a_{n+1}k_n + k_{n-1} = k_{n+1}.$$

This inequality and (7.9) imply that

$$\left|\xi - \frac{h_{n-1}}{k_{n-1}}\right| = \frac{1}{k_{n-1}(\xi_n k_{n-1} + k_{n-2})} > \frac{1}{k_{n-1}k_{n+1}}.$$

We multiply by k_{n-1} and use Theorem 7.11 to get

$$\xi k_{n-1} - h_{n-1} > \frac{1}{k_{n+1}} > |\xi k_n - h_n|.$$

The convergent h_n/k_n is the best approximation to ξ of all the rational fractions with denominator k_n or less. The following theorem states this in a different way.

Theorem 7.13 *If a/b is a rational number with positive denominator such that $|\xi - a/b| < |\xi - h_n/k_n|$ for some $n \geqq 1$, then $b > k_n$. In fact if $|\xi b - a| < |\xi k_n - h_n|$ for some $n \geqq 0$, then $b \geqq k_{n+1}$.*

Proof. First we show that the second part of the theorem implies the first. Suppose that the first part is false so that there is an a/b with

$$\left|\xi - \frac{a}{b}\right| < \left|\xi - \frac{h_n}{k_n}\right| \quad \text{and} \quad b \leqq k_n.$$

The product of these inequalities gives $|\xi b - a| < |\xi k_n - h_n|$. But the second part of the theorem says that this implies $b \geqq k_{n+1}$, so we have a contradiction, since $k_n < k_{n+1}$ for $n \geqq 1$.

To prove the second part of the theorem we proceed again by indirect argument, assuming that $|\xi b - a| < |\xi k_n - h_n|$ and $b < k_{n+1}$. Consider the linear equations in x and y,

$$xk_n + yk_{n+1} = b, \qquad xh_n + yh_{n+1} = a.$$

The determinant of coefficients is ± 1 by Theorem 7.5, and consequently these equations have an integral solution x, y. Moreover, neither x nor y is zero. For if $x = 0$ then $b = yk_{n+1}$, which implies that $y \neq 0$, in fact that $y > 0$ and $b \geqq k_{n+1}$, in contradiction to $b < k_{n+1}$. If $y = 0$ then $a = xh_n$, $b = xk_n$, and

$$|\xi b - a| = |\xi x k_n - xh_n| = |x|\,|\xi k_n - h_n| \geqq |k_n \xi - h_n|$$

since $|x| \geqq 1$, and again we have a contradiction.

Next we prove that x and y have opposite signs. First, if $y < 0$, then $xk_n = b - yk_{n+1}$ shows that $x > 0$. Second, if $y > 0$, then $b < k_{n+1}$ implies that $b < yk_{n+1}$, and so xk_n is negative, whence $x < 0$. Now it follows from Theorem 7.10 that $\xi k_n - h_n$ and $\xi k_{n+1} - h_{n+1}$ have opposite signs, and hence $x(\xi k_n - h_n)$ and $y(\xi k_{n+1} - h_{n+1})$ have the same sign. From the equations defining x and y we get $\xi b - a = x(\xi k_n - h_n) + y(\xi k_{n+1} - h_{n+1})$. Since the two terms on the right have the same sign, the absolute value of the whole equals the sum of the separate absolute values. Thus

$$\begin{aligned} |\xi b - a| &= |x(\xi k_n - h_n) + y(\xi k_{n+1} - h_{n+1})| \\ &= |x(\xi k_n - h_n)| + |y(\xi k_{n+1} - h_{n+1})| \\ &> |x(\xi k_n - h_n)| = |x|\,|\xi k_n - h_n| \geqq |\xi k_n - h_n|. \end{aligned}$$

This is a contradiction, and so the theorem is established.

Theorem 7.14 *Let ξ denote any irrational number. If there is a rational number a/b with $b \geqq 1$ such that*

$$\left|\xi - \frac{a}{b}\right| < \frac{1}{2b^2},$$

then a/b equals one of the convergents of the simple continued fraction expansion of ξ.

Proof. It suffices to prove the result in the case $(a, b) = 1$. Let the convergents of the simple continued fraction expansion of ξ be h_j/k_j, and suppose that a/b is not a convergent. The inequalities $k_n \leqq b < k_{n+1}$ determine an integer n. For this n, the inequality $|\xi b - a| < |\xi k_n - h_n|$ is impossible because of Theorem 7.13.

Therefore we have

$$|\xi k_n - h_n| \leqq |\xi b - a| < \frac{1}{2b},$$

$$\left|\xi - \frac{h_n}{k_n}\right| < \frac{1}{2bk_n}.$$

Using the facts that $a/b \neq h_n/k_n$ and that $bh_n - ak_n$ is an integer, we find that

$$\frac{1}{bk_n} \leqq \frac{|bh_n - ak_n|}{bk_n} = \left|\frac{h_n}{k_n} - \frac{a}{b}\right| \leqq \left|\xi - \frac{h_n}{k_n}\right| + \left|\xi - \frac{a}{b}\right| < \frac{1}{2bk_n} + \frac{1}{2b^2}.$$

This implies $b < k_n$ which is a contradiction.

Theorem 7.15 *The* nth *convergent of* $1/x$ *is the reciprocal of the* $(n - 1)$st *convergent of* x *if* x *is any real number* >1.

Proof. We have $x = \langle a_0, a_1, \cdots \rangle$ and $1/x = \langle 0, a_0, a_1, \cdots \rangle$. If h_n/k_n and h'_n/k'_n are the convergents for x and $1/x$ respectively, then

$$\begin{array}{llll} h'_0 = 0, & h'_1 = 1, & h'_2 = a_1, & h'_n = a_{n-1}h'_{n-1} + h'_{n-2} \\ & k_0 = 1, & k_1 = a_1, & k_{n-1} = a_{n-1}k_{n-2} + k_{n-3} \\ k'_0 = 1, & k'_1 = a_0, & k'_2 = a_0a_1 + 1, & k'_n = a_{n-1}k'_{n-1} + k'_{n-2} \\ & h_0 = a_0, & h_1 = a_0a_1 + 1, & h_{n-1} = a_{n-1}h_{n-2} + h_{n-3}. \end{array}$$

The theorem now follows by mathematical induction.

PROBLEMS

1. Prove that the first assertion in Theorem 7.13 holds in case $n = 0$ if $k_1 > 1$.
2. Prove that the first assertion in Theorem 7.13 becomes false if "$b > k_n$" is replaced by "$b \geqq k_{n+1}$." *Suggestion:* use $\xi = \pi^{-1}$ and $n = 1$.
3. Say that a rational number a/b with $b > 0$ is a "good approximation" to the irrational number ξ if

$$|\xi b - a| = \min_{\substack{\text{all } x \\ 0 < y \leqq b}} |\xi y - x|,$$

where, as indicated, the minimum on the right is to be taken over all integers x and all y satisfying $0 < y \leqq b$. Prove that every convergent to ξ is a "good approximation."

4. Prove that every "good approximation" to ξ is a convergent.

5. (*a*) Prove that if r/s lies between a/b and c/d, where the denominators of these rational fractions are positive, and if $ad - bc = \pm 1$, then $s > b$ and $s > d$.

(*b*) Let ξ be an irrational with convergents $\{h_n/k_n\}$. Prove that the sequence

$$\frac{h_{n-1}}{k_{n-1}}, \frac{h_{n-1} + h_n}{k_{n-1} + k_n}, \frac{h_{n-1} + 2h_n}{k_{n-1} + 2k_n}, \cdots, \frac{h_{n-1} + a_{n+1}h_n}{k_{n-1} + a_{n+1}k_n} = \frac{h_{n+1}}{k_{n+1}}$$

is increasing if n is odd, decreasing if n is even. If a/b and c/d denote any consecutive pair of this sequence, prove that $ad - bc = \pm 1$. The terms of this sequence, except the first and last, are called the *secondary convergents;* here n runs through all values $1, 2, \cdots$.

(*c*) Say that a rational number a/b is a "fair approximation" to ξ if $|\xi - a/b| = \min |\xi - x/y|$, the minimum being taken over all integers x and y with $0 < y \leqq b$. Prove that every good approximation is a fair approximation. Prove that every fair approximation is either a convergent or a secondary convergent to ξ.

(*d*) Prove that not every secondary convergent is a "fair approximation". *Suggestion:* consider $\xi = \sqrt{2}$.

(*e*) Say that an infinite sequence of rational numbers, $r_1, r_2, r_3, \cdots$ with limit ξ is an "approximating sequence" to an irrational number ξ if $|\xi - r_{j+1}| < |\xi - r_j|$, $j = 1, 2, 3, \cdots$, and if the positive denominators of the r_j are increasing with j. Prove that the "fair approximations" to ξ form an "approximating sequence."

(*f*) Let S_{n-1} denote the finite sequence of (*b*) with the first term deleted, so that S_{n-1} has a_{n+1} terms, the last term being h_{n+1}/k_{n+1}. Prove that the infinite sequence of rational numbers obtained by first taking the terms of S_0 in order, then the terms of S_2, then S_4, then $S_6, \cdots$, is also an "approximating sequence" to ξ. Prove also that this sequence is maximal in the sense that if any other rational number $< \xi$ is introduced into the sequence as a new member, we no longer have an approximating sequence.

(*g*) Establish analogous properties for the sequence obtained by taking the terms of $S_{-1}, S_1, S_3, S_5, \cdots$.

6. Let ξ be irrational, $\xi = \langle a_0, a_1, a_2, \cdots \rangle$. Verify that

$$-\xi = \langle -a_0 - 1, 1, a_1 - 1, a_2, a_3, \cdots \rangle \text{ if } a_1 > 1$$

and $-\xi = \langle -a_0 - 1, a_2 + 1, a_3, a_4, \cdots \rangle$ if $a_1 = 1$.

7.6 Best Possible Approximations

Theorem 7.11 provides another method of proving Theorem 6.9. For in the statement of Theorem 7.11 we can replace k_{n+1} by the smaller integer k_n to get the weaker, but still correct, inequality

$$\left|\xi - \frac{h_n}{k_n}\right| < \frac{1}{k_n^2}.$$

Moreover the process described in Section 7.4 enables us to determine for any given irrational ξ as many convergents h_n/k_n as we please. We can also use continued fractions in other proofs of Theorems 6.11 and 6.12. First we give a simple lemma.

Lemma 7.16 *If x is real, $x > 1$, and $x + x^{-1} < \sqrt{5}$, then $x < \frac{1}{2}(\sqrt{5} + 1)$ and $x^{-1} > \frac{1}{2}(\sqrt{5} - 1)$.*

Proof. For real $x \geqq 1$ we note that $x + x^{-1}$ increases with x, and $x + x^{-1} = \sqrt{5}$ if $x = \frac{1}{2}(\sqrt{5} + 1)$.

Theorem 7.17 (*Hurwitz*) *Given any irrational number ξ, there exist infinitely many rational numbers h/k such that*

$$\left|\xi - \frac{h}{k}\right| < \frac{1}{\sqrt{5}\,k^2}. \tag{7.13}$$

Proof. We will establish that, of every three consecutive convergents of the simple continued fraction expansion of ξ, at least one satisfies the inequality.

Let q_n denote k_n/k_{n-1}. We first prove that

$$q_j + q_j^{-1} < \sqrt{5} \tag{7.14}$$

if (7.13) is false for both $h/k = h_{j-1}/k_{j-1}$ and $h/k = h_j/k_j$. Suppose (7.13) is false for these two values of h/k. We have

$$\left|\xi - \frac{h_{j-1}}{k_{j-1}}\right| + \left|\xi - \frac{h_j}{k_j}\right| \geqq \frac{1}{\sqrt{5}\,k_{j-1}^2} + \frac{1}{\sqrt{5}\,k_j^2}.$$

But ξ lies between h_{j-1}/k_{j-1} and h_j/k_j and hence we find, using Theorem 7.5, that

$$\left|\xi - \frac{h_{j-1}}{k_{j-1}}\right| + \left|\xi - \frac{h_j}{k_j}\right| = \left|\frac{h_{j-1}}{k_{j-1}} - \frac{h_j}{k_j}\right| = \frac{1}{k_{j-1}k_j}.$$

Combining these results we get

$$\frac{k_j}{k_{j-1}} + \frac{k_{j-1}}{k_j} \leq \sqrt{5}.$$

Since the left side is rational we actually have a strict inequality, and (7.14) follows.

Now suppose (7.13) is false for $h/k = h_i/k_i$, $i = n - 1, n, n + 1$. We then have (7.14) for both $j = n$ and $j = n + 1$. By Lemma 7.16 we see that

$q_n^{-1} > \frac{1}{2}(\sqrt{5} - 1)$ and $q_{n+1} < \frac{1}{2}(\sqrt{5} + 1)$, and, by (7.6) we find $q_{n+1} = a_{n+1} + q_n^{-1}$. This gives us

$$\begin{aligned}\tfrac{1}{2}(\sqrt{5} + 1) > q_{n+1} = a_{n+1} + q_n^{-1} &> a_{n+1} + \tfrac{1}{2}(\sqrt{5} - 1)\\ &\geqq 1 + \tfrac{1}{2}(\sqrt{5} - 1) = \tfrac{1}{2}(\sqrt{5} + 1)\end{aligned}$$

and this is a contradiction.

Theorem 7.18 *The constant $\sqrt{5}$ in the preceding theorem is best possible. In other words Theorem* 7.17 *does not hold if $\sqrt{5}$ is replaced by any larger value.*

Proof. It suffices to exhibit an irrational number ξ for which $\sqrt{5}$ is the largest possible constant. Consider the irrational ξ whose continued fraction expansion is $\langle 1, 1, 1, \cdots \rangle$. We see that

$$\xi = 1 + \frac{1}{\langle 1, 1, \cdots \rangle} = 1 + \frac{1}{\xi}, \qquad \xi^2 = \xi + 1, \qquad \xi = \frac{1}{2}(\sqrt{5} + 1).$$

Using (7.7) we can prove by induction that $\xi_i = (\sqrt{5} + 1)/2$ for all $i \geqq 0$, for if $\xi_i = (\sqrt{5} + 1)/2$ then

$$\xi_{i+1} = (\xi_i - a_i)^{-1} = (\tfrac{1}{2}(\sqrt{5} + 1) - 1)^{-1} = \tfrac{1}{2}(\sqrt{5} + 1).$$

A simple calculation yields $h_0 = k_0 = k_1 = 1$, $h_1 = k_2 = 2$. Equations (7.6) become $h_i = h_{i-1} + h_{i-2}$, $k_i = k_{i-1} + k_{i-2}$, and so by mathematical induction $k_n = h_{n-1}$ for $n \geqq 1$. Hence we have

$$\lim_{n\to\infty} \frac{k_{n-1}}{k_n} = \lim_{n\to\infty} \frac{k_{n-1}}{h_{n-1}} = \frac{1}{\xi} = \frac{\sqrt{5} - 1}{2},$$

$$\lim_{n\to\infty} \left(\xi_{n+1} + \frac{k_{n-1}}{k_n}\right) = \frac{\sqrt{5} + 1}{2} + \frac{\sqrt{5} - 1}{2} = \sqrt{5}.$$

If c is any constant exceeding $\sqrt{5}$, then

$$\xi_{n+1} + \frac{k_{n-1}}{k_n} > c$$

holds for only a finite number of values of n. Thus, by (7.9),

$$\left| \xi - \frac{h_n}{k_n} \right| = \frac{1}{k_n^2(\xi_{n+1} + k_{n-1}/k_n)} < \frac{1}{ck_n^2}$$

holds for only a finite number of values of n. Thus there are only a finite number of rational numbers h/k satisfying $|\xi - h/k| < 1/(ck^2)$, because any such h/k is one of the convergents to ξ by Theorem 7.14.

PROBLEMS

1. Find two rational numbers a/b satisfying

$$\left|\sqrt{2} - \frac{a}{b}\right| < \frac{1}{\sqrt{5}\,b^2}.$$

2. Find two rational numbers a/b satisfying

$$\left|\pi - \frac{a}{b}\right| < \frac{1}{\sqrt{5}\,b^2}.$$

3. Prove that the following is false for any constant $c > 2$: Given any irrational number ξ, there exist infinitely many rational numbers h/k such that

$$\left|\xi - \frac{h}{k}\right| < \frac{1}{k^c}.$$

4. Given any constant c, prove that there exists an irrational number ξ and infinitely many rational numbers h/k such that

$$\left|\xi - \frac{h}{k}\right| < \frac{1}{k^c}.$$

5. Prove that of every two consecutive convergents h_n/k_n to ξ with $n \geqq 0$, at least one satisfies

$$\left|\xi - \frac{h}{k}\right| < \frac{1}{2k^2}.$$

Suggestion: Use the idea of Lemma 7.16.

7.7 Periodic Continued Fractions

An infinite simple continued fraction $\langle a_0, a_1, a_2, \cdots\rangle$ is said to be *periodic* if there is an integer n such that $a_r = a_{n+r}$ for all sufficiently large integers r. Thus a periodic continued fraction can be written in the form

$$(7.15)\quad \langle b_0, b_1, b_2, \cdots, b_j, a_0, a_1, \cdots, a_{n-1}, a_0, a_1, \cdots, a_{n-1}, \cdots\rangle$$
$$= \langle b_0, b_1, b_2, \cdots, b_j, \overline{a_0, a_1, \cdots, a_{n-1}}\rangle,$$

where the bar over the $a_0, a_1, \cdots, a_{n-1}$ indicates that this block of integers is repeated indefinitely. For example $\langle\overline{2, 3}\rangle$ denotes $\langle 2, 3, 2, 3, 2, 3, \cdots\rangle$ and its value is easily computed. Writing θ for $\langle\overline{2, 3}\rangle$ we have

$$\theta = 2 + \cfrac{1}{3 + \cfrac{1}{\theta}}.$$

This is a quadratic equation in θ, and we discard the negative root to get the value $\theta = (3 + \sqrt{15})/3$. As a second example consider $\langle 4, 1, \overline{2, 3}\rangle$. Calling this ξ, we have $\xi = \langle 4, 1, \theta\rangle$, with θ as above, and so

$$\xi = 4 + (1 + \theta^{-1})^{-1} = 4 + \frac{\theta}{\theta + 1} = \frac{29 + \sqrt{15}}{7}.$$

These two examples illustrate the following result.

Theorem 7.19 *Any periodic simple continued fraction is a quadratic irrational number, and conversely.*

Proof. Let us write ξ for the periodic continued fraction of (7.15) and θ for its purely periodic part, $\theta = \langle\overline{a_0, a_1, \cdots, a_{n-1}}\rangle = \langle a_0, a_1, \cdots, a_{n-1}, \theta\rangle$. Then equation (7.8) gives

$$\theta = \frac{\theta h_{n-1} + h_{n-2}}{\theta k_{n-1} + k_{n-2}},$$

and this is a quadratic equation in θ. Hence θ is either a quadratic irrational number or a rational number, but the latter is ruled out by Theorem 7.7. Now ξ can be written in terms of θ,

$$\xi = \langle b_0, b_1, \cdots, b_j, \theta\rangle = \frac{\theta m + m'}{\theta q + q'}$$

where m'/q' and m/q are the last two convergents to $\langle b_0, b_1, \cdots, b_j\rangle$. But θ is of the form $(a + \sqrt{b})/c$, and hence ξ is of similar form because, as with θ, we can rule out the possibility that ξ is rational.

To prove the converse, let us begin with any quadratic irrational ξ, or ξ_0, of the form $\xi = \xi_0 = (a + \sqrt{b})/c$, with integers $a, b, c, b > 0, c \neq 0$. The integer b is not a perfect square since ξ is irrational. We multiply numerator and denominator by $|c|$ to get

$$\xi_0 = \frac{ac + \sqrt{bc^2}}{c^2} \quad \text{or} \quad \xi_0 = \frac{-ac + \sqrt{bc^2}}{-c^2}$$

according as c is positive or negative. Thus we can write ξ in the form

$$\xi_0 = \frac{m_0 + \sqrt{d}}{q_0},$$

where $q_0 \mid (d - m_0^2)$, d, m_0 and q_0 are integers, $q_0 \neq 0$, d not a perfect square. By writing ξ_0 in this form we can get a simple formulation of its continued fraction expansion $\langle a_0, a_1, a_2, \cdots \rangle$. We shall prove that the equations

$$(7.16) \qquad a_i = [\xi_i], \qquad \xi_i = \frac{m_i + \sqrt{d}}{q_i},$$

$$m_{i+1} = a_i q_i - m_i, \qquad q_{i+1} = \frac{d - m_{i+1}^2}{q_i}$$

define infinite sequences of integers m_i, q_i, a_i, and irrationals ξ_i in such a way that equations (7.7) hold, and hence we will have the continued fraction expansion of ξ_0.

In the first place, we start with ξ_0, m_0, q_0 as determined above, and we let $a_0 = [\xi_0]$. If ξ_i, m_i, q_i, a_i are known, then we take $m_{i+1} = a_i q_i - m_i$, $q_{i+1} = (d - m_{i+1}^2)/q_i$, $\xi_{i+1} = (m_{i+1} + \sqrt{d})/q_{i+1}$, $a_{i+1} = [\xi_{i+1}]$. That is, (7.16) actually does determine sequences ξ_i, m_i, q_i, a_i that are at least real.

Now we use mathematical induction to prove that the m_i and q_i are integers such that $q_i \neq 0$ and $q_i \mid (d - m_i^2)$. This holds for $i = 0$. If it is true at the ith stage, we observe that $m_{i+1} = a_i q_i - m_i$ is an integer. Then the equation

$$q_{i+1} = \frac{d - m_{i+1}^2}{q_i} = \frac{d - m_i^2}{q_i} + 2a_i m_i - a_i^2 q_i$$

establishes that q_{i+1} is an integer. Moreover q_{i+1} cannot be zero, since if it were, we would have $d = m_{i+1}^2$, whereas d is not a perfect square. Finally, we have $q_i = (d - m_{i+1}^2)/q_{i+1}$, so that $q_{i+1} \mid (d - m_{i+1}^2)$.

Next we can verify that

$$\xi_i - a_i = \frac{-a_i q_i + m_i + \sqrt{d}}{q_i} = \frac{\sqrt{d} - m_{i+1}}{q_i} = \frac{d - m_{i+1}^2}{q_i(\sqrt{d} + m_{i+1})}$$

$$= \frac{q_{i+1}}{\sqrt{d} + m_{i+1}} = \frac{1}{\xi_{i+1}},$$

which verifies (7.7) and so we have proved that $\xi_0 = \langle a_0, a_1, a_2, \cdots \rangle$, with the a_i defined by (7.16).

By ξ_i' we denote the conjugate of ξ_i, that is, $\xi_i' = (m_i - \sqrt{d})/q_i$. Since the conjugate of a quotient equals the quotient of the conjugates, we get the equation

$$\xi_0' = \frac{\xi_n' h_{n-1} + h_{n-2}}{\xi_n' k_{n-1} + k_{n-2}}$$

by taking conjugates in (7.8). Solving for ξ_n' we have

$$\xi_n' = -\frac{k_{n-2}}{k_{n-1}}\left(\frac{\xi_0' - h_{n-2}/k_{n-2}}{\xi_0' - h_{n-1}/k_{n-1}}\right).$$

As n tends to infinity, both h_{n-1}/k_{n-1} and h_{n-2}/k_{n-2} tend to ξ_0, which is different from ξ_0', and hence the fraction in parentheses tends to 1. Thus for sufficiently large n, say $n > N$ where N is fixed, the fraction in parentheses is positive, and ξ_n' is negative. But ξ_n is positive for $n \geqq 1$ and hence $\xi_n - \xi_n' > 0$ for $n > N$. Applying (7.16) we see that this gives $2\sqrt{d}/q_n > 0$ and hence $q_n > 0$ for $n > N$.

It also follows from (7.16) that

$$q_n q_{n+1} = d - m_{n+1}^2 \leqq d, \qquad q_n \leqq q_n q_{n+1} \leqq d$$

$$m_{n+1}^2 < m_{n+1}^2 + q_n q_{n+1} = d, \qquad |m_{n+1}| < \sqrt{d},$$

for $n > N$. Since d is a fixed positive integer we conclude that q_n and m_{n+1} can assume only a fixed number of possible values for $n > N$. Hence the ordered pairs (m_n, q_n) can assume only a fixed number of possible pair values for $n > N$, and so there are distinct integers j and k such that $m_j = m_k$ and $q_j = q_k$. We can suppose we have chosen j and k so that $j < k$. By (7.16) this implies that $\xi_j = \xi_k$ and hence that

$$\xi_0 = \langle a_0, a_1, \cdots, a_{j-1}, \overline{a_j, a_{j+1}, \cdots, a_{k-1}}\rangle.$$

The proof of Theorem 7.19 is now complete.

Next we determine the subclass of real quadratic irrationals that have purely periodic continued fraction expansions, that is, expressions of the form $\langle \overline{a_0, a_1, \cdots, a_n}\rangle$.

Theorem 7.20 *The continued fraction expansion of the real quadratic irrational number ξ is purely periodic if and only if $\xi > 1$ and $-1 < \xi' < 0$, where ξ' denotes the conjugate of ξ.*

Proof. First we assume that $\xi > 1$ and $-1 < \xi' < 0$. As usual we write ξ_0 for ξ and take conjugates in (7.7) to obtain

$$\text{(7.17)} \qquad \frac{1}{\xi_{i+1}'} = \xi_i' - a_i.$$

Now $a_i \geqq 1$ for all i, even for $i = 0$, since $\xi_0 > 1$. Hence if $\xi_i' < 0$, then $1/\xi_{i+1}' < -1$, and we have $-1 < \xi_{i+1}' < 0$. Since $-1 < \xi_0' < 0$ we see, by mathematical induction, that $-1 < \xi_i' < 0$ holds for all $i \geqq 0$. Then, since $\xi_i' = a_i + 1/\xi_{i+1}'$ by (7.17), we have

$$0 < -\frac{1}{\xi_{i+1}'} - a_i < 1, \qquad a_i = \left[-\frac{1}{\xi_{i+1}'}\right].$$

Now ξ is a quadratic irrational, so $\xi_j = \xi_k$ for some integers j and k with $0 < j < k$. Then we have $\xi_j' = \xi_k'$ and

$$a_{j-1} = \left[-\frac{1}{\xi_j'}\right] = \left[-\frac{1}{\xi_k'}\right] = a_{k-1},$$

$$\xi_{j-1} = a_{j-1} + \frac{1}{\xi_j} = a_{k-1} + \frac{1}{\xi_k} = \xi_{k-1}.$$

Thus $\xi_j = \xi_k$ implies $\xi_{j-1} = \xi_{k-1}$. A j-fold iteration of this implication gives us $\xi_0 = \xi_{k-j}$, and we have

$$\xi = \xi_0 = \overline{\langle a_0, a_1, \cdots, a_{k-j-1}\rangle}.$$

To prove the converse, let us assume that ξ is purely periodic, say $\xi = \overline{\langle a_0, a_1, \cdots, a_{n-1}\rangle}$, where $a_0, a_1, \cdots, a_{n-1}$ are positive integers. Then $\xi > a_0 \geqq 1$. Also, by (7.8) we have

$$\xi = \langle a_0, a_1, \cdots, a_{n-1}, \xi_\delta\rangle = \frac{\xi h_{n-1} + h_{n-2}}{\xi k_{n-1} + k_{n-2}}.$$

Thus ξ satisfies the equation

$$f(x) = x^2 k_{n-1} + x(k_{n-2} - h_{n-1}) - h_{n-2} = 0.$$

This quadratic equation has two roots, ξ and its conjugate ξ'. Since $\xi > 1$, we need only prove that $f(x)$ has a root between -1 and 0 in order to establish that $-1 < \xi' < 0$. We will do this by showing that $f(-1)$ and $f(0)$ have opposite signs. First we observe that $f(0) = -h_{n-2} < 0$ by (7.6) since $a_i > 0$ for $i \geqq 0$. Next we see that for $n > 1$

$$\begin{aligned} f(-1) &= k_{n-1} - k_{n-2} + h_{n-1} - h_{n-2} \\ &= (k_{n-2} + h_{n-2})(a_{n-1} - 1) + k_{n-3} + h_{n-3} \\ &\geqq k_{n-3} + h_{n-3} > 0. \end{aligned}$$

Finally, if $n = 1$, we have $f(-1) = k_0 - k_{-1} + h_0 - h_{-1} = a_0 > 0$, and this completes the proof.

We now turn to the continued fraction expansion of $\sqrt{d}$ for a positive integer d not a perfect square. We get at this by considering the closely related irrational number $\sqrt{d} + [\sqrt{d}]$. This number satisfies the conditions of Theorem 7.20, and so its continued fraction is purely periodic,

$$(7.18)\qquad \sqrt{d} + [\sqrt{d}] = \overline{\langle a_0, a_1, \cdots, a_{r-1}\rangle} = \langle a_0, \overline{a_1, \cdots, a_{r-1}, a_0}\rangle.$$

We can suppose that we have chosen r to be the smallest integer for which $\sqrt{d} + [\sqrt{d}]$ has an expansion of the form (7.18). Now we note that

$\xi_i = \langle a_i, a_{i+1}, \cdots \rangle$ is purely periodic for all values of i, and that $\xi_0 = \xi_r = \xi_{2r} = \cdots$. Furthermore $\xi_1, \xi_2, \cdots, \xi_{r-1}$ are all different from ξ_0, since otherwise there would be a shorter period. Thus $\xi_i = \xi_0$ if and only if i is of the form mr.

Now we can start with $\xi_0 = \sqrt{d} + [\sqrt{d}]$, $q_0 = 1$, $m_0 = [\sqrt{d}]$ in (7.16) because $1 \mid (d - [\sqrt{d}]^2)$. Then, for all $j \geqq 0$,

$$(7.19) \qquad \frac{m_{jr} + \sqrt{d}}{q_{jr}} = \xi_{jr} = \xi_0 = \frac{m_0 + \sqrt{d}}{q_0} = [\sqrt{d}] + \sqrt{d},$$

$$m_{jr} - q_{jr}[\sqrt{d}] = (q_{jr} - 1)\sqrt{d},$$

and hence $q_{jr} = 1$ since the left side is rational and $\sqrt{d}$ is irrational. Moreover $q_i = 1$ for no other values of the subscript i. For $q_i = 1$ implies $\xi_i = m_i + \sqrt{d}$, but ξ_i has a purely periodic expansion so that, by Theorem 7.20 we have $-1 < m_i - \sqrt{d} < 0$, $\sqrt{d} - 1 < m_i < \sqrt{d}$, and hence $m_i = [\sqrt{d}]$. Thus $\xi_i = \xi_0$ and i is a multiple of r.

We also establish that $q_i = -1$ does not hold for any i. For $q_i = -1$ implies $\xi_i = -m_i - \sqrt{d}$ by (7.16), and by Theorem 7.20 we would have $-m_i - \sqrt{d} > 1$ and $-1 < -m_i + \sqrt{d} < 0$. But this implies $\sqrt{d} < m_i < -\sqrt{d} - 1$, which is impossible.

Noting that $a_0 = [\sqrt{d} + [\sqrt{d}]] = 2[\sqrt{d}]$, we can now turn to the case $\xi = \sqrt{d}$. Using (7.18) we have

$$\begin{aligned}\sqrt{d} &= -[\sqrt{d}] + (\sqrt{d} + [\sqrt{d}]) \\ &= -[\sqrt{d}] + \langle \overline{2[\sqrt{d}], a_1, a_2, \cdots, a_{r-1}, a_0} \rangle \\ &= \langle [\sqrt{d}], \overline{a_1, a_2, \cdots, a_{r-1}, a_0} \rangle\end{aligned}$$

with $a_0 = 2[\sqrt{d}]$.

When we apply (7.16) to $\sqrt{d} + [\sqrt{d}]$, $q_0 = 1$, $m_0 = [\sqrt{d}]$ we have $a_0 = 2[\sqrt{d}]$, $m_1 = [\sqrt{d}]$, $q_1 = d - [\sqrt{d}]^2$. But we can also apply (7.16) to $\sqrt{d}$ with $q_0 = 1$, $m_0 = 0$, and we find $a_0 = [\sqrt{d}]$, $m_1 = [\sqrt{d}]$, $q_1 = d - [\sqrt{d}]^2$. The value of a_0 is different, but the values of m_1, and of q_1, are the same in both cases. Since $\xi_i = (m_i + \sqrt{d})/q_i$ we see that further application of (7.16) yields the same values for the a_i, for the m_i, and for the q_i, in both cases. In other words, the expansions of $\sqrt{d} + [\sqrt{d}]$ and $\sqrt{d}$ differ only in the values of a_0 and m_0. Stating our results explicitly for the case $\sqrt{d}$ we have the following theorem.

Theorem 7.21 *If the positive integer d is not a perfect square, the simple continued fraction expansion of $\sqrt{d}$ has the form $\sqrt{d} = \langle a_0, \overline{a_1, a_2, \cdots, a_{r-1}, 2a_0} \rangle$*

with $a_0 = [\sqrt{d}]$. *Furthermore with* $\xi_0 = \sqrt{d}$, $q_0 = 1$, $m_0 = 0$, *in equations* (7.16), *we have* $q_i = 1$ *if and only if* $r \mid i$, *and* $q_i = -1$ *holds for no subscript* i. *Here* r *denotes the length of the shortest period in the expansion of* $\sqrt{d}$.

PROBLEM

1. For what positive integers c does the quadratic irrational $([\sqrt{d}] + \sqrt{d})/c$ have a purely periodic expansion?

7.8 Pell's Equation

The equation $x^2 - dy^2 = N$, with given integers d and N and unknowns x and y, is usually called *Pell's equation*. If d is negative, it can have only a finite number of solutions. If d is a perfect square, say $d = a^2$, the equation reduces to $(x - ay)(x + ay) = N$ and again there is only a finite number of solutions. The most interesting case of the equation arises when d is a positive integer not a perfect square. For this case, simple continued fractions are very useful.

We expand $\sqrt{d}$ into a continued fraction as in Theorem 7.21, with convergents h_n/k_n, and with q_n defined by equations (7.16) with $\xi_0 = \sqrt{d}$, $q_0 = 1$, $m_0 = 0$.

Theorem 7.22 *If* d *is a positive integer not a perfect square, then* $h_n^2 - dk_n^2 = (-1)^{n-1}q_{n+1}$ *for all integers* $n \geq -1$.

Proof. From equations (7.8) and (7.16) we have

$$\sqrt{d} = \xi_0 = \frac{\xi_{n+1}h_n + h_{n-1}}{\xi_{n+1}k_n + k_{n-1}} = \frac{(m_{n+1} + \sqrt{d})h_n + g_{n+1}h_{n-1}}{(m_{n+1} + \sqrt{d})k_n + q_{n+1}k_{n-1}}.$$

We simplify this equation and separate it into a rational and a purely irrational part much as we did in (7.19). Each part must be zero so we get two equations, and we can eliminate m_{n+1} from them. The final result is

$$h_n^2 - dk_n^2 = (h_nk_{n-1} - h_{n-1}k_n)q_{n+1} = (-1)^{n-1}q_{n+1},$$

where we used Theorem 7.5 in the last step.

Corollary 7.23 *Taking* r *as the length of the period of the expansion of* $\sqrt{d}$, *as in Theorem* 7.21, *we have for* $n \geq 0$,

$$h_{nr-1}^2 - dk_{nr-1}^2 = (-1)^{nr}q_{nr} = (-1)^{nr}.$$

It can be seen that Theorem 7.22 gives us solutions of Pell's equation for certain values of N. In particular, Corollary 7.23 gives infinitely many solutions of $x^2 - dy^2 = 1$ by the use of even values nr. Of course if r is even, all values of nr are even. If r is odd, Corollary 7.23 gives infinitely many solutions of $x^2 - dy^2 = -1$ by the use of odd integers $n \geqq 1$. The next theorem shows that every solution of $x^2 - dy^2 = \pm 1$ can be obtained from the continued fraction expansion of $\sqrt{d}$. But first we make this simple observation: apart from such trivial solutions as $x = \pm 1$, $y = 0$ of $x^2 - dy^2 = 1$, all solutions of $x^2 - dy^2 = N$ fall into sets of four by all combinations of signs $\pm x$, $\pm y$. Hence it is sufficient to discuss the positive solutions $x > 0$, $y > 0$.

Theorem 7.24 *Let d be a positive integer not a perfect square, and let the convergents to the continued fraction expansion of $\sqrt{d}$ be h_n/k_n. Let the integer N satisfy $|N| < \sqrt{d}$. Then any positive solution $x = s$, $y = t$ of $x^2 - dy^2 = N$ with $(s, t) = 1$ satisfies $s = h_n$, $t = k_n$ for some positive integer n.*

Proof. Let E and M be positive integers such that $(E, M) = 1$ and $E^2 - \rho M^2 = \sigma$, where $\sqrt{\rho}$ is irrational and $0 < \sigma < \sqrt{\rho}$. Here ρ and σ are real numbers, not necessarily integers. Then

$$\frac{E}{M} - \sqrt{\rho} = \frac{\sigma}{M(E + M\sqrt{\rho})},$$

and hence

$$0 < \frac{E}{M} - \sqrt{\rho} < \frac{\sqrt{\rho}}{M(E + M\sqrt{\rho})} = \frac{1}{M^2(E/(M\sqrt{\rho}) + 1)}.$$

Also $0 < E/M - \sqrt{\rho}$ implies $E/(M\sqrt{\rho}) > 1$, and therefore

$$\left| \frac{E}{M} - \sqrt{\rho} \right| < \frac{1}{2M^2}.$$

By Theorem 7.14, E/M is a convergent in the continued fraction expansion of $\sqrt{\rho}$.

If $N > 0$, we take $\sigma = N$, $\rho = d$, $E = s$, $M = t$, and the theorem holds in this case.

If $N < 0$, then $t^2 - (1/d)s^2 = -N/d$, and we take $\sigma = -N/d$, $\rho = 1/d$, $E = t$, $M = s$. We find that t/s is a convergent in the expansion of $1/\sqrt{d}$. Then Theorem 7.15 shows that s/t is a convergent in the expansion of $\sqrt{d}$.

Theorem 7.25 *All positive solutions of $x^2 - dy^2 = \pm 1$ are to be found among $x = h_n$, $y = k_n$, where h_n/k_n are the congruents of the expansion of $\sqrt{d}$. If r is the period of the expansion of $\sqrt{d}$, as in Theorem 7.21, and if r is* *even, then* $x^2 - dy^2 = -1$ *has no solutions, and all positive solutions of* $x^2 - dy^2 = 1$ *are given by* $x = h_{nr-1}$, $y = k_{nr-1}$ *for* $n = 1, 2, 3, \cdots$. *On the other hand if* r *is odd, then* $x = h_{nr-1}$, $y = k_{nr-1}$ *give all positive solutions of* $x^2 - dy^2 = -1$ *by use of* $n = 1, 3, 5, \cdots$, *and all positive solutions of* $x^2 - dy^2 = 1$ *by use of* $n = 2, 4, 6, \cdots$.

Proof. This result is a corollary of Theorems 7.21, 7.22, and 7.24.

The sequence of pairs $(h_0, k_0), (h_1, k_1), \cdots$ will include all positive solutions of $x^2 - dy^2 = 1$. Furthermore, $a_0 = [\sqrt{d}] > 0$ so the sequence $h_0, h_1, h_2, \cdots$ is strictly increasing. If we let x_1, y_1 denote the first solution that appears, then for every other solution x, y we will have $x > x_1$, and hence $y > y_1$ also. Having found this least positive solution by means of continued fractions, we can find all the remaining positive solutions by a simpler method.

Theorem 7.26 *Let* x_1, y_1 *be the least positive solution of* $x^2 - dy^2 = 1$, d *being a positive integer not a perfect square. Then all positive solutions are given by* x_n, y_n *for* $n = 1, 2, 3, \cdots$ *where* x_n *and* y_n *are the integers defined by* $x_n + y_n\sqrt{d} = (x_1 + y_1\sqrt{d})^n$.

The values of x_n and y_n are determined by expanding the power and equating the rational parts, and the purely irrational parts. For example, $x_3 + y_3\sqrt{d} = (x_1 + y_1\sqrt{d})^3$ so that $x_3 = x_1^3 + 3x_1y_1^2d$ and $y_3 = 3x_1^2y_1 + y_1^3d$.

Proof. First we establish that x_n, y_n is a solution. We have $x_n - y_n\sqrt{d} = (x_1 - y_1\sqrt{d})^n$, since the conjugate of a product is the product of the conjugates. Hence we can write

$$\begin{aligned} x_n^2 - y_n^2d &= (x_n - y_n\sqrt{d})(x_n + y_n\sqrt{d}) \\ &= (x_1 - y_1\sqrt{d})^n(x_1 + y_1\sqrt{d})^n = (x_1^2 - y_1^2d)^n = 1. \end{aligned}$$

Next we show that every positive solution can be obtained. Suppose there is a positive solution s, t that is not in the collection $\{x_n, y_n\}$. Since both $x_1 + y_1\sqrt{d}$ and $s + t\sqrt{d}$ are greater than 1, there must be some integer m such that $(x_1 + y_1\sqrt{d})^m \leqq s + t\sqrt{d} < (x_1 + y_1\sqrt{d})^{m+1}$. We cannot have $(x_1 + y_1\sqrt{d})^m = s + t\sqrt{d}$, for this would imply $x_m + y_m\sqrt{d} = s + t\sqrt{d}$, and hence $s = x_m$, $t = y_m$. Now $(x_1 - y_1\sqrt{d})^m = (x_1 + y_1\sqrt{d})^{-m}$, and we can multiply the above inequality by $(x_1 - y_1\sqrt{d})^m$ to obtain

$$1 < (s + t\sqrt{d})(x_1 - y_1\sqrt{d})^m < x_1 + y_1\sqrt{d}.$$

Defining integers a and b by $a + b\sqrt{d} = (s + t\sqrt{d})(x_1 - y_1\sqrt{d})^m$ we have

$$a^2 - b^2d = (s^2 - t^2d)(x_1^2 - y_1^2d)^m = 1$$

so a, b is a solution of $x^2 - dy^2 = 1$ such that $1 < a + b\sqrt{d} < x_1 + y_1\sqrt{d}$. But then $0 < (a + b\sqrt{d})^{-1} < 1$, and hence $0 < a - b\sqrt{d} < 1$. Now we have

$$a = \tfrac{1}{2}(a + b\sqrt{d}) + \tfrac{1}{2}(a - b\sqrt{d}) > \tfrac{1}{2} + 0 > 0,$$
$$b\sqrt{d} = \tfrac{1}{2}(a + b\sqrt{d}) - \tfrac{1}{2}(a - b\sqrt{d}) > \tfrac{1}{2} - \tfrac{1}{2} = 0,$$

so a, b is a positive solution. Therefore $a > x_1$, $b > y_1$, but this contradicts $a + b\sqrt{d} < x_1 + y_1\sqrt{d}$, and hence our supposition was false. All positive solutions are given by $x_n, y_n, n = 1, 2, 3, \cdots$.

It may be noted that the definition of x_n, y_n can be extended to zero and negative n. They then give nonpositive solutions.

For N different from 1 there are certain results that can be proved, but they are not as complete as what we have shown to be true in the case $N = 1$. For example, if x_1, y_1 is the smallest positive solution of $x^2 - dy^2 = 1$, and if $r_0^2 - ds_0^2 = N$, then integers r_n, s_n can be defined by $r_n + s_n\sqrt{d} = (r_0 + s_0\sqrt{d})(x_1 + y_1\sqrt{d})^n$, and it is easy to show that r_n, s_n are solutions of $x^2 - dy^2 = N$. However, there is no assurance that all positive solutions can be obtained in this way starting from a fixed r_0, s_0.

PROBLEMS

The symbol d denotes a positive integer, not a perfect square.

1. Assuming that $x^2 - dy^2 = -1$ is solvable, let x_1, y_1 be the smallest positive solution. Prove that x_2, y_2, defined by $x_2 + y_2\sqrt{d} = (x_1 + y_1\sqrt{d})^2$ is the smallest positive solution of $x^2 - dy^2 = 1$. Also prove that all solutions of $x^2 - dy^2 = -1$ are given by x_n, y_n, where $x_n + y_n\sqrt{d} = (x_1 + y_1\sqrt{d})^n$, with $n = 1, 3, 5, 7, \cdots$, and that all solutions of $x^2 - dy^2 = 1$ are given by x_n, y_n with $n = 2, 4, 6, 8, \cdots$.

2. Prove that if $x^2 - dy^2 = N$ has one solution, it has infinitely many. *Suggestion:* use the identity $(x_1^2 - dy_1^2)(x_2^2 - dy_2^2) = (x_1x_2 - dy_1y_2)^2 - d(x_1y_2 - x_2y_1)^2$.

3. Prove that $x^2 - dy^2 = -1$ has no solution if $d \equiv 3 \pmod 4$.

4. Let d be a positive integer, not a perfect square. If k is any positive integer, prove that there are infinitely many solutions in integers of $x^2 - dy^2 = 1$ with $k \mid y$.

7.9 Numerical Computation

The numerical computations involved in finding a simple continued fraction can be rather lengthy. In general the algorithm (7.7) must be used. However if ξ_0 is a quadratic irrational the work can be simplified. It is probably best to use (7.16) in a slightly altered form. From (7.16) we have

$$q_{i+1} = \frac{d - m_{i+1}^2}{q_i} = \frac{d - (a_i q_i - m_i)^2}{q_i} = \frac{d - m_i^2}{q_i} - a_i^2 q_i + 2a_i m_i$$

$$= q_{i-1} - a_i(a_i q_i - m_i) + a_i m_i = q_{i-1} + a_i(m_i - m_{i+1}).$$

Starting with $\xi_0 = (m_0 + \sqrt{d})/q_0$, $q_0 \mid (d - m_0^2)$, we obtain, in turn,

$$a_0 = \left[\frac{m_0 + \sqrt{d}}{q_0}\right], \qquad m_1 = a_0 q_0 - m_0, \qquad q_1 = \frac{d - m_1^2}{q_0},$$

$$a_1 = \left[\frac{m_1 + \sqrt{d}}{q_1}\right], \qquad m_2 = a_1 q_1 - m_1, \qquad q_2 = q_0 + a_1(m_1 - m_2),$$

. . . .

$$a_{i-1} = \left[\frac{m_{i-1} + \sqrt{d}}{q_{i-1}}\right], \qquad m_i = a_{i-1} q_{i-1} - m_{i-1},$$

$$q_i = q_{i-2} + a_{i-1}(m_{i-1} - m_i), \qquad i \geqq 1.$$

The formula $q_i q_{i+1} = d - m_{i+1}^2$ serves as a good check. Even for large numbers, this procedure is fairly simple to carry out.

For quite small numbers it is often easiest to obtain the expansion directly. For example for $\sqrt{3}$ we can compute as follows:

$$\xi_0 = \sqrt{3} = 1 + \frac{1}{\xi_1},$$

$$\xi_1 = \frac{1}{\sqrt{3} - 1} = \frac{\sqrt{3} + 1}{2} = 1 + \frac{1}{\xi_2},$$

$$\xi_2 = \frac{2}{\sqrt{3} - 1} = \sqrt{3} + 1 = 2 + \frac{1}{\xi_3},$$

$$\xi_3 = \frac{1}{\sqrt{3} - 1}.$$

In this case $\xi_3 = \xi_1$ so we stop. We have

$$a_0 = 1, \qquad a_1 = 1, \qquad a_2 = 2, \qquad a_3 = a_1 = 1, \cdots, \sqrt{3} = \langle 1, \overline{1, 2} \rangle.$$

When a continued fraction is known the convergents can be obtained from (7.6). The work can be systematized. The following example, for $\sqrt{3}$, demonstrates a convenient method.

	0	1
	1	0
1	1	1
1	2	1
2	5	3
1	7	4
2	19	11
.	.	.
.	.	.
.	.	.

$h_0 = 1, k_0 = 1$
$h_1 = 2, k_1 = 1$
$h_2 = 5, k_2 = 3$
. . . .

NOTES ON CHAPTER 7

A completely different approach to continued fractions, specifically with the continued fractions arising naturally out of the approximations rather than the other way about, can be found (for example) in Chapter 1 of the book by J. W. S. Cassels listed in the General References, page 269.

8

Elementary Remarks on the Distribution of Primes

8.1 The Function $\pi(x)$

The discussion in Section 1.3 makes it clear that the primes are distributed among the natural numbers in a very irregular way. Theorem 1.18 shows that there are arbitrarily large gaps in the sequence of primes. The proof of Theorem 1.17 not only shows that there are infinitely many primes but also that the rth prime p_r is no greater than $\prod_{j=1}^{r-1} p_j + 1$, the product of the first $r - 1$ primes plus 1. A minor change in the proof shows that $p_r \leqq \prod_{j=1}^{r-1} p_j - 1$ if $r > 2$.

In this chapter we abandon our convention that letters of the roman alphabet represent integers.

Definition 8.1 *For real x we let $\pi(x)$ denote the number of primes that do not exceed x. Thus for example*

$$\pi(-1) = \pi(1) = 0, \qquad \pi(2) = \pi(5/2) = 1.$$

Theorem 8.1 (*Tschebyschef*) *There exist positive constants a and b such that*

$$a \frac{x}{\log x} < \pi(x) < b \frac{x}{\log x}$$

for $x \geqq 2$.

Proof. For n a positive integer and p a prime let p^{μ_p} be the largest power of p that divides the binomial coefficient $\binom{2n}{n}$. According to Theorem 4.2

$$\mu_p = \sum_{j\geqq 1}\left(\left[\frac{2n}{p^j}\right] - 2\left[\frac{n}{p^j}\right]\right). \tag{8.1}$$

We define the integer ν_p by the inequalities $p^{\nu_p} \leqq 2n < p^{1+\nu_p}$. Clearly ν_p exists and is unique. Then $\left(\frac{2n}{p^j}\right) - 2\left(\frac{n}{p^j}\right) = 0 - 0 = 0$ for $j > \nu_p$. We also have for every j,

$$\left[\frac{2n}{p^j}\right] - 2\left[\frac{n}{p^j}\right] < \frac{2n}{p^j} - 2\left(\frac{n}{p^j} - 1\right) = 2$$

and hence $[2n/p^j] - 2[n/p^j] \leqq 1$ for all $j \geqq 1$. Using this in (8.1) we obtain

$$\mu_p \leqq \sum_{j=1}^{\nu_p} 1 = \nu_p \tag{8.2}$$

and therefore

$$\binom{2n}{n} \Bigm| \prod_{p\leqq 2n} p^{\nu_p}. \tag{8.3}$$

On the other hand if $n < p \leqq 2n$ then $p \mid (2n)!$ and $p \nmid n!$ so we have $\prod_{n<p\leqq 2n} p \Bigm| \binom{2n}{n}$, which, along with (8.3), gives us

$$\prod_{n<p\leqq 2n} p \leqq \binom{2n}{n} \leqq \prod_{p\leqq 2n} p^{\nu_p} \leqq \prod_{p\leqq 2n} 2n$$

and hence, changing each p in the first product to n,

$$n^{\pi(2n)-\pi(n)} \leqq \binom{2n}{n} \leqq (2n)^{\pi(2n)}. \tag{8.4}$$

But $\binom{2n}{n} \leqq (1 + 1)^{2n} = 2^{2n}$ and

$$\binom{2n}{n} = \frac{(2n)(2n-1)\cdots(n+1)}{n!} = \prod_{j=1}^{n} \frac{n+j}{j} \geqq \prod_{j=1}^{n} 2 = 2^n.$$

Using these inequalities in (8.4) and taking logarithms, we obtain

$$\pi(2n) - \pi(n) \leqq \frac{2n \log 2}{\log n}, \qquad \pi(2n) \geqq \frac{n \log 2}{\log (2n)}, \tag{8.5}$$

where we must assume $n > 1$ in the first inequality.

For any real $x \geqq 2$ we let $2n$ be the greatest even integer that does not exceed x. Then we have $x \geqq 2n$, $n \geqq 1$, $2n + 2 > x$, and hence

$$\pi(x) \geqq \pi(2n) \geqq \frac{n \log 2}{\log (2n)} \geqq \frac{n \log 2}{\log x} \geqq \frac{(2n+2)\log 2}{4 \log x} > \frac{\log 2}{4} \frac{x}{\log x}.$$

To get the other half of the inequality in the theorem, in the first part of (8.5) we take $2n$ to be an exact power of 2, say 2^r with $r \geqq 3$, to get

$$\pi(2^r) - \pi(2^{r-1}) \leqq 2^r/(r-1).$$

Replace r by $2j, 2j-1, 2j-2, \cdots, 3$, and add these inequalities to the obvious result $\pi(2^2) < 2^2$ to obtain, for $j \geqq 2$,

$$\pi(2^{2j}) < \sum_{r=2}^{2j} 2^r/(r-1) = \sum_{r=j+1}^{2j} 2^r/(r-1) + \sum_{r=2}^{j} 2^r/(r-1).$$

In the final two sums we replace $r - 1$ by j and 1 respectively to get

$$\pi(2^{2j}) < \sum_{r=j+1}^{2j} 2^r/j + \sum_{r=2}^{j} 2^r < 2^{2j+1}/j + 2^{j+1}.$$

Now $2^{j+1} < 2^{2j+1}/j$ since $j < 2^j$, and hence we have for $j \geqq 2$

$$\pi(2^{2j}) < 2 \cdot 2^{2j+1}/j \quad \text{or} \quad \frac{\pi(2^{2j})}{2^{2j}} < \frac{4}{j}. \tag{8.6}$$

Note that this holds also for $j = 1$. Now for any real $x \geqq 2$ there is an integer $j \geqq 1$ such that $2^{2j-2} < x \leqq 2^{2j}$ and we can write

$$\frac{\pi(x)}{x} \leqq \frac{\pi(2^{2j})}{2^{2j-2}} = \frac{4\pi(2^{2j})}{2^{2j}}, \quad \text{and} \quad 2j \geqq \frac{\log x}{\log 2}, \quad \frac{4}{j} \leqq \frac{8 \log 2}{\log x}.$$

Substituting these in (8.6) we get

$$\frac{\pi(x)}{x} < \frac{16}{j} \leqq \frac{32 \log 2}{\log x}.$$

Therefore we can take $a = (\log 2)/4$, $b = 32 \log 2$, and the theorem is proved. We have found values of a and b that suffice, but they are by no means the best possible values.

The theorem we have just proved tells us something about how numerously and how scarcely the primes are distributed. Since there are infinitely many primes, we cannot say that there are more natural numbers than there are primes. However the ratio $\pi(n)/n$ represents the proportion of primes in the first n natural numbers. Since $\pi(n)/n < b/\log n$ tends to zero as n increases, we are led to say that the primes are scarcer than the natural numbers.

This is often stated as: "almost all positive integers are composite." Of course the theorem tells us a good deal more. Since

$$a < \frac{\pi(x)}{x/\log x} < b,$$

the function $\pi(x)$ is of order $x/\log x$. The primes are neither too numerous nor too scarce. Since

$$\frac{\sqrt{x}}{\pi(x)} < \frac{\log x}{a\sqrt{x}} \to 0 \qquad \text{as } x \to \infty,$$

the sequences of squares is scarcer than the sequence of primes. The prime number theorem mentioned in Section 1.3 is a refinement of the present theorem.

PROBLEM

1. Prove that ν_p, as defined following (8.1), is equal to $\left[\dfrac{\log 2n}{\log p}\right]$.

8.2 The Sequence of Primes

Results concerning the size of the rth prime, p_r, can also be used to describe how numerous the primes are. Our first result is essentially a corollary to Theorem 8.1.

Theorem 8.2 *There exist positive constants c and d such that $cr \log r < p_r < dr \log r$ for $r \geqq 2$.*

Proof. Using Theorem 8.1 and the fact that $p_r \geqq r$ we have

$$r = \pi(p_r) < b\frac{p_r}{\log p_r}, \qquad p_r > \frac{r \log p_r}{b} \geqq \frac{1}{b} r \log r.$$

The other way around we have $r = \pi(p_r) > ap_r/\log p_r$. If r is large so is p_r, and there is a constant k such that $\log p_r/\sqrt{p_r} < a$ if $r \geqq k$. Then, for $r \geqq k$,

$$r\frac{\log p_r}{p_r} > a > \frac{\log p_r}{\sqrt{p_r}},$$

hence $r > \sqrt{p_r}$, $\log p_r < 2 \log r$ and therefore $ap_r < r \log p_r < 2r \log r$. If d is larger than the largest number among

$$\frac{2}{a}, \quad \frac{p_2}{2 \log 2}, \quad \frac{p_3}{3 \log 3}, \ldots, \frac{p_{k-1}}{(k-1) \log (k-1)},$$

then $p_r < dr \log r$ for $r \geqq 2$.

Theorem 8.3 *The series* $\sum_{r=1}^{\infty} \frac{1}{p_r}$ *diverges.*

Proof. For $r > 1$ we have

$$\frac{1}{p_r} > \frac{1}{dr \log r}$$

and the series $\sum_{r=2}^{\infty} 1/(r \log r)$ diverges.

Alternate proof. Whereas the proof just given is based on Theorem 8.2 and 8.1, it is possible to establish Theorem 8.3 in terms of simpler concepts, as follows. First we prove that $\sum' 1/k$ diverges, where $\sum'$ denotes the sum over all positive integers that are square-free. Given any positive integer m, let r^2 be the largest integer square that divides m, so that $r \geqq 1$, and m/r^2 is square-free. Thus every positive integer is uniquely expressible as a product of a perfect square and a square-free positive integer. It follows that for any positive integer n

$$\left(\sum_{j<n} 1/j^2\right)\left(\sum_{k<n}{}' 1/k\right) \geqq \sum_{m<n} 1/m$$

because when the two sums on the left are multiplied together the result is a collection of reciprocals of integers $1/m$, including all integers $m < n$. Now as n tends to infinity the sum on the right is unbounded because the infinite sum $\sum 1/m$ is a divergent series. But the first sum on the left is bounded as n tends to infinity because the infinite sum $\sum 1/j^2$ is a convergent series. Hence the second sum on the left is unbounded, and this proves that $\sum' 1/k$ is a divergent series.

Next suppose, contrary to what we want to prove, that $\sum 1/p$ converges to a value β, where the sum is over all primes p. In the next step of the argument we use the fact that by dropping all terms beyond x in the power series expansion of e^x or $\exp(x)$, the inequality $\exp(x) > 1 + x$ holds for all positive real numbers x. Hence for any positive integer n

$$\exp(\beta) > \exp\left(\sum_{p<n} 1/p\right) = \prod_{p<n} \exp(1/p) > \prod_{p<n} (1 + 1/p) \geqq \sum_{k<n}{}' (1/k)$$

where the last step follows from multiplying out the factors $(1 + 1/p)$. Now the last sum is unbounded as n tends to infinity, whereas $\exp(\beta)$ is a fixed real number. Thus we have a contradiction, and the proof is complete.

If a series $\sum_{r=1}^{\infty} a_r^k$ of positive terms converges for all values of $k > f$ and diverges for all values of $k < f$, f being fixed, then f is called the *exponent of convergence* of the sequence a_r. The sequence $1/p_r$ has exponent of convergence 1. If $k > 1$, then $1/p_r^k < 1/r^k$ and $\sum_{r=1}^{\infty} 1/r^k$ converges. If $0 < k \leqq 1$, it follows from Theorem 8.3 that $\sum_{r=1}^{\infty} 1/p_r^k$ diverges.

The sequence $a_r = 1/r^2$ has exponent of convergence $\frac{1}{2}$. In fact, if $a_r = 1/[r^{1+\varepsilon}]$, $\varepsilon > 0$, then $a_r < 1/(r^{1+\varepsilon} - 1) < 2/r^{1+\varepsilon}$ if $r \geqq 2$ and the sequence a_r has exponent of convergence $(1 + \varepsilon)^{-1} < 1$. In a sense the primes are denser than the sequence consisting of $[r^{1+\varepsilon}]$.

Theorem 8.4 *There is a constant k such that*

$$\sum_{2<p\leqq x} \frac{1}{p} < k \log\log x \qquad \text{if } x \geqq 3.$$

Proof. By Theorem 8.2

$$\sum_{2<p\leqq x} \frac{1}{p} < \sum_{r=2}^{\pi(x)} \frac{1}{cr\log r} \leqq \frac{1}{c}\sum_{r=2}^{[x]} \frac{1}{r\log r} = \frac{1}{c}\left(\frac{1}{2\log 2} + \sum_{r=3}^{[x]} \int_{r-1}^{r} \frac{dt}{r\log r}\right)$$

$$\leqq \frac{1}{2c\log 2} + \frac{1}{c}\sum_{r=3}^{[x]} \int_{r-1}^{r} \frac{dt}{t\log t} \leqq \frac{1}{2c\log 2} + \frac{1}{c}\int_{2}^{x} \frac{dt}{t\log t}$$

$$= \frac{1}{2c\log 2} + \frac{1}{c}\log\log x - \frac{1}{c}\log\log 2,$$

and this is less than $k \log\log x$ for $x \geqq 3$, if k is large enough.

The method used in this proof occurs frequently in number theory. It is the method ordinarily found in proofs of the integral test in the theory of series. For a monotonic function $f(x)$, we compare $\sum_{n=M}^{N} f(n)$ with $\int_{M-1}^{N} f(x)\,dx$. Geometrically, $\int_{M-1}^{N} f(x)\,dx$ is the area under the curve $y = f(x)$, whereas $\sum_{n=M}^{N} f(n)$ represents the area covered by rectangles having unit bases and altitudes $f(n)$, $n = M, M+1, \cdots, N$.

In contrast with Theorem 8.3, Theorem 8.4 shows that the primes are not too numerous.

Theorem 8.5 *If $x \geqq 2$ then $\prod_{p\leqq x} p < 4^x$.*

Proof. This theorem is obviously true for $2 \leqq x < 3$. If it is true when x is an odd integer $n \geqq 3$, then it is true for $n \leqq x < n + 2$ since $\prod_{p\leqq x} p = \prod_{p\leqq n} p < 4^n \leqq 4^x$. Therefore we need to consider only odd integers n with

$n \geqq 3$. The proof is now by induction on the odd integer n. Noting that the theorem holds for $n = 3$, we assume the result for all odd integers greater than 1 that are less than some odd integer $n \geqq 5$. We define $k = (n \pm 1)/2$ where the sign is chosen so that k is odd. Then $k \geqq 3$. Now

$$\binom{n}{k} = \frac{n!}{k!(n-k)!} \tag{8.7}$$

and $n - k$ is even and $n - k = 2k \mp 1 - k \leqq k + 1$. If p is a prime such that $k < p \leqq n$, then p is odd and $p \mid n!$, $p \nmid k!$, $p \nmid (n - k)!$. Hence from (8.7) we see that the product of all such primes divides $\binom{n}{k}$, and so

$$\prod_{k<p\leqq n} p \leqq \binom{n}{k}.$$

But $\binom{n}{k} = \binom{n}{n-k}$ and both these binomial coefficients appear in the expansion of $(1 + 1)^n$. This implies that $\binom{n}{k} < 2^{n-1}$. Using this and the induction hypothesis we have

$$\prod_{p\leqq n} p = \prod_{p\leqq k} p \cdot \prod_{k<p\leqq n} p < 4^k \cdot 2^{n-1} = 2^{n+2k-1} \leqq 2^{2n} = 4^n$$

because $n \geqq 2k - 1$.

PROBLEMS

1. Find the exponent of convergence of the sequence of the reciprocals of the increasing positive integers that lack the digit 9 when written in ordinary decimal notation.

2. Give an independent proof of Theorem 1.18 ("There are arbitrarily large gaps in the series of primes") by using Theorem 8.2.

3. Write s_r for the sum of the first r primes. Prove that there are positive constants a_1 and b_1 such that for $r > 1$

$$a_1 r^2 \log r < s_r < b_1 r^2 \log r.$$

4. (*a*) Let $\{a_j\}$, $\{b_j\}$, $\{c_j\}$ be increasing sequences of real numbers, each with limit infinity. Say that a_j is asymptotic to b_j, written $a_j \sim b_j$, if and only if $\lim a_j/b_j = 1$. Prove that $a_j \sim b_j$ implies $\log a_j \sim \log b_j$, but that the converse is false.

(*b*) If $a_j \sim b_j$ prove that $b_j \sim a_j$; if $a_j \sim b_j$ and $c_j \sim d_j$ prove that $a_j c_j \sim b_j d_j$.

(*c*) Prove that $\lim (\log a_j)/a_j = 0$.

(*d*) If $\lim c_j/a_j = 0$, prove that $a_j \sim b_j$ if and only if $a_j \sim b_j + c_j$.

(*e*) The prime number theorem states that $\pi(n) \sim n/\log n$. A sketch is now given that this is implied by $p_n \sim n \log n$. Verify the steps in this proof and prove the converse result. Note that $p_{n+1} \sim (n+1)\log(n+1) \sim n \log n \sim p_n$. For integers $k > 1$ define n by $p_n \leqq k < p_{n+1}$, so that n is a function of k, and $k \sim p_n$. Also $n = \pi(k)$ and hence $k \sim n \log n = \pi(k) \log \pi(k)$ and $\log k \sim \log \pi(k) + \log \log \pi(k) \sim \log \pi(k)$. It follows that $k \sim \pi(k) \log k$ or $\pi(k) \sim k/\log k$.

8.3 Bertrand's Postulate

The following theorem was conjectured by Bertrand, but proved by Tschebyschef.

Theorem 8.6 *For every positive integer $n > 1$ there is a prime p such that $n < p < 2n$.*

Proof. For $n \geqq 128$ the result will be established by a general argument. To verify the theorem for the smaller values of n we take $p = 3, 5$ for $n = 2, 3$ respectively, and

$$\begin{array}{rcr} p = 7 & \text{for} & 4 \leqq n \leqq 6 \\ 13 & & 7 \leqq n \leqq 12 \\ 23 & & 13 \leqq n \leqq 22 \\ 43 & & 23 \leqq n \leqq 42 \\ 83 & & 43 \leqq n \leqq 82 \\ 131 & & 83 \leqq n \leqq 127. \end{array}$$

Next, suppose the result is false for some integer $n \geqq 128$. By the definition of μ_p and (8.2), with this supposition we have

$$\binom{2n}{n} = \prod_{p \leqq 2n} p^{\mu_p} = \prod_{p \leqq n} p^{\mu_p}, \qquad \mu_p \leqq \nu_p. \tag{8.8}$$

For any prime p in the range $2n/3 < p \leqq n$ we have

$$p \geqq 3, \qquad p^2 > \frac{2}{3} np \geqq 2n, \qquad 1 \leqq \frac{n}{p} < \frac{3}{2}, \qquad 2 \leqq \frac{2n}{p} < 3$$

and so by (8.1)

$$\mu_p = \left[\frac{2n}{p}\right] - 2\left[\frac{n}{p}\right] = 2 - 2 = 0.$$

For any prime p in the range $\sqrt{2n} < p \leqq 2n/3$ we have $p^2 > 2n$, hence $\nu_p = 1$ and $\mu_p \leqq 1$. For any prime p satisfying $p \leqq \sqrt{2n}$ we have $p^{\mu_2} \leqq$

$p^{\nu_p} \leqq 2n$. Using these facts in (8.8) we obtain

$$\binom{2n}{n} = \prod_{p \leqq \sqrt{2n}} p^{\mu_p} \prod_{\sqrt{2n} < p \leqq 2n/3} p^{\mu_p} \prod_{2n/3 < p \leqq n} p^{\mu_p}$$

$$\leqq \prod_{p \leqq \sqrt{2n}} 2n \prod_{p \leqq 2n/3} p.$$

If $y > 3$, the number of primes $\leqq y$ is at most the number of positive odd integers $\leqq y$, and so $\pi(y) \leqq (1 + y)/2$. Applying this simple observation to the first product in the last expression we see that the number of factors is at most $(1 + \sqrt{2n})/2$. We apply Theorem 8.5 to the second product and then we have

$$\binom{2n}{n} \leqq (2n)^{(1+\sqrt{2n})/2} 4^{2n/3}. \tag{8.9}$$

Now $\binom{2n}{n}$ is the largest of $2n + 1$ terms in the binomial expansion of $(1 + 1)^{2n}$, or the largest of $2n$ terms if we combine the first and last terms into a single term 2, so we have

$$2n\binom{2n}{n} > 2^{2n}, \qquad \binom{2n}{n} > (2n)^{-1} 2^{2n}.$$

This with (8.9) implies

$$(2n)^{-1} 2^{2n} < (2n)^{(1+\sqrt{2n})/2} 4^{2n/3}, \qquad 2^{2n/3} < (2n)^{(3+\sqrt{2n})/2}.$$

Also $(3 + \sqrt{2n})/2 < (2\sqrt{2n})/3$ for $n \geqq 128$, and hence

$$2^{2n/3} < (2n)^{(2\sqrt{2n})/3}, \qquad 2^n < (2n)^{\sqrt{2n}}.$$

Taking logarithms and then dividing by $\sqrt{n}$, we get

$$\sqrt{n} \log 2 < \sqrt{2} \log (2n). \tag{8.10}$$

But $\sqrt{n} \log 2 - \sqrt{2} \log (2n)$ is zero for $n = 128 = 2^7$ and

$$\frac{d}{dn} \{\sqrt{n} \log 2 - \sqrt{2} \log (2n)\} = \frac{\log 2}{2\sqrt{n}} - \frac{\sqrt{2}}{n},$$

which is positive for $n \geqq 128$ because $\log 2 > 1/4$. Hence (8.10) is false for $n \geqq 128$, and so the Theorem holds for $n \geqq 128$.

This proof is fairly typical of a great many proofs in number theory. In the proof one uses inequalities and estimates the sizes of various expressions.

Often these estimates are good enough to prove the theorem for large values of n, say, but are too crude to yield the desired result for smaller n. We are then forced to take care of these smaller n by more special methods.

PROBLEMS

1. Prove that for every positive real number $x > 1$ there is a prime p such that $x < p < 2x$.
2. Prove that $n! = m^k$ is impossible in integers $m, n > 1, k > 1$.
3. Let k and r be positive integers, $k > 1, r > 1$. Prove that there is a prime whose digital representation to base r has exactly k digits.
4. For this problem include 1 as a prime. Prove that every positive integer can be represented as a sum of one or more distinct primes.
5. Prove that the following three properties of a positive integer n are equivalent: (i) all primes $\leqq \sqrt{n}$ are divisors of n; (ii) all positive integers $<n$ and prime to n are primes or 1; (iii) every composite integer $<n$ has a factor in common with n. Furthermore, prove that only a finite number of positive integers have these properties, and find them. *Suggestion:* if n is large enough there are distinct primes p_1, p_2, p_3, p_4 such that $\frac{\sqrt{n}}{2^j} < p_j < \frac{\sqrt{n}}{2^{j-1}}$ for $j = 1, 2, 3, 4$.

NOTES ON CHAPTER 8

The final part of the proof of Theorem 8.1 has been improved by use of an arrangement by E. L. Spitznagel, Jr., "An elementary proof that primes are scarce," *Amer. Math. Monthly*, **77**, 396–397 (1970).

9

Algebraic Numbers

9.1 Polynomials

Algebraic numbers are the roots of certain types of polynomials, so it is natural to begin our discussion with this topic. Our plan in this chapter is to proceed from the most general results about algebraic numbers to stronger specific results about special classes of algebraic numbers. In this process of proving more and more about less and less, we have selected material of a number theoretic aspect as contrasted with the more "algebraic" parts of the theory. In other words, we are concerned with such questions as divisibility, uniqueness of factorization, and prime numbers, rather than questions concerning the algebraic structure of the groups, rings, and fields arising in the theory.

The polynomials that we will consider will have rational numbers for coefficients. Such polynomials are called polynomials over Q, where Q denotes the field of rational numbers. This collection of polynomials in one variable x is often denoted by $Q[x]$, just as all polynomials in x with integral coefficients is denoted by $Z[x]$, and the set of all polynomials in x with coefficients in any set of numbers F is denoted by $F[x]$. That the set of rational numbers forms a field can be verified from the postulates in Section 2.10. In a polynomial such as

$$f(x) = a_0x^n + a_1x^{n-1} + \cdots + a_n, \qquad a_0 \neq 0,$$

the non-negative integer n is called the *degree* of the polynomial, and a_0 is called the *leading* coefficient. If $a_0 = 1$, the polynomial is called "*monic*." Since we assign no degree to the zero polynomial, we can assert without exception that the degree of the product of two polynomials is the sum of the degrees of the polynomials.

A polynomial $f(x)$ is said to be *divisible* by a polynomial $g(x)$, not identically zero, if there exists a polynomial $q(x)$ such that $f(x) = g(x)q(x)$ and we write

$$g(x) \mid f(x).$$

Also, $g(x)$ is said to be a divisor or factor of $f(x)$. The degree of $g(x)$ here does not exceed that of $f(x)$, unless $f(x)$ is identically zero, written $f(x) \equiv 0$. This concept of divisibility is not the same as the divisibility that we have considered earlier. In fact $3 \mid 7$ holds if 3 and 7 are thought of as polynomials of degree zero, whereas it is not true that the integer 3 divides the integer 7.

Theorem 9.1 *To any polynomials $f(x)$ and $g(x)$ over Q with $g(x) \not\equiv 0$, there correspond unique polynomials $q(x)$ and $r(x)$ such that $f(x) = g(x)q(x) + r(x)$, where either $r(x) \equiv 0$ or $r(x)$ is of lower degree than $g(x)$.*

Proof. In case $f(x) \equiv 0$ or $f(x)$ has lower degree than $g(x)$, define $q(x) \equiv 0$ and $r(x) = f(x)$. Otherwise divide $g(x)$ into $f(x)$ to get a quotient $q(x)$ and a remainder $r(x)$. Clearly $q(x)$ and $r(x)$ are polynomials over Q, and either $r(x) \equiv 0$ or the degree of $r(x)$ is less than the degree of $g(x)$ if the division has been carried to completion. If there were another pair, $q_1(x)$ and $r_1(x)$, then we would have

$$f(x) = g(x)q_1(x) + r_1(x), \qquad r(x) - r_1(x) = g(x)\{q_1(x) - q(x)\}.$$

Thus $g(x)$ would be a divisor of the polynomial $r(x) - r_1(x)$ which, unless identically zero, has lower degree than $g(x)$. Hence $r(x) - r_1(x) \equiv 0$, and it follows that $q(x) = q_1(x)$.

Theorem 9.2 *Any polynomials $f(x)$ and $g(x)$, not both identically zero, have a common divisor $h(x)$ which is a linear combination of $f(x)$ and $g(x)$. Thus $h(x) \mid f(x)$, $h(x) \mid g(x)$, and*

$$h(x) = f(x)F(x) + g(x)G(x) \tag{9.1}$$

for some polynomials $F(x)$ and $G(x)$.

Proof. From all the polynomials of the form (9.1) that are not identically zero, choose any one of least degree and designate it by $h(x)$. If $h(x)$ were not a divisor of $f(x)$, Theorem 9.1 would give us $f(x) = h(x)q(x) + r(x)$ with $r(x) \not\equiv 0$ and $r(x)$ of degree lower than $h(x)$. But then $r(x) = f(x) - h(x)q(x) = f(x)\{1 - F(x)q(x)\} - g(x)\{G(x)q(x)\}$ which is of the form (9.1) in contradiction with the choice of $h(x)$. Thus $h(x) \mid f(x)$ and similarly $h(x) \mid g(x)$.

Theorem 9.3 *To any polynomials $f(x)$ and $g(x)$, not both identically zero, there corresponds a unique monic polynomial $d(x)$ having the properties*

(1) $d(x) \mid f(x)$, $d(x) \mid g(x)$;

(2) *$d(x)$ is a linear combination of $f(x)$ and $g(x)$, as in* (9.1);

(3) *any common divisor of $f(x)$ and $g(x)$ is a divisor of $d(x)$, and thus there is no common divisor having higher degree than that of $d(x)$.*

Proof. Define $d(x) = c^{-1}h(x)$, where c is the leading coefficient of $h(x)$, so that $d(x)$ is monic. Properties (1) and (2) are inherited from $h(x)$ by $d(x)$. Equation (9.1) implies $d(x) = c^{-1}f(x)F(x) + c^{-1}g(x)G(x)$, and this equation shows that if $m(x)$ is a common divisor of $f(x)$ and $g(x)$, then $m(x) \mid d(x)$. Finally, to prove that $d(x)$ is unique, suppose that $d(x)$ and $d_1(x)$ both satisfy properties (1), (2), (3). We then have $d(x) \mid d_1(x)$ and $d_1(x) \mid d(x)$, hence $d_1(x) = q(x)d(x)$ and $d(x) = q_1(x)d_1(x)$ for some polynomials $q(x)$ and $q_1(x)$. This implies $q(x)q_1(x) = 1$, from which we see that $q(x)$ and $q_1(x)$ are of degree zero. Since both $d(x)$ and $d_1(x)$ are monic, we have $q(x) = 1$, $d_1(x) = d(x)$.

Definition 9.1 *The polynomial $d(x)$ is called the greatest common divisor of $f(x)$ and $g(x)$. We write $(f(x), g(x)) = d(x)$.*

Definition 9.2 *A polynomial $f(x)$, not identically zero, is irreducible, or prime, over Q if there is no factoring, $f(x) = g(x)h(x)$, of $f(x)$ into two polynomials $g(x)$ and $h(x)$ of positive degrees over Q.*

For example $x^2 - 2$ is irreducible over Q. It has the factoring $(x - \sqrt{2})(x + \sqrt{2})$ over the field of real numbers, but it has no factoring over R.

Theorem 9.4 *If an irreducible polynomial $p(x)$ divides a product $f(x)g(x)$, then $p(x)$ divides at least one of the polynomials $f(x)$ and $g(x)$.*

Proof. If $f(x) \equiv 0$ or $g(x) \equiv 0$ the result is obvious. If neither is identically zero, let us assume that $p(x) \nmid f(x)$ and prove that $p(x) \mid g(x)$. The assumption that $p(x) \nmid f(x)$ implies that $(p(x), f(x)) = 1$, and hence by Theorem 9.3 there exist polynomials $F(x)$ and $G(x)$ such that $1 = p(x)F(x) + f(x)G(x)$. Multiplying by $g(x)$ we get

$$g(x) = p(x)g(x)F(x) + f(x)g(x)G(x).$$

Now $p(x)$ is a divisor of the right member of this equation because $p(x) \mid f(x)g(x)$, and hence $p(x) \mid g(x)$.

Theorem 9.5 *Any polynomial $f(x)$ over Q of positive degree can be factored into a product $f(x) = cp_1(x)p_2(x) \cdots p_k(x)$ where the $p_j(x)$ are irreducible monic polynomials over Q. This factoring is unique apart from order.*

Proof. Clearly $f(x)$ can be factored repeatedly until it becomes a product of irreducible polynomials, and the constant c can be adjusted to make all the factors monic. We must prove uniqueness. Let us consider another factoring, $f(x) = cq_1(x)q_2(x) \cdots q_j(x)$, into irreducible monic polynomials. According to Theorem 9.4, $p_1(x)$ divides some $q_i(x)$, and we can reorder the $q_m(x)$ to make $p_1(x) \mid q_1(x)$. Since $p_1(x)$ and $q_1(x)$ are irreducible and monic, we have $p_1(x) = q_1(x)$. A repetition of this argument yields

$$p_2(x) = q_2(x), \qquad p_3(x) = q_3(x), \cdots \quad \text{and} \quad k = j.$$

Definition 9.3 *A polynomial* $f(x) = a_0x^n + \cdots + a_n$ *with integral coefficients* a_j *is said to be primitive if the greatest common divisor of its coefficients is* 1. *Obviously, here we mean the greatest common divisor of integers as defined in Definition* 1.2.

Theorem 9.6 *The product of two primitive polynomials is primitive.*

Proof. Let $a_0x^n + \cdots + a_n$ and $b_0x^m + \cdots + b_m$ be primitive polynomials and denote their product by $c_0x^{n+m} + \cdots + c_{n+m}$. Suppose that this product polynomial is not primitive, so that there is a prime p which divides every coefficient c_k. Since $a_0x^n + \cdots + a_n$ is primitive, at least one of its coefficients is not divisible by p. Let a_i denote the first such coefficient and let b_j denote the first coefficient of $b_0x^m + \cdots + b_m$, not divisible by p. Then the coefficient of $x^{n+m-i-j}$ in the product polynomial is

$$c_{i+j} = \sum a_k b_{i+j-k} \tag{9.2}$$

summed over all k such that $0 \leqq k \leqq n$, $0 \leqq i + j - k \leqq m$. In this sum, any term with $k < i$ is a multiple of p. Any term with $k > i$ that appears in the sum will have the factor b_{i+j-k} with $i + j - k < j$ and will also be a multiple of p. The term a_ib_j, for $k = i$, appears in the sum, and we have $c_{i+j} \equiv a_ib_j \pmod p$. But this is in contradiction with $p \mid c_{i+j}$, $p \nmid a_i$, $p \nmid b_j$.

Theorem 9.7 *Gauss' Lemma. If a monic polynomial* $f(x)$ *with integral coefficients factors into two monic polynomials with rational coefficients, say* $f(x) = g(x)h(x)$, *then* $g(x)$ *and* $h(x)$ *have integral coefficients.*

Proof. Let c be the least positive integer such that $cg(x)$ has integral coefficients; if $g(x)$ has integral coefficients take $c = 1$. Then $cg(x)$ is a primitive polynomial, because if p is a divisor of its coefficients, then $p \mid c$ because c is the leading coefficient, and $(c/p)g(x)$ would have integral coefficients contrary to the minimal property of c. Similarly let c_1 be least positive integer such that $c_1h(x)$ has integral coefficients, and hence $c_1h(x)$ is also primitive. Then by Theorem 9.6 the product $\{cg(x)\}\{c_1h(x)\} = cc_1f(x)$ is primitive. But since $f(x)$ has integral coefficients, it follows that $cc_1 = 1$ and $c = c_1 = 1$.

PROBLEMS

1. If $f(x) \mid g(x)$ and $g(x) \mid f(x)$, prove that there is a rational number c such that $g(x) = cf(x)$.
2. If $f(x) \mid g(x)$ and $g(x) \mid h(x)$, prove that $f(x) \mid h(x)$.
3. If $p(x)$ is irreducible and $g(x) \mid p(x)$, prove that either $g(x)$ is a constant or $g(x) = cp(x)$ for some rational number c.
4. If $p(x)$ is irreducible, prove that $cp(x)$ is irreducible for any rational $c \neq 0$.
5. If a polynomial $f(x)$ with integral coefficients factors into a product $g(x)h(x)$ of two polynomials with coefficients in Q, prove that there is a factoring $g_1(x)h_1(x)$ with integral coefficients.
6. If $f(x)$ and $g(x)$ are primitive polynomials, and if $f(x) \mid g(x)$ and $g(x) \mid f(x)$, prove that $f(x) = \pm g(x)$.

9.2 Algebraic Numbers

Definition 9.4 *A complex number ξ is called an algebraic number if it satisfies some polynomial equation $f(x) = 0$ where $f(x)$ is a polynomial over Q.*

Every rational number r is an algebraic number because $f(x)$ can be taken as $x - r$ in this case.

Theorem 9.8 *An algebraic number ξ satisfies a unique irreducible monic polynomial equation $g(x) = 0$ over Q. Furthermore every polynomial equation over Q satisfied by ξ is divisible by $g(x)$.*

Proof. From all polynomial equations over Q satisfied by ξ, choose one of lowest degree, say $G(x) = 0$. If the leading coefficient of $G(x)$ is c, define $g(x) = c^{-1}G(x)$, so that $g(\xi) = 0$ and $g(x)$ is monic. The polynomial $g(x)$ is irreducible, for if $g(x) = h_1(x)h_2(x)$, then one at least of $h_1(\xi) = 0$ and $h_2(\xi) = 0$ would hold, contrary to the fact that $G(x) = 0$ and $g(x) = 0$ are polynomial equations over Q of least degree satisfied by ξ.

Next let $f(x) = 0$ be any polynomial equation over Q having ξ as a root. Applying Theorem 9.1, we get $f(x) = g(x)q(x) + r(x)$. The remainder $r(x)$ must be identically zero, for otherwise the degree of $r(x)$ would be less than that of $g(x)$, and ξ would be a root of $r(x)$ since $f(\xi) = g(\xi) = 0$. Hence $g(x)$ is a divisor of $f(x)$.

Finally to prove that $g(x)$ is unique, suppose that $g_1(x)$ is an irreducible monic polynomial such that $g_1(\xi) = 0$. Then $g(x) \mid g_1(x)$ by the argument above, say $g_1(x) = g(x)q(x)$. But the irreducibility of $g_1(x)$ then implies that $q(x)$ is a constant, in fact $q(x) = 1$ since $g_1(x)$ and $g(x)$ are monic. Thus we have $g_1(x) = g(x)$.

Definition 9.5 *The minimal equation of an algebraic number ξ is the equation $g(x) = 0$ described in Theorem 9.8. The minimal polynomial of ξ is $g(x)$. The degree of an algebraic number is the degree of its minimal polynomial.*

Definition 9.6 *An algebraic number ξ is an algebraic integer if it satisfies some monic polynomial equation*

$$f(x) = x^n + b_1 x^{n-1} + \cdots + b_n = 0 \tag{9.3}$$

with integral coefficients.

Theorem 9.9 *Among the rational numbers, the only ones that are algebraic integers are the integers $0, \pm 1, \pm 2, \cdots$.*

Proof. Any integer m is an algebraic integer because $f(x)$ can be taken as $x - m$. On the other hand, if any rational number m/q is an algebraic integer, then we may suppose $(m, q) = 1$, and we have

$$\left(\frac{m}{q}\right)^n + b_1\left(\frac{m}{q}\right)^{n-1} + \cdots + b_n = 0,$$

$$m^n + b_1 q m^{n-1} + \cdots + b_n q^n = 0.$$

Thus $q \mid m^n$, so that $q = \pm 1$, and m/q is an integer.

The word "integer" in Definition 9.6 is thus simply a generalization of our previous usage. In algebraic number theory, $0, \pm 1, \pm 2, \cdots$ are often referred to as "rational integers" to distinguish them from the other algebraic integers, that are not rational. For example, $\sqrt{2}$ is an algebraic integer but not a rational integer.

Theorem 9.10 *The minimal equation of an algebraic integer is monic with integral coefficients.*

Proof. The equation is monic by definition, so we need prove only that the coefficients are integers. Let the algebraic integer ξ satisfy $f(x) = 0$ as in (9.3), and let its minimal equation be $g(x) = 0$, monic and irreducible over Q. By Theorem 9.8, $g(x)$ is a divisor of $f(x)$, say $f(x) = g(x)h(x)$, and the quotient $h(x)$, like $f(x)$ and $g(x)$, is monic and has coefficients in Q. Applying Theorem 9.7, we see that $g(x)$ has integral coefficients.

Theorem 9.11 *Let n be a positive rational integer and ξ a complex number. Suppose that the complex numbers $\theta_1, \theta_2, \cdots, \theta_n$, not all zero, satisfy the equations*

$$\xi\theta_j = a_{j,1}\theta_1 + a_{j,2}\theta_2 + \cdots + a_{j,n}\theta_n, \qquad j = 1, 2, \cdots, n, \tag{9.4}$$

where the n^2 coefficients $a_{j,i}$ are rational. Then ξ is an algebraic number. Moreover, if the $a_{j,i}$ are rational integers, ξ is an algebraic integer.

Proof. Equations (9.4) can be thought of as a system of homogeneous linear equations in $\theta_1, \theta_2, \cdots, \theta_n$. Since the θ_i are not all zero, the determinant of coefficients must vanish:

$$\begin{vmatrix} \xi - a_{1,1} & -a_{1,2} & \cdots & -a_{1,n} \\ -a_{2,1} & \xi - a_{2,2} & \cdots & -a_{2,n} \\ \cdots & \cdots & \cdots & \cdots \\ -a_{n,1} & -a_{n,2} & \cdots & \xi - a_{n,n} \end{vmatrix} = 0.$$

Expansion of this determinant gives an equation $\xi^n + b_1\xi^{n-1} + \cdots + b_n = 0$, where the b_i are polynomials in the $a_{j,k}$. Thus the b_i are rational, and they are rational integers if the $a_{j,k}$ are.

Theorem 9.12 *If α and β are algebraic numbers, so are $\alpha + \beta$ and $\alpha\beta$. If α and β are algebraic integers, so are $\alpha + \beta$ and $\alpha\beta$.*

Proof. Suppose that α and β satisfy

$$\alpha^m + a_1\alpha^{m-1} + \cdots + a_m = 0,$$
$$\beta^r + b_1\beta^{r-1} + \cdots + b_r = 0$$

with rational coefficients a_i and b_j. Let $n = mr$, and define the complex numbers $\theta_1, \cdots, \theta_n$ as the numbers

$$\begin{array}{ccccc} 1, & \alpha, & \alpha^2, & \cdots, & \alpha^{m-1}, \\ \beta, & \alpha\beta, & \alpha^2\beta, & \cdots, & \alpha^{m-1}\beta, \\ \cdots & \cdots & \cdots & \cdots & \cdots \\ \beta^{r-1}, & \alpha\beta^{r-1}, & \alpha^2\beta^{r-1}, & \cdots, & \alpha^{m-1}\beta^{r-1}, \end{array}$$

in any order. Thus $\theta_1, \cdots, \theta_n$ are the numbers $\alpha^s\beta^t$ with $s = 0, 1, \cdots, m-1$ and $t = 0, 1, \cdots, r-1$. Hence for any θ_j,

$$\alpha\theta_j = \alpha^{s+1}\beta^t = \begin{cases} \text{some } \theta_k & \text{if } s+1 \leqq m-1 \\ (-a_1\alpha^{m-1} - a_2\alpha^{m-2} - \cdots - a_m)\beta^t & \text{if } s+1 = m. \end{cases}$$

In either case we see that there are rational constants $h_{j,1}, \cdots, h_{j,n}$ such that $\alpha\theta_j = h_{j,1}\theta_1 + \cdots + h_{j,n}\theta_n$. Similarly there are rational constants $k_{j,1}, \cdots, k_{j,n}$ such that $\beta\theta_j = k_{j,1}\theta_1 + \cdots + k_{j,n}\theta_n$, and hence $(\alpha + \beta)\theta_j = (h_{j,1} + k_{j,1})\theta_1 + \cdots + (h_{j,n} + k_{j,n})\theta_n$. These equations are of the form (9.4), and so we conclude that $\alpha + \beta$ is algebraic. Furthermore, if α and β are algebraic integers, then the a_j, b_j, $h_{j,i}$, $k_{j,i}$ are all rational integers, and $\alpha + \beta$ is an algebraic integer.

We also have $\alpha\beta\theta_j = \alpha(k_{j,1}\theta_1 + \cdots + k_{j,n}\theta_n) = k_{j,1}\alpha\theta_1 + \cdots + k_{j,n}\alpha\theta_n$ from which we find $\alpha\beta\theta_j = c_{j,1}\theta_1 + \cdots + c_{j,n}\theta_n$ where $c_{j,i} = k_{j,1}h_{1,i} + k_{j,2}h_{2,i} + \cdots + k_{j,n}h_{n,i}$. Again we apply Theorem 9.11 to conclude that $\alpha\beta$ is algebraic, and that it is an algebraic integer if α and β are.

This theorem states that the set of algebraic numbers is closed under addition and multiplication, and likewise for the set of algebraic integers. The following result states a little more.

Theorem 9.13 *The set of all algebraic numbers forms a field. The class of all algebraic integers forms a ring.*

Proof. Rings and fields are defined in Definition 2.12. The rational numbers 0 and 1 serve as the zero and unit for the system. Most of the postulates are easily seen to be satisfied if we remember that algebraic numbers are complex numbers whose properties we are familiar with. The only place where any difficulty arises is in proving the existence of additive and multiplicative inverses. If $\alpha \neq 0$ is a solution of

$$a_0x^n + a_1x^{n-1} + \cdots + a_n = 0$$

then $-\alpha$ and α^{-1} are solutions of

$$a_0x^n - a_1x^{n-1} + a_2x^{n-2} - \cdots + (-1)^n a_n = 0$$

and

$$a_0 + a_1x + a_2x^2 + \cdots + a_nx^n = 0,$$

respectively. Therefore if α is an algebraic number, then so are $-\alpha$ and α^{-1}. If α is an algebraic integer, then so is $-\alpha$, but not necessarily α^{-1}. Therefore the algebraic numbers form a field, the algebraic integers a ring.

PROBLEMS

1. Find the minimal polynomial of each of the following algebraic numbers: 7, $\sqrt[3]{7}$, $(1 + \sqrt[3]{7})/2$, $1 + \sqrt{2} + \sqrt{3}$. Which of these are algebraic integers?

2. Prove that if α is algebraic of degree n, then $-\alpha$, α^{-1}, and $\alpha - 1$ are also of degree n, assuming $\alpha \neq 0$ in the case of α^{-1}.

3. Prove that if α is algebraic of degree n, and β is algebraic of degree m, then $\alpha + \beta$ is of degree $\leqq mn$. Prove a similar result for $\alpha\beta$.

4. Prove that the set of all real algebraic numbers (i.e., algebraic numbers that are real) forms a field, and the set of all real algebraic integers forms a ring.

9.3 Algebraic Number Fields

The field discussed in Theorem 9.13 contains the totality of algebraic numbers. In general, an *algebraic number field* is any subset of this total collection that is a field itself. For example, if ξ is an algebraic number, then it can be readily verified that the collection of all numbers of the form $f(\xi)/h(\xi)$, $h(\xi) \neq 0$, f and h polynomials over Q, constitutes a field. This field is denoted by $Q(\xi)$, and it is called the extension of Q by ξ.

Theorem 9.14 *If ξ is an algebraic number of degree n, then every number in $Q(\xi)$ can be written uniquely in the form*

$$a_0 + a_1\xi + \cdots + a_{n-1}\xi^{n-1} \tag{9.5}$$

where the a_i are rational numbers.

Proof. Consider any number $f(\xi)/h(\xi)$ of $Q(\xi)$. If the minimal polynomial of ξ is $g(x)$, then $g(x) \nmid h(x)$ since $h(\xi) \neq 0$. But $g(x)$ is irreducible, so the greatest common polynomial divisor of $g(x)$ and $h(x)$ is 1, and so by Theorem 9.3 there exist polynomials $G(x)$ and $H(x)$ such that $1 = g(x)G(x) + h(x)H(x)$. Replacing x by ξ and using the fact that $g(\xi) = 0$, we get $1/h(\xi) = H(\xi)$ and $f(\xi)/h(\xi) = f(\xi)H(\xi)$. Let $k(x) = f(x)H(x)$ so that $f(\xi)/h(\xi) = k(\xi)$. Dividing $k(x)$ by $g(x)$, we get $k(x) = g(x)q(x) + r(x)$, and hence $f(\xi)/h(\xi) = k(\xi) = r(\xi)$ where $r(\xi)$ is of the form (9.5).

To prove that the form (9.5) is unique, suppose $r(\xi)$ and $r_1(\xi)$ are expressions of the form (9.5). If $r(x) - r_1(x)$ is not identically zero, then it is a polynomial of degree less than n. Since the minimal polynomial of ξ has degree n, we have $r(\xi) - r_1(\xi) \neq 0$, $r(\xi) \neq r_1(\xi)$, unless $r(x)$ and $r_1(x)$ are the same polynomial.

The field $Q(\xi)$ can be looked at in a different way, by consideration of congruences modulo the polynomial $g(x)$. That is, in analogy with Definition 2.1, for any polynomial $G(x)$ of degree at least one we will write

$$f_1(x) \equiv f_2(x) \pmod{G(x)}$$

if $G(x) \mid (f_1(x) - f_2(x))$. Ultimately, in order to get back to $Q(\xi)$ we will take the minimal polynomial $g(x)$ of ξ for $G(x)$. However, the theory of congruences is more general, and we start with the polynomial $G(x)$ over Q, irreducible or not. The properties of congruences in Theorem 2.1 can be extended at once to the polynomial case. For example, part (*c*) of the theorem has the analogue: If $f_1(x) \equiv f_2(x) \pmod{G(x)}$ and $h_1(x) \equiv h_2(x) \pmod{G(x)}$, then $f_1(x)h_1(x) \equiv f_2(x)h_2(x) \pmod{G(x)}$.

By the division algorithm Theorem 9.1, any polynomial $f(x)$ over Q is mapped by division by $G(x)$ onto a unique polynomial $r(x)$ modulo $G(x)$;

$$f(x) = G(x)q(x) + r(x), \qquad f(x) \equiv r(x) \pmod{G(x)}.$$

Thus the set of polynomials $r(x)$ consisting of 0 and all polynomials over Q of degree less than n constitute a "complete residue system modulo $G(x)$" in the sense of Definition 2.2. Of course the present residue system has infinitely many members, whereas the residue system modulo m contained precisely m elements.

Theorem 9.15 *Let $G(x)$ be a polynomial over Q of degree $n \geqq 1$. The totality of polynomials*

$$r(x) = a_0 + a_1x + \cdots + a_{n-1}x^{n-1} \tag{9.6}$$

with coefficients in Q, and with addition and multiplication modulo $G(x)$, forms a ring.

Proof. This theorem is the analogue of the first part of Theorem 2.33, and its proof is virtually the same. First we note that the polynomials (9.6) form a group under addition, with identity element 0, the additive inverse of $r(x)$ being $-r(x)$. Next, the polynomials (9.6) are closed under multiplication modulo $G(x)$, and the associative property of multiplication comes from the corresponding property for polynomials over Q with ordinary multiplication, that is

$$\{r_1(x)r_2(x)\}r_3(x) = r_1(x)\{r_2(x)r_3(x)\}$$

implies

$$\{r_1(x)r_2(x)\}r_3(x) \equiv r_1(x)\{r_2(x)r_3(x)\} \pmod{G(x)}.$$

Similarly, the distributive property modulo $G(x)$ is inherited from the distributive property of polynomials over Q.

Before stating the next theorem, we extend Definition 2.10 to the concept of isomorphism between fields. Two fields F and F' are isomorphic if there is a one-to-one correspondence between the elements of F and the elements of F' such that if a and b in F correspond respectively to a' and b' in F', then $a + b$ and ab in F correspond respectively to $a' + b'$ and $a'b'$ in F'. A virtually identical definition is used for the concept of isomorphism between rings. The following result is a direct analogue of the second part of Theorem 2.33.

Theorem 9.16 *The ring of polynomials modulo $G(x)$ described in Theorem* 9.15 *is a field if and only if $G(x)$ is an irreducible polynomial. If $G(x)$ is the minimal polynomial of the algebraic number ξ, then this field is isomorphic to $Q(\xi)$.*

Proof. If the polynomial $G(x)$ is reducible over Q, say $G(x) = G_1(x)G_2(x)$ where $G_1(x)$ and $G_2(x)$ have degrees between 1 and $n - 1$, then $G_1(x)$ and $G_2(x)$ are of the form (9.6). But then $G_1(x)$ has no multiplicative inverse modulo $G(x)$ since $G_1(x)f(x) \equiv 1 \pmod{G(x)}$ implies

$$G(x) \mid \{G_1(x)f(x) - 1\},\ G_1(x) \mid \{G_1(x)f(x) - 1\},\ G_1(x) \mid 1.$$

Hence the ring of polynomials modulo $G(x)$ is not a field.

On the other hand, if $G(x)$ is irreducible over Q, then every polynomial $r(x)$ of the form (9.6) has a unique multiplicative inverse $r_1(x)$ modulo $G(x)$, of the form (9.6). To show this we note that the greatest common divisor of $G(x)$ and $r(x)$ is 1, and so by Theorem 9.3 there exist polynomials $f(x)$ and $h(x)$ such that

$$1 = r(x)f(x) + G(x)h(x). \tag{9.7}$$

Applying Theorem 9.1 to $f(x)$ and $G(x)$ we get $f(x) = G(x)q(x) + r_1(x)$ where $r_1(x)$ is of the form (9.6). Thus (9.7) can be written

$$1 = r(x)r_1(x) + G(x)\{h(x) + r(x)q(x)\}, \qquad r(x)r_1(x) \equiv 1 \pmod{G(x)},$$

so $r_1(x)$ is a multiplicative inverse of $r(x)$ of the form (9.6). This inverse is unique because if $r(x)r_2(x) \equiv 1 \pmod{G(x)}$ then

$$r(x)r_1(x) \equiv r(x)r_2(x) \pmod{G(x)}, \ G(x) \mid r(x)\{r_1(x) - r_2(x)\}.$$

Since $G(x) \nmid r(x)$ we have $G(x) \mid \{r_1(x) - r_2(x)\}$ by Theorem 9.4. But the polynomial $r_1(x) - r_2(x)$ is either identically zero or is of degree less than n, the degree of $G(x)$. Hence $r_1(x) - r_2(x) = 0$, $r_1(x) = r_2(x)$.

Finally, if $G(x)$ is the minimal polynomial $g(x)$ of the algebraic number ξ, we must show that the field is isomorphic to $Q(\xi)$. To each $r(x)$ of the form (9.6) we let correspond the number $r(\xi)$ of $Q(\xi)$. Theorem 9.14 shows that this correspondence is one-to-one. If

$$r_1(x)r_2(x) \equiv r_3(x), \qquad r_1(x) + r_2(x) \equiv r_4(x) \pmod{G(x)}$$

then

$$r_1(x)r_2(x) = r_3(x) + q_1(x)G(x), \qquad r_1(x) + r_2(x) = r_4(x) + q_2(x)G(x),$$

and hence

$$r_1(\xi)r_2(\xi) = r_3(\xi), \qquad r_1(\xi) + r_3(\xi) = r_4(\xi),$$

since $G(\xi) = 0$. Therefore the correspondence preserves multiplication and addition.

The theorem we have just proved is significant in that it makes possible the development of the theory of algebraic numbers from the consideration of polynomials without any reference to the roots of the polynomials. The *fundamental theorem of algebra* states that every polynomial of positive degree over Q has a root that is a complex number. Therefore the algebraic number fields obtained by means of Theorem 9.16 are essentially the same—isomorphic to—the fields $R(\xi)$ of Theorem 9.14, but one does not need a knowledge of the fundamental theorem of algebra to use the method of Theorem 9.16.

The fundamental theorem of algebra implies, and is sometimes stated in the form, that every polynomial $f(x)$ of degree n over Q has n complex roots. If $f(x)$ is irreducible over Q, then the n roots, say $\xi_1, \cdots, \xi_n$, are called *conjugate algebraic numbers*, and the conjugates of any one of them are simply all the others. Now Theorem 9.16 does not make any distinction between conjugates, whereas Theorem 9.14 allows for such a distinction. For example, let $g(x)$ be the irreducible polynomial $x^3 - 2$. In Theorem 9.14 we can take ξ to be any one of the three algebraic numbers which are solutions of $x^3 - 2 = 0$, namely $\sqrt[3]{2}$, $\omega\sqrt[3]{2}$, $\omega^2\sqrt[3]{2}$ where $\omega = (-1 + i\sqrt{3})/2$. Thus there are three fields

$$R(\sqrt[3]{2}), \qquad R(\omega\sqrt[3]{2}), \qquad R(\omega^2\sqrt[3]{2}). \tag{9.8}$$

The first of these consists of real numbers, whereas the other two contain nonreal elements. Therefore, the first is certainly a different field from the others. It is not so apparent, but can be proved, that the last two differ from each other. On the other hand, if we apply Theorem 9.16 to the polynomial $x^3 - 2$, we obtain a single field consisting of all polynomials $a_0 + a_1x + a_2x^2$ over Q modulo $x^3 - 2$. According to Theorem 9.16, this field is isomorphic to each of the fields (9.8). Since isomorphism is a transitive property, the fields (9.8) are isomorphic to each other. They differ in that they contain different elements, but they are essentially the same except for the names of their elements.

PROBLEMS

1. Prove that the fields of (9.8), although isomorphic, are distinct. *Suggestion:* to prove that $Q(\omega^2\sqrt[3]{2})$ is different from $Q(\omega\sqrt[3]{2})$, assume that $\omega^2\sqrt[3]{2}$ is an element of the latter field, that is, assume that there are rationals a, b, c such that $\omega^2\sqrt[3]{2} = a + b\omega\sqrt[3]{2} + c(\omega\sqrt[3]{2})^2$. Prove that no such rationals exist.

2. Prove that the field $Q(i)$, where $i^2 = -1$, is isomorphic to the field of all polynomials $a + bx$ with a and b in Q, taken modulo $x^2 + 1$.

3. Prove that any algebraic number field contains Q as a subfield.

4. Assuming the fundamental theorem of algebra, prove Theorem 9.10 by the following procedure. Let the algebraic integer ξ satisfy some monic polynomial equation $f(x) = 0$ with integral coefficients. Then we can factor $f(x)$ in the field of complex numbers, say

$$f(x) = (x - \xi)(x - \xi_2)(x - \xi_3) \cdots (x - \xi_n).$$

If $g(x)$ is the minimal polynomial of ξ, then $g(x) \mid f(x)$ by Theorem 9.8, and so

$$g(x) = (x - \xi)(x - \theta_2) \cdots (x - \theta_r)$$

where $\theta_2, \cdots, \theta_r$ are a subset of $\xi_2, \cdots, \xi_n$. Thus $\xi, \theta_2, \cdots, \theta_r$ are algebraic integers, and by Theorem 9.12 the coefficients of $g(x)$ are algebraic integers. Then apply Theorem 9.9.

9.4 Algebraic Integers

Any algebraic number field contains the elements 0 and 1, and so, by the postulates for a field, must contain all the rational numbers. Thus any algebraic number field contains at least some algebraic integers, the rational integers $0, \pm 1, \pm 2, \cdots$. The following result shows that, in general, an algebraic number field also contains other algebraic integers.

Theorem 9.17 *If α is any algebraic number, there is a rational integer b such that $b\alpha$ is an algebraic integer.*

Proof. Let $f(x)$ be a polynomial over Q such that $f(\alpha) = 0$. We may presume that the coefficients of $f(x)$ are rational integers, since we can multiply by the least common multiple of the denominators of the coefficients. Thus we can take $f(x)$ in the form

$$f(x) = bx^n + a_1x^{n-1} + \cdots + a_n = bx^n + \sum_{j=1}^{n} a_jx^{n-j}$$

with rational integers b and a_j. Then $b\alpha$ is a zero of

$$b^{n-1}f\left(\frac{x}{b}\right) = x^n + \sum_{j=1}^{n} a_jb^{j-1}x^{n-j},$$

and hence $b\alpha$ is an algebraic integer.

Theorem 9.18 *The integers of any algebraic number field form a ring.*

Proof. If α and β are integers in such a field F, then $\alpha + \beta$ and $\alpha\beta$ are in F since F is a field. But by Theorems 9.12 and 9.13, $\alpha + \beta$, $\alpha\beta$, and $-\alpha$ are algebraic integers. Thus the integers of F form a ring with 0 and 1 as the identity elements of addition and multiplication.

Definition 9.7 *In any algebraic number field F an integer $\alpha \neq 0$ is said to be a divisor of an integer β if there exists an integer γ such that $\beta = \alpha\gamma$. In this case we write $\alpha \mid \beta$. Any divisor of the integer 1 is called a unit of F. Nonzero integers α and β are called associates if α/β is a unit.*

This definition of associates does not appear to be symmetrical in α and β, but we shall establish that the property really is symmetric.

Theorem 9.19 *The reciprocal of a unit is a unit. The units of an algebraic number field form a multiplicative group.*

Proof. If ε_1 is a unit, then there exists an integer ε_2 such that $\varepsilon_1\varepsilon_2 = 1$. Hence ε_2 is also a unit, and it is the reciprocal of ε_1. If, similarly, ε_3 is any unit with reciprocal ε_4, then the product $\varepsilon_1\varepsilon_3$ is a unit because $(\varepsilon_1\varepsilon_3)(\varepsilon_2\varepsilon_4) = 1$. Hence the units of an algebraic number field form a multiplicative group where the identity element is 1, and the inverse of ε is the reciprocal of ε.

If α and β are associates, then α/β is a unit by definition, and by the above theorem β/α is also a unit. Hence the definition of associates is symmetric: if α and β are associates, then so are β and α.

PROBLEMS

1. Prove that the units of the rational number field Q are ± 1, and that integers α and β are associates in this field if and only if $\alpha = \pm\beta$.
2. For any algebraic number α, define m as the smallest positive rational integer such that $m\alpha$ is an algebraic integer. Prove that if $b\alpha$ is an algebraic integer, where b is a rational integer, then $m \mid b$.
3. Let $\alpha = \alpha_1 + \alpha_2 i$ be an algebraic number, where α_1 and α_2 are real. Does it follow that α_1 and α_2 are algebraic numbers? If α is an algebraic integer, would α_1 and α_2 necessarily be algebraic integers?

9.5 Quadratic Fields

A quadratic field is one of the form $Q(\xi)$ where ξ is a root of an irreducible quadratic polynomial over Q. By Theorem 9.14 the elements of such a field are the totality of numbers of the form $a_0 + a_1\xi$, where a_0 and a_1 are rational numbers. Since ξ is of the form $(a + b\sqrt{m})/c$ where a, b, c, m are integers, we see that

$$Q(\xi) = Q\left(\frac{a + b\sqrt{m}}{c}\right) = Q(a + b\sqrt{m}) = Q(b\sqrt{m}) = Q(\sqrt{m}).$$

Here we have presumed that $c \neq 0$ and that m is square-free, $m \neq 1$. On the other hand, if m and n are two different square-free rational integers, neither of which is 1, then $Q(\sqrt{m}) \neq Q(\sqrt{n})$ since $\sqrt{m}$ is not in $Q(\sqrt{n})$. That is, it is impossible to find rational numbers a and b such that $\sqrt{m} = a + b\sqrt{n}$.

Theorem 9.20 *Every quadratic field is of the form $Q(\sqrt{m})$ where m is a square-free rational integer, positive or negative but not equal to 1. Numbers of the form $a + b\sqrt{m}$ with rational integers a and b are integers of $Q(\sqrt{m})$. These are the only integers of $Q(\sqrt{m})$ if $m \equiv 2$ or $3 \pmod 4$. If $m \equiv 1 \pmod 4$, the numbers $(a + b\sqrt{m})/2$, with odd rational integers a and b, are also integers of $Q(\sqrt{m})$, and there are no further integers.*

Proof. We have already proved the first part of the theorem. All that remains is to identify the algebraic integers. Any number in $Q(\sqrt{m})$ is of the form $\alpha = (a + b\sqrt{m})/c$ where a, b, c are rational integers with $c > 0$. There is no loss in generality in assuming that $(a, b, c) = 1$ so that α is in its lowest terms. If $b = 0$, then α is rational and, by Theorem 9.9, is an algebraic integer if and only if it is a rational integer, that is $c = 1$. If $b \neq 0$, then α

is not rational, and its minimal equation is quadratic,

$$\left(x - \frac{a + b\sqrt{m}}{c}\right)\left(x - \frac{a - b\sqrt{m}}{c}\right) = x^2 - \frac{2a}{c}x + \frac{a^2 - b^2m}{c^2} = 0.$$

According to Theorem 9.10, α will then be an algebraic integer if and only if this equation is monic with integral coefficients. Thus α is an algebraic integer if and only if

$$c \mid 2a \quad \text{and} \quad c^2 \mid (a^2 - b^2m), \tag{9.9}$$

and this includes the case $b = 0$, since $(a, b, c) = 1$. If $(a, c) > 1$ and $c \mid 2a$, then a and c have some common prime factor, say p, and $p \nmid b$ since $(a, b, c) = 1$. Then $p^2 \mid a^2$ and $p^2 \mid c^2$, and if $c^2 \mid (a^2 - b^2m)$, we would have $p^2 \mid b^2m$, $p^2 \mid m$, which is impossible since m is square-free. Therefore (9.9) can hold only if $(a, c) = 1$. If $c \mid 2a$ and $c > 2$ then $(a, c) > 1$, so that (9.9) can hold only if $c = 1$ or $c = 2$. It is obvious that (9.9) holds for $c = 1$. For $c = 2$ condition (9.9) becomes $a^2 \equiv b^2m \pmod 4$ and we also have a odd since $(a, c) = 1$. Then (9.9) becomes $b^2m \equiv a^2 \equiv 1 \pmod 4$, which requires that b be odd, and then reduces to $m \equiv b^2m \equiv 1 \pmod 4$. To sum up: (9.9) is satisfied if and only if either $c = 1$ or $c = 2$, a odd, b odd, $m \equiv 1 \pmod 4$, and this completes the proof.

Definition 9.8 *The norm $N(\alpha)$ of a number $\alpha = (a + b\sqrt{m})/c$ in $Q(\sqrt{m})$ is the product of α and its conjugate, $\bar{\alpha} = (a - b\sqrt{m})/c$,*

$$N(\alpha) = \alpha\bar{\alpha} = \frac{a + b\sqrt{m}}{c}\,\frac{a - b\sqrt{m}}{c} = \frac{a^2 - b^2m}{c^2}.$$

Note that by Theorem 9.20 the number α is an integer in $Q(\sqrt{m})$ if and only if its conjugate $\bar{\alpha}$ is an integer, and that if α is a rational number then $\bar{\alpha} = \alpha$.

Theorem 9.21 *The norm of a product equals the product of the norms, $N(\alpha\beta) = N(\alpha)N(\beta)$. $N(\alpha) = 0$ if and only if $\alpha = 0$. The norm of an integer in $Q(\sqrt{m})$ is a rational integer. If γ is an integer in $Q(\sqrt{m})$, then $N(\gamma) = \pm 1$ if and only if γ is a unit.*

Proof. For α and β in $Q(\sqrt{m})$ it is easy to verify that $\overline{(\alpha\beta)} = \bar{\alpha}\bar{\beta}$. Then we have $N(\alpha\beta) = \alpha\beta\bar{\alpha}\bar{\beta} = \alpha\bar{\alpha}\beta\bar{\beta} = N(\alpha)N(\beta)$. If $\alpha = 0$, then $\bar{\alpha} = 0$ and $N(\alpha) = 0$. Conversely if $N(\alpha) = 0$, then $\alpha\bar{\alpha} = 0$ so that $\alpha = 0$ or $\bar{\alpha} = 0$; but $\bar{\alpha} = 0$ implies $\alpha = 0$.

Next, if γ is an algebraic integer in $Q(\sqrt{m})$, it has degree either 1 or 2. If it has degree 1, then γ is a rational integer by Theorem 9.9, and $N(\gamma) = \gamma\bar{\gamma} = \gamma^2$ so that $N(\gamma)$ is a rational integer. If γ is of degree 2, then the minimal equation of γ, $x^2 - (\gamma + \bar{\gamma})x + \gamma\bar{\gamma} = 0$, has rational integer coefficients, and again $N(\gamma) = \gamma\bar{\gamma}$ is a rational integer.

If $N(\gamma) = \pm 1$ and γ is an integer, then $\gamma\bar{\gamma} = \pm 1$, $\gamma \mid 1$, so that γ is a unit. To prove the converse, let γ be a unit. Then there is an integer ε such that $\gamma\varepsilon = 1$. This implies $N(\gamma)N(\varepsilon) = N(1) = 1$, so that $N(\gamma) = \pm 1$ since $N(\gamma)$ and $N(\varepsilon)$ are rational integers.

Remark. The integers of $Q(i)$ are often called *Gaussian integers.*

PROBLEMS

1. If an integer α in $Q(\sqrt{m})$ is neither zero nor a unit, prove that $|N(\alpha)| > 1$.
2. If $m \equiv 1 \pmod 4$, prove that the integers of $Q(\sqrt{m})$ are all numbers of the form

$$a + b\frac{1 + \sqrt{m}}{2},$$

where a and b are rational integers.
3. If α is any integer, and ε any unit, in $Q(\sqrt{m})$, prove that $\varepsilon \mid \alpha$.
4. If α and $\beta \neq 0$ are integers in $Q(\sqrt{m})$, and if $\alpha \mid \beta$, prove that $\bar{\alpha} \mid \bar{\beta}$ and $N(\alpha) \mid N(\beta)$.
5. If α is an algebraic number in $Q(\sqrt{m})$ with $m < 0$, prove that $N(\alpha) \geqq 0$. Show that this is false if $m > 0$.
6. Prove that the following assertion is false in $Q(i)$: If $N(\alpha)$ is a rational integer, then α is an algebraic integer.
7. Prove that the assertion of the preceding problem is false in every quadratic field. *Suggestion:* Define $\alpha = (x - 2\sqrt{m})/y$, so that $N(\alpha)$ is certainly an integer if x and y satisfy $x^2 - y^2 = 4m$. Choose $x = m + 1$, $y = m - 1$ so that α is not an integer if $|m - 1| > 4$. The cases $|m - 1| \leqq 4$ can be treated specially.

9.6 Units in Quadratic Fields

A quadratic field $Q(\sqrt{m})$ is called *imaginary* if $m < 0$, and it is called *real* if $m > 1$. There are striking differences between these two sorts of quadratic fields. We shall see that an imaginary quadratic field has only a finite number of units; in fact for most of these fields ± 1 are the only units. On the other hand, every real quadratic field has infinitely many units.

Theorem 9.22 *Let m be a negative square-free rational integer. The field $Q(\sqrt{m})$ has units ± 1, and these are the only units except in the cases $m = -1$, and $m = -3$. The units for $Q(i)$ are ± 1 and $\pm i$. The units for $Q(\sqrt{-3})$ are ± 1, $(1 \pm \sqrt{-3})/2$, and $(-1 \pm \sqrt{-3})/2$.*

Proof. Taking note of Theorem 9.21, we look for all integers α in $Q(\sqrt{m})$ such that $N(\alpha) = \pm 1$. According to Theorem 9.20 we can write α in one of the two forms $x + y\sqrt{m}$ and $(x + y\sqrt{m})/2$ where x and y are rational integers and where, in the second form, x and y are odd and $m \equiv 1 \pmod 4$. Then $N(\alpha) = x^2 - my^2$ or $N(\alpha) = (x^2 - my^2)/4$ respectively. Since m is negative we have $x^2 - my^2 \geqq 0$ so there are no α with $N(\alpha) = -1$. For $m < -1$ we have $x^2 - my^2 \geqq -my^2 \geqq 2y^2$ and the only solutions of $x^2 - my^2 = 1$ are $y = 0$, $x = \pm 1$ in this case. For $m = -1$, the equation $x^2 - my^2 = 1$ has the solutions $x = 0, y = \pm 1$, and $x = \pm 1, y = 0$ and no others. For $m \equiv 1 \pmod 4$, $m < -3$ there are no solutions of $(x^2 - my^2)/4 = 1$ with odd x and y since $x^2 - my^2 \geqq 1 - m > 4$. Finally, for $m = -3$, we see that the solutions of the equation $(x^2 + 3y^2)/4 = 1$ with odd x and y are just $x = 1, y = \pm 1$, and $x = -1, y = \pm 1$. These solutions give exactly the units described in the theorem.

Theorem 9.23 *There are infinitely many units in any real quadratic field.*

Proof. The numbers $\alpha = x + y\sqrt{m}$ with integers x, y are integers in $Q(\sqrt{m})$ with norms $N(\alpha) = x^2 - my^2$. If $x^2 - my^2 = 1$, then α is a unit. But the equation $x^2 - my^2 = 1$, $m > 1$, was treated in Theorems 7.25 and 7.26 where it was proved that it has infinitely many solutions.

PROBLEM

1. Prove that the units of $Q(\sqrt{2})$ are $\pm(1 + \sqrt{2})^n$ where n ranges over all integers.

9.7 Primes in Quadratic Fields

Definition 9.9 *An algebraic integer α, not a unit, in a quadratic field $Q(\sqrt{m})$ is called a prime if it is divisible only by its associates and the units of the field.*

This definition is almost the same as the definition of primes among the rational integers. There is this difference, however. In Q all primes are positive, whereas in $Q(\sqrt{m})$ no such property is required. Thus if π is a prime and ε is a unit in $Q(\sqrt{m})$, then $\varepsilon\pi$ is an associated prime in $Q(\sqrt{m})$. For example, $-\pi$ is an associated prime of π.

Theorem 9.24 *If the norm of an integer α in $Q(\sqrt{m})$ is $\pm p$, where p is a rational prime, then α is a prime.*

Proof. Suppose that $\alpha = \beta\gamma$ where β and γ are integers in $Q(\sqrt{m})$. By Theorem 9.21 we have $N(\alpha) = N(\beta)N(\gamma) = \pm p$. Then since $N(\beta)$ and $N(\gamma)$

are rational integers, one of them must be ± 1, so that either β or γ is a unit and the other an associate of α. Thus α is a prime.

Theorem 9.25 *Every integer in $Q(\sqrt{m})$, not zero or a unit, can be factored into a product of primes.*

Proof. If α is not a prime, it can be factored into a product $\beta\gamma$ where neither β nor γ is a unit. Repeating the procedure, we factor β and γ if they are not primes. The process of factoring must stop since otherwise we could get α in the form $\beta_1\beta_2 \cdots \beta_n$ with n arbitrarily large, and no factor β_j a unit. But this would imply that

$$N(\alpha) = \prod_{j=1}^{n} N(\beta_j), \qquad |N(\alpha)| = \prod_{j=1}^{n} |N(\beta_j)| \geqq 2^n, \qquad n \text{ arbitrary,}$$

since $|N(\beta_j)|$ is an integer >1.

Although we have established that there is factorization into primes, this factorization may not be unique. In fact, we showed in Section 1.3 that factorization in the field $Q(\sqrt{-6})$ is not unique. In the next section we prove that factorization is unique in the field $Q(i)$. The general question of the values of m for which $Q(\sqrt{m})$ has the unique factorization property is an unsolved problem. There is, however, a close connection between unique factorization and the Euclidean algorithm, as we now show.

Just as in the case of the rational field, a unique factorization theorem will have to disregard the order in which the various prime factors appear. But now a new ambiguity arises due to the existence of associated primes. The two factorings

$$\alpha = \pi_1\pi_2 \cdots \pi_r = (\varepsilon_1\pi_1)(\varepsilon_2\pi_2) \cdots (\varepsilon_r\pi_r)$$

where the ε_j are units with product 1, will have to be considered as being the same.

Definition 9.10 *A quadratic field $Q(\sqrt{m})$ is said to have the unique factorization property if every integer α in $Q(\sqrt{m})$, not zero or a unit, can be factored into primes uniquely, apart from the order of the primes and ambiguities between associated primes.*

Definition 9.11 *A quadratic field $Q(\sqrt{m})$ is said to be Euclidean if the integers of $Q(\sqrt{m})$ satisfy a Euclidean algorithm, that is, if α and β are integers of $Q(\sqrt{m})$ with $\beta \neq 0$, there exist integers γ and δ of $Q(\sqrt{m})$ such that $\alpha = \beta\gamma + \delta$, $|N(\delta)| < |N(\beta)|$.*

Theorem 9.26 *Every Euclidean quadratic field has the unique factorization property.*

Proof. The proof of this theorem is similar to the procedure used in establishing the fundamental theorem of arithmetic, Theorem 1.16. First we establish that if α and β are any two integers of $Q(\sqrt{m})$ having no common factors except units, then there exist integers λ_0 and μ_0 in $Q(\sqrt{m})$ such that $\alpha\lambda_0 + \beta\mu_0 = 1$. Let S denote the set of integers of the form $\alpha\lambda + \beta\mu$ where λ and μ range over all integers of $Q(\sqrt{m})$. The norm $N(\alpha\lambda + \beta\mu)$ of any integer in S is a rational integer, so we can choose an integer, $\alpha\lambda_1 + \beta\mu_1 = \varepsilon$ say, such that $|N(\varepsilon)|$ is the least positive value taken on by $|N(\alpha\lambda + \beta\mu)|$. Applying the Euclidean algorithm to α and ε we get

$$\alpha = \varepsilon\gamma + \delta, \qquad |N(\delta)| < |N(\varepsilon)|.$$

Then we have

$$\delta = \alpha - \varepsilon\gamma = \alpha - \gamma(\alpha\lambda_1 + \beta\mu_1) = \alpha(1 - \gamma\lambda_1) + \beta(-\gamma\mu_1)$$

so that δ is an integer in S. Now this requires $|N(\delta)| = 0$ by the definition of ε, and we have $\delta = 0$ by Theorem 9.21. Thus $\alpha = \varepsilon\gamma$ and hence $\varepsilon \mid \alpha$. Similarly we find $\varepsilon \mid \beta$, and therefore ε is a unit. Then ε^{-1} is also a unit by Theorem 9.19, and we have

$$1 = \varepsilon^{-1}\varepsilon = \varepsilon^{-1}(\alpha\lambda_1 + \beta\mu_1) = \alpha(\varepsilon^{-1}\lambda_1) + \beta(\varepsilon^{-1}\mu_1) = \alpha\lambda_0 + \beta\mu_0,$$

say.

Next we prove that if π is a prime in $Q(\sqrt{m})$ and if $\pi \mid \alpha\beta$, then $\pi \mid \alpha$ or $\pi \mid \beta$. For if $\pi \nmid \alpha$, then π and α have no common factors except units, and hence there exist integers λ_0 and μ_0 such that $1 = \pi\lambda_0 + \alpha\mu_0$. Then $\beta = \pi\beta\lambda_0 + \alpha\beta\mu_0$ and $\pi \mid \beta$ because $\pi \mid \alpha\beta$. This can be extended by mathematical induction to prove that if $\pi \mid (\alpha_1\alpha_2 \cdots \alpha_n)$, then π divides at least one factor α_j of the product.

From this point on the proof is identical with the first proof of Theorem 1.16, and there is no need to repeat the details.

PROBLEMS

1. If π is a prime and ε a unit in $Q(\sqrt{m})$, prove that $\varepsilon\pi$ is a prime.
2. Prove that $1 + i$ is a prime in $Q(i)$.
3. Prove that $11 + 2\sqrt{6}$ is a prime in $Q(\sqrt{6})$.
4. Prove that 3 is a prime in $Q(i)$, but not a prime in $Q(\sqrt{6})$.
5. Prove that there are infinitely many primes in any quadratic field $Q(\sqrt{m})$.

9.8 Unique Factorization

In this section we shall apply Theorem 9.26 to various quadratic fields, namely $Q(i)$, $Q(\sqrt{-2})$, $Q(\sqrt{-3})$, $Q(\sqrt{-7})$, $Q(\sqrt{2})$, $Q(\sqrt{3})$. We shall show that these fields have the unique factorization property by proving that they are Euclidean fields. There are other Euclidean quadratic fields, but we focus our attention on these few for which the Euclidean algorithm is easily established.

Theorem 9.27 *The fields $Q(\sqrt{m})$ for $m = -1, -2, -3, -7, 2, 3$, are Euclidean and so have the unique factorization property.*

Proof. Consider any integers α and β of $Q(\sqrt{m})$ with $\beta \neq 0$. Then $\alpha/\beta = u + v\sqrt{m}$ where u and v are rational numbers, and we choose rational integers x and y that are closest to u and v, that is, so that

$$0 \leqq |u - x| \leqq \tfrac{1}{2}, \qquad 0 \leqq |v - y| \leqq \tfrac{1}{2}. \tag{9.10}$$

If we denote $x + y\sqrt{m}$ by γ and $\alpha - \beta\gamma$ by δ, then γ and δ are integers in $Q(\sqrt{m})$ and $N(\delta) = N(\alpha - \beta\gamma) = N(\beta)N(\alpha/\beta - \gamma) = N(\beta)N((u - x) + (v - y)\sqrt{m}) = N(\beta)\{(u - x)^2 - m(v - y)^2\}$,

$$|N(\delta)| = |N(\beta)|\,|(u - x)^2 - m(v - y)^2|. \tag{9.11}$$

By Equations (9.10) we have

$$-\frac{m}{4} \leqq (u - x)^2 - m(v - y)^2 \leqq \frac{1}{4} \quad \text{if} \quad m > 0,$$

$$0 \leqq (u - x)^2 - m(v - y)^2 \leqq \frac{1}{4} + \frac{1}{4}(-m) \quad \text{if} \quad m < 0,$$

and hence, by (9.11), $|N(\delta)| < |N(\beta)|$ if $m = 2, 3, -1, -2$. Therefore $Q(\sqrt{m})$ is Euclidean for these values of m.

For the cases $m = -3$ and $m = -7$ we must choose γ in a different way. With u and v defined as above, we choose a rational integer s as close to $2v$ as possible and then choose a rational integer r, such that $r \equiv s \pmod 2$, as close to $2u$ as possible. Then we have $|2v - s| \leqq \frac{1}{2}$ and $|2u - r| \leqq 1$, and the number $\gamma = (r + s\sqrt{m})/2$ is an integer of $Q(\sqrt{m})$ by Theorem 9.20, since $m \equiv 1 \pmod 4$ in the cases under discussion. As before, $\delta = \alpha - \beta\gamma$ is an integer in $Q(\sqrt{m})$ and

$$N(\delta) = N(\beta)N\left(\frac{\alpha}{\beta} - \gamma\right) = N(\beta)\left\{\left(u - \frac{r}{2}\right)^2 - m\left(v - \frac{s}{2}\right)^2\right\},$$

$$|N(\delta)| \leqq |N(\beta)|\left\{\frac{1}{4} + \frac{1}{16}(-m)\right\} < |N(\beta)|,$$

for $m = -3$ and $m = -7$.

PROBLEMS

1. Prove that $Q(\sqrt{-11})$ has the unique factorization property.
2. Prove that $Q(\sqrt{5})$ has the unique factorization property.
3. Prove that in $Q(i)$ the quotient γ and remainder δ obtained in the proof of Theorem 9.27 are not necessarily unique. That is, prove that in $Q(i)$ there exist integers $\alpha, \beta, \gamma, \delta, \gamma_1, \delta_1$ such that

$$\alpha = \beta\gamma + \delta = \beta\gamma_1 + \delta_1, \qquad N(\delta) < N(\beta),$$
$$N(\delta_1) < N(\beta), \qquad \gamma \neq \gamma_1, \qquad \delta \neq \delta_1.$$

4. If α and β are integers of $Q(i)$, not both zero, say that γ is a greatest common divisor of α and β if $N(\gamma)$ is greatest among norms of all common divisors of α and β. Prove that there are exactly four greatest common divisors of any fixed pair α, β, and that each of the four is divisible by any common divisor.

9.9 Primes in Quadratic Fields Having the Unique Factorization Property

If a field $Q(\sqrt{m})$ has the unique factorization property, we can say much more about the primes of the field than we did in Section 9.7.

Theorem 9.28 *Let $Q(\sqrt{m})$ have the unique factorization property. Then to any prime π in $Q(\sqrt{m})$ there corresponds one and only one rational prime p such that $\pi \mid p$.*

Proof. The prime π is a divisor of the rational integer $N(\pi)$, and hence there exist positive rational integers divisible by π. Let n be the least of these. Then n is a rational prime. For otherwise $n = n_1n_2$, and we have, by the unique factorization property, $\pi \mid n$, $\pi \mid (n_1n_2)$, $\pi \mid n_1$ or $\pi \mid n_2$, a contradiction since $0 < n_1 < n$, $0 < n_2 < n$. Hence n is a rational prime, call it p. And, if π were a divisor of another rational prime q, we could find rational integers by Theorem 1.3 such that $1 = px + qy$. Since $\pi \mid (px + qy)$ this implies $\pi \mid 1$, which is false, and hence the prime p is unique.

Theorem 9.29 *Let $Q(\sqrt{m})$ have the unique factorization property. Then:*

(1) *Any rational prime p is either a prime π of the field or a product $\pi_1\pi_2$ of two primes, not necessarily distinct, of $Q(\sqrt{m})$.*
(2) *The totality of primes π, π_1, π_2 obtained by applying part 1 to all rational primes, together with their associates, constitute the set of all primes of $Q(\sqrt{m})$.*

(3) *An odd rational prime p satisfying $(p, m) = 1$ is a product $\pi_1\pi_2$ of two primes in $Q(\sqrt{m})$ if and only if $\left(\frac{m}{p}\right) = 1$. Furthermore if $p = \pi_1\pi_2$, the product of two primes, then π_1 and π_2 are not associates, but π_1 and $\bar{\pi}_2$ are, and π_2 and $\bar{\pi}_1$ are.*

(4) *If $(2, m) = 1$, then 2 is the associate of a square of a prime if $m \equiv 3 \pmod 4$; 2 is a prime if $m \equiv 5 \pmod 8$; and 2 is the product of two distinct primes if $m \equiv 1 \pmod 8$.*

(5) *Any rational prime p that divides m is the associate of the square of a prime in $Q(\sqrt{m})$.*

Proof. (1) If the rational prime p is not a prime in $Q(\sqrt{m})$, then $p = \pi\beta$ for some prime π and some integer β of $Q(\sqrt{m})$. Then we have $N(\pi)N(\beta) = N(p) = p^2$. Since $N(\pi) \neq \pm 1$, we must have either $N(\beta) = \pm 1$ or $N(\beta) = \pm p$. If $N(\beta) = \pm 1$, then β is a unit by Theorem 9.21, and π is an associate of p, which then must be a prime in $Q(\sqrt{m})$. If $N(\beta) = \pm p$ then β is a prime by Theorem 9.24, and so p is a product $\pi\beta$ of two primes in $Q(\sqrt{m})$.

(2) The statement (2) now follows directly from Theorem 9.28 and statement (1).

(3) If p is an odd rational prime such that $(p, m) = 1$ and $\left(\frac{m}{p}\right) = 1$, there exists a rational integer x satisfying

$$x^2 \equiv m \pmod p,\ p \mid (x^2 - m),\ p \mid (x - \sqrt{m})(x + \sqrt{m}).$$

If p were a prime of $Q(\sqrt{m})$, it would divide one of the factors $x - \sqrt{m}$ and $x + \sqrt{m}$, so that one of

$$\frac{x}{p} - \frac{\sqrt{m}}{p}, \qquad \frac{x}{p} + \frac{\sqrt{m}}{p}$$

would be an integer in $Q(\sqrt{m})$. But this is impossible by Theorem 9.20, and hence p is not a prime in $Q(\sqrt{m})$. Therefore, by statement (1), $p = \pi_1\pi_2$ if $\left(\frac{m}{p}\right) = 1$.

Now suppose that p is an odd rational prime, that $(p, m) = 1$, and that p is not a prime in $Q(\sqrt{m})$. Then from the proof of statement (1) we see that $p = \pi\beta$, $N(\beta) = \pm p$, and $N(\pi) = \pm p$. We can write $\pi = a + b\sqrt{m}$ where a and b are rational integers or, if $m \equiv 1 \pmod 4$, halves of odd rational integers. Then $a^2 - mb^2 = N(\pi) = \pm p$, and we have $(2a)^2 - m(2b)^2 = \pm 4p$, $(2a)^2 \equiv m(2b)^2 \pmod p$. Here $2a$ and $2b$ are rational integers and neither is a multiple of p, for if p divided either one it would divide the other and we would have $p^2 \mid 4a^2$, $p^2 \mid 4b^2$, $p^2 \mid (4a^2 - 4mb^2)$, $p^2 \mid 4p$. Therefore $(2b, p) = 1$, and there is a rational integer w such that $2bw \equiv 1 \pmod{p}$, $(2aw)^2 \equiv m(2bw)^2 \equiv m \pmod{p}$, and we have $\left(\frac{m}{p}\right) = 1$.

Furthermore, with the notation of the preceding paragraph we prove that π and β are not associates, but π and $\bar{\beta}$ are, and $\bar{\pi}$ and β are. From $p = \pi\beta$ and $N(\pi) = a^2 - mb^2 = \pm p$ we have

$$\beta = \frac{p}{\pi} = \frac{p}{a + b\sqrt{m}} = \pm(a - b\sqrt{m}), \qquad \bar{\beta} = \pm(a + b\sqrt{m}),$$

so π and $\bar{\beta}$ are associates. On the other hand we note that

$$\frac{\pi}{\beta} = \pm\frac{a + b\sqrt{m}}{a - b\sqrt{m}} = \frac{(2a)^2 + m(2b)^2}{4p} + \frac{8ab\sqrt{m}}{4p},$$

and this is not an integer, and so not a unit, because p does not divide $8ab$. Thus π and β are not associates.

(4) If $m \equiv 3 \pmod{4}$, then

$$m^2 - m = 2\frac{m^2 - m}{2} = (m + \sqrt{m})(m - \sqrt{m}),$$

and $2 \nmid (m \pm \sqrt{m})$, so 2 cannot be a prime of $Q(\sqrt{m})$. Hence 2 is divisible by a prime $x + y\sqrt{m}$ and this prime must have norm ± 2. Therefore $x^2 - my^2 = \pm 2$. But this implies that

$$\pm\frac{x - y\sqrt{m}}{x + y\sqrt{m}} = \pm\frac{x^2 + my^2 - 2xy\sqrt{m}}{x^2 - my^2} = \frac{x^2 + my^2}{2} - xy\sqrt{m},$$

and, similarly,

$$\pm\frac{x + y\sqrt{m}}{x - y\sqrt{m}} = \frac{x^2 + my^2}{2} + xy\sqrt{m},$$

and therefore $(x - y\sqrt{m})(x + y\sqrt{m})^{-1}$ and its inverse are integers of $Q(\sqrt{m})$. Hence $(x - y\sqrt{m})(x + y\sqrt{m})^{-1}$ is a unit, and $x - y\sqrt{m}$ and $x + y\sqrt{m}$ are associates.

If $m \equiv 1 \pmod{4}$ and if 2 is not a prime in $Q(\sqrt{m})$ then 2 is divisible by a prime $\frac{1}{2}(x + y\sqrt{m})$ having norm ± 2. This would mean that there are rational integers x and y, both even or both odd such that

$$x^2 - my^2 = \pm 8. \tag{9.12}$$

If x and y are even, say $x = 2x_0$, $y = 2y_0$, then (9.12) would require $x_0^2 - my_0^2 = \pm 2$. But, since $m \equiv 1 \pmod{4}$, $x_0^2 - my_0^2$ is either odd or a multiple of 4. Thus (9.12) can have solutions only with odd x and y. Then $x^2 \equiv y^2 \equiv 1 \pmod 8$, and (9.12) implies

$$x^2 - my^2 \equiv 1 - m \equiv 0, \qquad m \equiv 1 \pmod 8.$$

It follows that 2 is a prime in $Q(\sqrt{m})$ if $m \equiv 5 \pmod 8$.

Now if $m \equiv 1 \pmod 8$ we observe that

$$\frac{1-m}{4} = 2\,\frac{1-m}{8} = \frac{1-\sqrt{m}}{2}\cdot\frac{1+\sqrt{m}}{2},$$

and $2 \nmid (1 \pm \sqrt{m})/2$, so 2 cannot be a prime in $Q(\sqrt{m})$. Hence (9.12) has solutions in odd integers x and y. Now the primes $\frac{1}{2}(x + y\sqrt{m})$ and $\frac{1}{2}(x - y\sqrt{m})$ are not associates in $Q(\sqrt{m})$ because their quotient is not a unit. In fact their quotient is

$$\frac{x + y\sqrt{m}}{x - y\sqrt{m}} = \pm\,\frac{x^2 + my^2}{8} \pm \frac{xy\sqrt{m}}{4}$$

which is not even an integer in $Q(\sqrt{m})$.

(5) Let p be a rational prime divisor of m. If $p = |m|$ then $p = \pm\sqrt{m}\cdot\sqrt{m}$ and hence p is the associate of the square of a prime in $Q(\sqrt{m})$ by Theorem 9.24. If $p < |m|$, we note that

$$m = p\,\frac{m}{p} = \sqrt{m}\cdot\sqrt{m}. \tag{9.13}$$

But p is not a divisor of $\sqrt{m}$ in $Q(\sqrt{m})$ by Theorem 9.20 and hence p is not a prime in $Q(\sqrt{m})$. Therefore p is divisible by a prime π, with $N(\pi) = \pm p$, and hence π is not a divisor of m/p. But, by (9.13), π is also a divisor of $\sqrt{m}$, π^2 is a divisor of m, and hence π^2 is a divisor of p.

The theorem we have just proved provides a method for determining the primes of a quadratic field having the unique factorization property. For such $Q(\sqrt{m})$ we look at all the rational primes p. Those p for which $(p, 2m) = 1$ and $\left(\frac{m}{p}\right) = -1$, together with all their associates in $Q(\sqrt{m})$, are primes in $Q(\sqrt{m})$. Those p for which $(p, 2m) = 1$ and $\left(\frac{m}{p}\right) = +1$ will factor into $p = \pi_1\pi_2$, a product of two primes of $Q(\sqrt{m})$, with $N(\pi_1) = N(\pi_2) = \pm p$. Any other factoring of p will merely replace π_1 and π_2 by associates. The primes p for which $(p, 2m) > 1$ will either be primes of $Q(\sqrt{m})$ or products of two primes of $Q(\sqrt{m})$.

Suppose that α is an integer in $Q(\sqrt{m})$ and that $N(\alpha) = \pm p$, p a rational prime. Then $\bar{\alpha}$ is also an integer in $Q(\sqrt{m})$ and $\alpha\bar{\alpha} = N(\alpha) = \pm p$, and this necessitates that $\bar{\alpha}$ be a prime in $Q(\sqrt{m})$. If $m \not\equiv 1 \pmod 4$, we can write $\alpha = x + y\sqrt{m}$, $N(\alpha) = x^2 - my^2$, with integers x and y. If $m \equiv 1 \pmod 4$, we can write $\alpha = (x + y\sqrt{m})/2$, $4N(\alpha) = x^2 - my^2$, with x and y integers, both odd or both even.

Combining these facts we have the following. Let $Q(\sqrt{m})$ have the unique factorization property, and let p be a rational prime such that $(p, 2m) = 1$, $\left(\frac{m}{p}\right) = +1$. Then if $m \not\equiv 1 \pmod 4$, one at least of the two equations $x^2 - my^2 = \pm p$ has a solution. Let $x = a$, $y = b$ be such a solution. Then the numbers $\alpha = a + b\sqrt{m}$, $\bar{\alpha} = a - b\sqrt{m}$, and the associates of α and $\bar{\alpha}$ are primes in $Q(\sqrt{m})$, and these are the only primes in $Q(\sqrt{m})$ that divide p. On the other hand, if $m \equiv 1 \pmod 4$, one at least of the two equations $x^2 - my^2 = \pm 4p$ has a solution with x and y both odd or both even. Again denoting such a solution by $x = a$, $y = b$, we can say that the numbers $\alpha = (a + b\sqrt{m})/2$, $\bar{\alpha} = (a - b\sqrt{m})/2$, and their associates are primes in $Q(\sqrt{m})$, and these are the only primes in $Q(\sqrt{m})$ that divide p. It is worth noting that our consideration of algebraic number fields has thus given us information concerning diophantine equations.

It must be remembered that these results apply only to those $Q(\sqrt{m})$ that have the unique factorization property.

Example. $m = -1$. Gaussian primes. The field is $Q(i)$ and we have

$$2m = -2, \qquad 1^2 + 1^2 = 2, \qquad \overline{1 + i} = 1 - i,$$

$$\left(\frac{m}{p}\right) = \begin{cases} +1 \text{ if } p = 4k + 1 \\ -1 \text{ if } p = 4k + 3. \end{cases}$$

For each rational prime p of the form $4k + 1$ the equation $x^2 + y^2 = p$ has a solution since $x^2 + y^2 = -p$ is clearly impossible. For each such p choose a solution $x = a_p$, $y = b_p$.

The primes in $Q(i)$ are $1 + i$, all rational primes $p = 4k + 3$, all $a_p + ib_p$, all $a_p - ib_p$, together with all their associates. Note that $1 - i = \overline{1 + i}$ has not been included since $1 - i = -i(1 + i)$, i is a unit of $Q(i)$, and hence $1 - i$ is an associate of $1 + i$.

Example. $m = -3$. The field is $Q(\sqrt{-3})$ and we have

$$2m = -6, \qquad x^2 + 3y^2 = \pm 4 \cdot 2 \text{ has no solution},$$

$$3^2 + 3 \cdot 1^2 = 4 \cdot 3, \qquad \overline{\frac{3 + \sqrt{-3}}{2}} = \frac{3 - \sqrt{-3}}{2},$$

$$\left(\frac{m}{p}\right) = \begin{cases} +1 \text{ if } p = 3k + 1, (p, 6) = 1 \\ -1 \text{ if } p = 3k + 2, (p, 6) = 1. \end{cases}$$

For each odd $p = 3k + 1$, choose a_p, b_p such that $a_p^2 + 3b_p^2 = 4p$.

The primes in $Q(\sqrt{-3})$ are 2, $(3 + \sqrt{-3})/2$, all odd rational primes $p = 3k + 2$, all $(a_p + b_p\sqrt{-3})/2$, all $(a_p - b_p\sqrt{-3})/2$, together with all their associates. Here, again, we omit $(3 - \sqrt{-3})/2$ because it can be shown to be an associate of $(3 + \sqrt{-3})/2$. We could have included 2 among the $p = 3k + 2$ by just omitting the word "odd."

PROBLEMS

1. In the second example, where $m = -3$, we know from the theory that if p is any prime of the form $3k + 1$, then there are integers x and y such that $x^2 + 3y^2 = 4p$. Let $x = 2u - y$ and establish that any such prime can be expressed in the form $u^2 - uy + y^2$.

2. The rational prime 13 can be factored in two ways in $Q(\sqrt{-3})$,

$$13 = \frac{7 + \sqrt{-3}}{2} \cdot \frac{7 - \sqrt{-3}}{2} = (1 + 2\sqrt{-3})(1 - 2\sqrt{-3}).$$

Prove that this is not in conflict with the fact that $Q(\sqrt{-3})$ has the unique factorization property.

3. Prove that $\sqrt{3} - 1$ and $\sqrt{3} + 1$ are associates in $Q(\sqrt{3})$.

4. Prove that the primes of $Q(\sqrt{3})$ are $\sqrt{3} - 1$, $\sqrt{3}$, all rational primes $p \equiv \pm 5 \pmod{12}$, all factors $a + b\sqrt{3}$ of rational primes $p \equiv \pm 1 \pmod{12}$, and all associates of these primes.

5. Prove that the primes of $Q(\sqrt{2})$ are $\sqrt{2}$, all rational primes of the form $8k \pm 3$, and all factors $a + b\sqrt{2}$ of rational primes of the form $8k \pm 1$, and all associates of these primes.

6. Prove that if m is square-free, $m < -1$, $|m|$ not a prime, then $Q(\sqrt{m})$ does not have the unique factorization property. *Suggestion:* use part (5) of Theorem 9.29.

9.10 The Equation $x^3 + y^3 = z^3$

We shall prove that $x^3 + y^3 = z^3$ has no solutions in positive rational integers x, y, z. Even more, it will be established that $\alpha^3 + \beta^3 + \gamma^3 = 0$ has no solutions in nonzero integers in the quadratic field $Q(\sqrt{-3})$. Note that this amounts to proving that $\alpha^3 + \beta^3 = \gamma^3$ has no solutions in nonzero integers of $Q(\sqrt{-3})$, because this equation can be written as $\alpha^3 + \beta^3 + (-\gamma)^3 = 0$.

For convenience throughout this discussion we denote $(-1 + \sqrt{-3})/2$ by ω, which satisfies the equations $\omega^2 + \omega + 1 = 0$ and $\omega^3 = 1$. In this notation the units of $Q(\sqrt{-3})$ are ± 1, $\pm\omega$, $\pm\omega^2$, as given in Theorem 9.22. Also, in this field the integer $\sqrt{-3}$ is a prime, by Theorem 9.24. Because this prime plays a central role in the discussion we denote it by θ. Multiplying θ by the six units, we observe that the associates of θ are

$$\pm(1 - \omega), \pm(1 - \omega^2), \pm(\omega - \omega^2) = \pm\theta = \pm\sqrt{-3}. \tag{9.14}$$

Lemma 9.30 *Every integer in* $Q(\sqrt{-3})$ *is congruent to exactly one of* 0, $+1$, -1 *modulo* θ.

Proof. Consider any integer $(a + b\theta)/2$ in $Q(\sqrt{-3})$, where a and b are rational integers, both even or both odd. Then $(b + a\theta)/2$ is also an integer, and so

$$\tfrac{1}{2}(a + b\theta) = \tfrac{1}{2}(b + a\theta)\theta + 2a \equiv 2a \pmod{\theta}.$$

Now the rational integer $2a$ is congruent to 0, 1, or -1 modulo 3, and $\theta \mid 3$, so the lemma is proved.

Lemma 9.31 *Let* ξ *and* η *be integers of* $Q(\sqrt{-3})$, *not divisible by* θ. *If* $\xi \equiv 1 \pmod{\theta}$ *then* $\xi^3 \equiv 1 \pmod{\theta^4}$. *If* $\xi \equiv -1 \pmod{\theta}$ *then* $\xi^3 \equiv -1 \pmod{\theta^4}$. *If* $\xi^3 + \eta^3 \equiv 0 \pmod{\theta}$ *then* $\xi^3 + \eta^3 \equiv 0 \pmod{\theta^4}$. *Finally if* $\xi^3 - \eta^3 \equiv 0 \pmod{\theta}$ *then* $\xi^3 - \eta^3 \equiv 0 \pmod{\theta^4}$.

Proof. From Lemma 9.30 it follows that $\xi \equiv \pm 1 \pmod{\theta}$. First if $\xi \equiv +1 \pmod{\theta}$ then $\xi = 1 + \beta\theta$ for some integer β. Then

$$\xi^3 = (1 + \beta\theta)^3 = 1 + 3\beta\theta - 9\beta^2 + \beta^3\theta^3 \equiv 1 + 3\beta\theta + \beta^3\theta^3 \pmod{\theta^4}$$

because $\theta^4 = 9$. Also we note that

$$3\beta\theta + \beta^3\theta^3 = \theta^3(\beta^3 - \beta) = \theta^3(\beta)(\beta - 1)(\beta + 1).$$

But θ is a divisor of $\beta(\beta - 1)(\beta + 1)$ by Lemma 9.30 and hence $\xi^3 \equiv 1 \pmod{\theta^4}$. Second if $\xi \equiv -1 \pmod{\theta}$ then $(-\xi) \equiv 1 \pmod{\theta}$, $(-\xi)^3 \equiv 1 \pmod{\theta^4}$ and $\xi^3 \equiv -1 \pmod{\theta^4}$.

Now $\xi^3 \equiv \xi \pmod{\theta}$ because θ is a divisor of $\xi(\xi - 1)(\xi + 1)$, and so $\xi^3 + \eta^3 \equiv 0 \pmod{\theta}$ implies $\xi + \eta \equiv 0 \pmod{\theta}$. If $\xi \equiv 1 \pmod{\theta}$ then $\eta \equiv -1 \pmod{\theta}$ and hence $\xi^3 + \eta^3 \equiv 1 - 1 \equiv 0 \pmod{\theta^4}$. Finally if $\xi^3 - \eta^3 \equiv 0 \pmod{\theta}$ then $\xi^3 + (-\eta)^3 \equiv 0 \pmod{\theta}$ and so $\xi^3 + (-\eta)^3 \equiv 0 \pmod{\theta^4}$.

Lemma 9.32 *Suppose there are integers* α, β, γ *of* $Q(\sqrt{-3})$ *such that* $\alpha^3 + \beta^3 + \gamma^3 = 0$. *If g.c.d.* $(\alpha, \beta, \gamma) = 1$ *then* θ *divides one and only one of* α, β, γ.

Proof. Suppose that θ divides none of α, β, γ. Then by Lemma 9.31,

$$0 = \alpha^3 + \beta^3 + \gamma^3 \equiv \pm 1 \pm 1 \pm 1 \pmod{\theta^4}.$$

Considering all possible combinations of signs, we conclude that θ^4 is a divisor of $3, 1, -1$, or -3. But $\theta^4 = 9$, and hence we conclude that θ divides at least one of α, β, γ.

Furthermore if θ divides any two of them, it must divide the third, contrary to hypothesis.

Lemma 9.33 *Suppose there are nonzero integers α, β, γ of $Q(\sqrt{-3})$, with $\theta \nmid \alpha\beta\gamma$, and units $\varepsilon_1, \varepsilon_2$, and a positive rational integer r such that*

$$\alpha^3 + \varepsilon_1\beta^3 + \varepsilon_2(\theta^r\gamma)^3 = 0.$$

Then $\varepsilon_1 = \pm 1$ and $r \geqq 2$.

Proof. Since $r > 0$ we see that $\alpha^3 + \varepsilon_1\beta^3 \equiv 0 \pmod{\theta^3}$. Using Lemma 9.31 we see that $\alpha^3 + \varepsilon_1\beta^3 \equiv \pm 1 + \varepsilon_1(\pm 1) \equiv 0 \pmod{\theta^3}$. The unit ε_1 is one of $\pm 1, \pm\omega, \pm\omega^2$, and so $\pm 1 + \varepsilon_1(\pm 1)$ is one of $2, 0, -2, \pm(1 \pm \omega)$, $\pm(1 \pm \omega^2)$ with all possible combinations of signs. But θ^3 divides none of these except 0, because $1 - \omega$ and $1 - \omega^2$ are associates of θ, $1 + \omega = -\omega^2$ and $1 + \omega^2 = -\omega$ are units, and $N(\pm 2) = 4$ whereas $N(\theta^3) = 27$. It follows that $\pm 1 + \varepsilon_1(\pm 1) = 0$, so $\varepsilon_1 = \pm 1$.

By Lemma 9.31, $\alpha^3 + \varepsilon_1\beta^3 \equiv 0 \pmod{\theta^3}$ implies $\alpha^3 + \varepsilon_1\beta^3 \equiv 0 \pmod{\theta^4}$. From this it follows that θ^4 is a divisor of $\varepsilon_2(\theta^r\gamma)^3$ and $r \geqq 2$.

Lemma 9.34 *There do not exist nonzero integers α, β, γ in $Q(\sqrt{-3})$, a unit ε, and a rational integer $r \geqq 2$ such that*

$$\alpha^3 + \beta^3 + \varepsilon(\theta^r\gamma)^3 = 0. \tag{9.15}$$

Proof. We may presume that g.c.d. $(\alpha, \beta, \theta^r\gamma) = 1$, and that $\theta \nmid \gamma$. Furthermore, θ does not divide both α and β, and so, interchanging α and β if necessary, we may presume that $\theta \nmid \beta$. If there are integers satisfying (9.15) select a set such that

$$N(\alpha^3\beta^3\theta^{3r}\gamma^3) \tag{9.16}$$

is a minimum. This can be done because every norm in $Q(\sqrt{-3})$ is a non-negative integer. Note that ε in (9.15) is omitted in (9.16) because $N(\varepsilon) = +1$. We now construct a solution of (9.15) with a smaller norm in (9.16), and this will establish the lemma.

Since $r \geqq 2$, we have $\alpha^3 + \beta^3 \equiv 0 \pmod{\theta^6}$. Also

$$\alpha^3 + \beta^3 = (\alpha + \beta)(\alpha + \omega\beta)(\alpha + \omega^2\beta). \tag{9.17}$$

We first prove that if any prime π divides any two of $\alpha + \beta$, $\alpha + \omega\beta$, and $\alpha + \omega^2\beta$, it must be an associate of θ. First if $\pi \mid (\alpha + \beta)$ and $\pi \mid (\alpha + \omega\beta)$ then $\pi \mid \beta(1 - \omega)$ and $\pi \mid \alpha(1 - \omega)$. But g.c.d. $(\alpha, \beta) = 1$ and $1 - \omega$ is an associate of θ by (9.14). Second if $\pi \mid (\alpha + \beta)$ and $\pi \mid (\alpha + \omega^2\beta)$ then $\pi \mid \beta(1 - \omega^2)$ and $\pi \mid \alpha(1 - \omega^2)$. Again we see that $\pi \mid (1 - \omega^2)$ and so $\pi \mid \theta$ by (9.14). Third if $\pi \mid (\alpha + \omega\beta)$ and $\pi \mid (\alpha + \omega^2\beta)$ then $\pi \mid \beta(\omega - \omega^2)$ and $\pi \mid \alpha(\omega - \omega^2)$, and again by (9.14) we get $\pi \mid \theta$.

Furthermore, because of (9.14) and the fact that $\theta \nmid \beta$, we notice that the differences between $\alpha + \beta$, $\alpha + \omega\beta$, and $\alpha + \omega^2\beta$ are divisible by θ, but not by θ^2. The product of these three is divisible by θ^6, as in (9.17). Hence if θ^a, θ^b, θ^c are the highest powers of θ dividing $\alpha + \beta$, $\alpha + \omega\beta$, and $\alpha + \omega^2\beta$ respectively, then from this argument and (9.15) we conclude that a, b, c are 1, 1, $3r - 2$ in some order, and

$$\frac{\alpha + \beta}{\theta^a}, \qquad \frac{\alpha + \omega\beta}{\theta^b}, \qquad \frac{\alpha + \omega^2\beta}{\theta^c}$$

are integers with no common prime factor in $Q(\sqrt{-3})$. And (9.15) can be written as

$$\text{(9.18)} \qquad \frac{\alpha + \beta}{\theta^a} \cdot \frac{\alpha + \omega\beta}{\theta^b} \cdot \frac{\alpha + \omega^2\beta}{\theta^c} = -\varepsilon\gamma^3,$$

so each of the factors on the left is an associate of the cube of an integer, say

$$\text{(9.19)} \quad \alpha + \beta = \varepsilon_1\theta^a\lambda_1^3, \qquad \alpha + \omega\beta = \varepsilon_2\theta^b\lambda_2^3, \qquad \alpha + \omega^2\beta = \varepsilon_3\theta^c\lambda_3^3,$$

where ε_1, ε_2, ε_3 are units. Also we note that

$$(\alpha + \beta) + \omega(\alpha + \omega\beta) + \omega^2(\alpha + \omega^2\beta) = (\alpha + \beta)(1 + \omega + \omega^2) = 0,$$

and so

$$\text{(9.20)} \qquad \varepsilon_1\theta^a\lambda_1^3 + \varepsilon_4\theta^b\lambda_2^3 + \varepsilon_5\theta^c\lambda_3^3 = 0,$$

where $\varepsilon_4 = \omega\varepsilon_2$ and $\varepsilon_5 = \omega^2\varepsilon_3$.

Thus ε_4 and ε_5 are units, and (9.20) is symmetric in the three terms on the left side of the equation. Thus we can assign the values 1, 1, $3r - 2$ to a, b, c in any order, say $a = 1$, $b = 1$, $c = 3r - 2$. Substituting these values in (9.20) and dividing by $\varepsilon_1\theta$ we get

$$\text{(9.21)} \qquad \lambda_1^3 + \varepsilon_6\lambda_2^3 + \varepsilon_7(\theta^{r-1}\lambda_3)^3 = 0,$$

where ε_6 and ε_7 are the units $\varepsilon_4/\varepsilon_1$ and $\varepsilon_5/\varepsilon_1$. Since $\gamma \neq 0$ we see that $\lambda_1\lambda_2\lambda_3 \neq 0$ from (9.18) and (9.19). Also $\theta \nmid (\lambda_1\lambda_2\lambda_3)$ so by Lemma 9.33 we conclude that $\varepsilon_6 = \pm 1$ and $r - 1 \geq 2$. But (9.21) is of the form (9.15) because $\varepsilon_6\lambda_2^3$ is either λ_2^3 or $(-\lambda_2)^3$. Taking the norm analogous to (9.16) we have by (9.19),

(9.18), and $a + b + c = 3r$,

$$N(\lambda_1^3\lambda_2^3\theta^{3r-3}\lambda_3^3) = N(\theta^{-3}(\alpha + \beta)(\alpha + \omega\beta)(\alpha + \omega^2\beta))$$
$$= N(\theta^{3r-3}\gamma^3) < N(\alpha^3\beta^3\theta^{3r}\gamma^3),$$

because $N(\theta) = 3$ and $N(\alpha) \geqq 1$, $N(\beta) \geqq 1$.

This completes the proof of Lemma 9.34.

Theorem 9.35 *There are no nonzero integers α, β, γ in $Q(\sqrt{-3})$ such that $\alpha^3 + \beta^3 + \gamma^3 = 0$. There are no positive rational integers x, y, z such that $x^3 + y^3 = z^3$.*

Proof. The second assertion follows from the first. To prove the first, suppose there are nonzero integers α, β, γ such that $\alpha^3 + \beta^3 + \gamma^3 = 0$. We may presume that g.c.d. $(\alpha, \beta, \gamma) = 1$. Then by Lemma 9.32, θ divides exactly one of α, β, γ, say $\theta \mid \gamma$. Let θ^r be the highest power of θ dividing γ, say $\gamma = \theta^r\gamma_1$ where $\theta \nmid \gamma_1$. Then by Lemma 9.33 we conclude that $r \geqq 2$, and

$$\alpha^3 + \beta^3 + (\theta^r\gamma_1)^3 = 0.$$

But this contradicts Lemma 9.34.

PROBLEMS

1. Suppose there are nonzero integers α, β, γ in $Q(\sqrt{-3})$ and units ε_1, ε_2, ε_3 such that $\varepsilon_1\alpha^3 + \varepsilon_2\beta^3 + \varepsilon_3\gamma^3 = 0$. Since $\varepsilon_1\alpha^3$ can be written $-\varepsilon_1(-\alpha)^3$ we may presume that $\varepsilon_1 = 1$, ω, or ω^2. Likewise for ε_2 and ε_3. Prove that ε_1, ε_2, ε_3 are 1, ω, ω^2 in some order.

2. Prove that there *are* nonzero integers and units as in Problem 1 such that $\varepsilon_1\alpha^3 + \varepsilon_2\beta^3 + \varepsilon_3\gamma^3 = 0$.

NOTES ON CHAPTER 9

It can be noted that after Sections 9.1–9.4 on algebraic numbers in general, we turned our attention to quadratic fields. Many of our theorems can be extended to fields of algebraic numbers of higher degree, but of course it is not possible to obtain results as detailed as those for quadratic fields. Our brief survey of algebraic numbers has omitted not only these generalizations but also many other aspects of algebraic number theory that have been investigated.

§9.2. A complex number is said to be *nonalgebraic* or *transcendental* if it is not algebraic. The basic mathematical constants π and e are transcendental numbers; proofs are given in the books by G. H. Hardy and E. M. Wright, W. J. LeVeque, and Ivan Niven, listed in the General References.

§9.8. The only fields $(Q\sqrt{m})$ with $m < 0$ having unique factorization are the cases $m = -1, -2, -3, -7, -11, -19, -43, -67, -163$. This result was only settled completely quite recently. Harold M. Stark, who contributed to establishing this result, gives an excellent account of the history and status of this and related problems in Chapter 8 of his book, "An Introduction to Number Theory" listed in the General References on page 270; see especially his Theorems 8.21, 8.22, and 8.23.

An alternative proof to that given in Theorem 9.27 of the unique factorization property of Gaussian integers is given in Special Topics, page 268. The Diophantine problem of expressing Gaussian integers as sums of squares can be found on page 267. The Eisenstein irreducibility criterion for polynomials is set forth on page 268.

For further reading on the topic of this chapter, see Chapter 8 of the book by Harold M. Stark listed in the General References, page 270, and:

Ethan D. Bolker, *Elementary Number Theory*, Chapter 6, New York, W. A. Benjamin, 1970;

Harry Pollard, *The Theory of Algebraic Numbers*, Carus Monograph 9, New York, John Wiley and Sons, 1950;

Abraham Robinson, *Numbers and Ideals*, San Francisco, Holden-Day, 1965.

10

The Partition Function

10.1 Partitions

Definition 10.1 *The partition function $p(n)$ is defined as the number of ways that the positive integer n can be written as a sum of positive integers. Two partitions are not considered to be different if they differ only in the order of their summands. It is convenient to define $p(0) = 1$.*

For example $5 = 5 = 4 + 1 = 3 + 2 = 3 + 1 + 1 = 2 + 2 + 1 = 2 + 1 + 1 + 1 = 1 + 1 + 1 + 1 + 1$, and $p(5) = 7$. Similarly, $p(1) = 1$, $p(2) = 2$, $p(3) = 3$, $p(4) = 5$.

Other partition functions can be defined for which the summands must satisfy certain restrictions. We will make use of some of these.

Definition 10.2

$p_m(n)$ = *the number of partitions of n into summands less than or equal to m.*
$p^o(n)$ = *the number of partitions of n into odd summands.*
$p^d(n)$ = *the number of partitions of n into distinct summands.*
$q^e(n)$ = *the number of partitions of n into an even number of distinct summands.*
$q^o(n)$ = *the number of partitions of n into an odd number of distinct summands.*

We make the convention $p_m(0) = p^o(0) = p^d(0) = q^e(0) = 1$, $q^o(0) = 0$.

Since $5 = 2 + 2 + 1 = 2 + 1 + 1 + 1 = 1 + 1 + 1 + 1 + 1$ we have $p_2(5) = 3$. Also $5 = 5 = 3 + 1 + 1 = 1 + 1 + 1 + 1 + 1$, and $5 = 5 = 4 + 1 = 3 + 2$, and $5 = 4 + 1 = 3 + 2$, and $5 = 5$, so that $p^o(5) = 3$, $p^d(5) = 3$, $q^e(5) = 2$, $q^o(5) = 1$.

Theorem 10.1 *We have*

(*a*) $p_m(n) = p(n)$ *if* $n \leqq m$,
(*b*) $p_m(n) \leqq p(n)$ *for all* $n \geqq 0$,
(*c*) $p_m(n) = p_{m-1}(n) + p_m(n - m)$ *if* $n \geqq m > 1$,
(*d*) $p^d(n) = q^e(n) + q^o(n)$.

Proof. With the possible exception of (*c*), these are all obvious from the definitions. To prove (*c*) we note that each partition of n counted by $p_m(n)$ either has or does not have a summand equal to m. The partitions of the second sort are counted by $p_{m-1}(n)$. The partitions of the first sort are obtained by adding a summand m to each partition of $n - m$ into summands less than or equal to m, and hence are $p_m(n - m)$ in number. If $n = m$, the term $p_m(n - m) = 1$ counts the single partition $n = m$.

Theorem 10.2 *For* $n \geqq 1$ *we have* $p^d(n) = p^o(n)$.

Proof. Consider any partition counted by $p^o(n)$. It will consist of, say, r_1 summands a_1, r_2 summands $a_2, \cdots, r_s$ summands a_s, where the a_i are distinct odd integers and $\sum_{i=1}^{s} r_i a_i = n$. Now we can write each r_i in the form $r_i = \sum_j b_j^{(i)} 2^j$, $b_j^{(i)} = 0$ or 1. Then $n = \sum_{i=1}^{s} \sum_j b_j^{(i)} 2^j a_i$ gives us a partition of n whose summands are all the integers $2^j a_i$ for which $b_j^{(i)} = 1$. Also, since the a_i are distinct and odd, the summands $2^j a_i$ are distinct, and this new partition is counted by $p^d(n)$.

We can reverse the process. If $n = \sum_{k=1}^{t} c_k$ and the c_k are distinct, we write each c_k as $2^{e_k} d_k$, d_k odd. Then we let $a_1, a_2, \cdots, a_s$ be all the different integers to be found among the $d_1, d_2, \cdots, d_t$, and we let $b_{(j)}^{i} = 1$ if $2^j a_i$ is some c_k, $b_{(j)}^{i} = 0$ otherwise. Then $\sum_{i=1}^{s} \sum_j b_j^{(i)} 2^j a_i = \sum_{k=1}^{t} c_k = n$, and we get back the partition of n into r_1 summands a_1, r_2 summands $a_2, \cdots, r_s$ summands a_s where $r_i = \sum_j b_j^{(i)} 2^j$. We have found a one-to-one correspondence between the partitions counted by $p^o(n)$ and those counted by $p^d(n)$ and hence we have $p^d(n) = p^o(n)$.

10.2 Graphs

A partition of n can be represented geometrically. If $n = a_1 + a_2 + \cdots + a_r$ is a partition, we can arrange the summands a_i in such a way that $a_1 \geqq a_2 \geqq \cdots \geqq a_r$. Then the graph of this partition is the array of points having a_1 points in the top row, a_2 in the next row, and so on down to a_r in the bottom row.

$$\begin{array}{llllll} \bullet & \bullet & \bullet & \bullet & \bullet & \bullet \\ \bullet & \bullet & \bullet & \bullet & \bullet & \\ \bullet & \bullet & \bullet & \bullet & \bullet & \\ \bullet & \bullet & & & & \\ \bullet & & & & & \end{array}$$

$$19 = 6 + 5 + 5 + 2 + 1.$$

If we read the graph vertically instead of horizontally, we obtain a possibly different partition. For example, from $19 = 6 + 5 + 5 + 2 + 1$ we get $19 = 5 + 4 + 3 + 3 + 3 + 1$. From a partition $n = a_1 + a_2 + \cdots + a_r$ consisting of r summands with largest summand a_1, we obtain a partition of n into a_1 summands with largest summand r. Since this correspondence is reversible we have the following theorem.

Theorem 10.3 *The number of partitions of n into m summands is the same as the number of partitions of n having largest summand m. The number of partitions of n into at most m summands is* $p_m(n)$.

Theorem 10.4 *If* $n \geqq 0$ *then*

$$q^e(n) - q^o(n) = \begin{cases} (-1)^j & \text{if } n = (3j^2 \pm j)/2 \text{ for some } j = 0, 1, 2, \cdots \\ 0 & \text{otherwise.} \end{cases}$$

Proof. For $n = 0$ we have $j = 0$ and $q^e(0) - q^o(0) = 1$. We now suppose $n \geqq 1$ and consider a partition $n = a_1 + a_2 + \cdots + a_r$ into distinct summands. In the graph of this partition we let A_1 denote the point farthest to the right in the first row. Since the summands are distinct there will be no point directly below A_1. If $a_2 = a_1 - 1$, there will be a point A_2 directly below the point that is immediately to the left of A_1. If $a_2 < a_1 - 1$, there will be no such point A_2. If $a_3 = a_1 - 2$, then $a_2 = a_1 - 1$ and there will be a point A_3 directly below the point that is immediately to the left of A_2. If $a_2 = a_1 - 1$ and $a_3 < a_2 - 1$, there will be no point A_3. We continue this process as far as possible, thus obtaining a set of points $A_1, A_2, \cdots, A_s$, $s \geqq 1$, lying on a line through A_1 with slope 1. We also label the points of the bottom row $B_1, B_2, \cdots, B_t$, $t = a_r$. Notice that B_t and A_s may be the same point.

$$\begin{array}{llllllll} \bullet & \bullet & \bullet & \bullet & \bullet & \bullet & \bullet & A_1 \\ \bullet & \bullet & \bullet & \bullet & \bullet & \bullet & A_2 & \\ \bullet & \bullet & \bullet & \bullet & \bullet & A_3 & & \\ \bullet & \bullet & \bullet & & & & & \\ \bullet & \bullet & & & & & & \\ B_1 & B_2 & & & & & & \end{array}$$

Now we wish to change the graph into the graph of another partition of n into distinct summands. First, we try taking the points $B_1, B_2, \cdots, B_t$ and placing them to the right of $A_1, A_2, \cdots, A_t$; B_1 to the right of A_1, B_2 to the right of A_2, etc. It is obvious that we cannot do this if $t > s$ or if $t = s$ and $B_t = A_s$. However we can do it if $t < s$ or if $t = s$ and $B_t \neq A_s$, and we obtain a graph of a partition into distinct summands. Second, we try the reverse process, putting $A_1, A_2, \cdots, A_s$ underneath $B_1, B_2, \cdots, B_s$. This will give a proper graph if and only if $s < t - 1$ or $s = t - 1$ and $B_t \neq A_s$.

In other words, we can move the B_i if $t < s$ or if $t = s$ and $B_t \neq A_s$. We can move the A_i if $s < t - 1$ or if $s = t - 1$ and $B_t \neq A_s$. There is no graph in which both the A_i and the B_i can be moved. The graphs in which neither is possible are those for which $t = s$ and $B_t = A_s$ and those for which $s = t - 1$ and $B_t = A_s$. Starting with a partition P for which the A_i can be moved, we obtain a partition P' having the s points that were just moved in the bottom row. These points are in the B_i position and, of course, can be moved back to return to the partition P. A similar situation arises if the partition P is such that the B_i can be moved. In both cases the number of summands in P' differs from that in P by 1. This pairs off all the partitions of n for which either the A_i or the B_i can be moved, and in any pair P, P', one has an even number of summands, the other an odd number.

Let us consider the exceptional partitions for which neither the A_i nor the B_i can be moved, those for which $B_t = A_s$, $s = t$ or $t - 1$. Since $B_t = A_s$ the graph consists of s rows. The bottom row has t points and the partition is $n = a_1 + a_2 + \cdots + a_s$, $a_s = t$, $a_{s-1} = t + 1$, $a_1 = t + s - 1$. Therefore $n = st + (s - 1)s/2$. If $s = t$, this is $n = (3s^2 - s)/2$; and if $s = t - 1$, it is $n = (3s^2 + s)/2$. It is easy to verify that the integers $(3s^2 \pm s)/2$, $s = 1, 2, \cdots$, are all distinct. Therefore we have paired off the partitions counted by $q^e(n)$ with those counted by $q^o(n)$ except for a single partition into s summands if $n = (3s^2 \pm s)/2$. This means

$$q^e(n) - q^o(n) = \begin{cases} (-1) & \text{if } n = (3s^2 \pm s)/2 \text{ for some } s = 1, 2, \cdots \\ 0 & \text{otherwise.} \end{cases}$$

PROBLEMS

1. Let $p'(n)$ denote the number of partitions, $n = a_1 + a_2 + \cdots + a_r$, of n into summands $a_1 \geqq a_2 \geqq a_3 \geqq \cdots \geqq a_r = 1$ such that consecutive a_i differ by at most 1. Read the graphs of such partitions vertically in order to prove that, $p'(n) = p^d(n)$.

2. Consider n dots in a row, with a separator between adjacent dots, so $n - 1$ separators in all. By choosing $j - 1$ separators to be left in place while the others are removed, and then counting the number of dots between adjacent separators, prove that the equation

$$x_1 + x_2 + \cdots + x_j = n$$

has $\binom{n-1}{j-1}$ solutions in positive integers, where two solutions $x_1, x_2, \cdots, x_j$ and $x_1', x_2', \cdots, x_j'$ are counted as distinct if $x_k \neq x_k'$ for at least one subscript k. (Note that the order of summands is taken into account here, so these are *not* partitions of n.)

3. By taking $j = 1, 2, 3, \cdots$ in the preceding problem, prove that the number of ways of writing n as a sum of positive integers is 2^{n-1}, where again the order of summands is taken into account. For example, if $n = 4$ the sums being counted are

$$1 + 1 + 1 + 1,\ 1 + 1 + 2,\ 1 + 2 + 1,\ 2 + 1 + 1,\ 1 + 3,\ 3 + 1,\ 2 + 2,\ 4.$$

10.3 Formal Power Series and Euler's Identity

Many results concerning the partition function depend on the theory of analytic functions and are beyond the scope of this book. In the sections following this one we will obtain some interesting results without the use of very much analysis. In this section we get several results on partition functions by using power series in a formal way, so that no theorems on analysis are needed.

The power series that we use are of the form $a_0 + a_1x + a_2x^2 + a_3x^3 + \cdots$, where $a_0 = 1$ in most cases. Such a power series is treated "formally" if no numerical values are ever substituted for x. Thus x is a dummy variable, and the power series is really just a sequence of constants $a_0, a_1, a_2, a_3, \cdots$. However, it is very convenient to retain the x for easy identification of the general term. Two power series $\sum a_jx^j$ and $\sum b_jx^j$ are said to be equal if $a_j = b_j$ for all subscripts j. The product of these two power series is defined to be

$$a_0b_0 + (a_0b_1 + a_1b_0)x + (a_0b_2 + a_1b_1 + a_2b_0)x^2 + \cdots + \left(\sum_{j=0}^{n} a_jb_{n-j}\right)x^n + \cdots.$$

With these definitions of equality and multiplication of formal power series, the set of all power series with real coefficients with $a_0 \neq 0$ (and $b_0 \neq 0$) forms an abelian group. The associative property is easy to prove; in fact it follows from the associative property for polynomials in x because the coefficient of x^n in any product is determined by the terms up to x^n so that all terms in higher powers of x can be discarded in all power series in proving that the coefficients of x^n are identical.

The identity element of the group is 1 or $1 + 0x + 0x^2 + \cdots$. The inverse of any given power series $\sum a_j x^j$ is a power series $\sum b_j x^j$ (if it exists) such that

$$\left(\sum_{j=0}^{\infty} a_j x^j\right)\left(\sum_{j=0}^{\infty} b_j x^j\right) = 1$$

holds. The definition of multiplication of formal power series suggests at once that the coefficients $b_0, b_1, b_2, \cdots$ can be calculated in turn from the sequence of equations

$$a_0 b_0 = 1, \qquad a_0 b_1 + a_1 b_0 = 0,$$

$$a_0 b_2 + a_1 b_1 + a_2 b_0 = 0, \quad \cdots, \quad \sum_{j=0}^{n} a_j b_{n-j} = 0, \quad \cdots,$$

because $a_0 \neq 0$. Thus the inverse exists. Finally, the group is abelian because of the symmetry of the definition of multiplication.

The inverse of $1 - x$ is readily calculated to be $1 + x + x^2 + x^3 + \cdots$. As in analysis, this is called the power series expansion of $(1 - x)^{-1}$.

Under suitable circumstances an infinite number of power series can be multiplied. An illustration of this is

$$(1 + x)(1 + x^2)(1 + x^3)(1 + x^4) \cdots = \prod_{n=1}^{\infty} (1 + x^n),$$

a product that will be used in what follows. The reason that this infinite product is well-defined is that the coefficient of x^m for any positive integer m depends on only a finite number of factors, in fact it depends on

$$(1 + x)(1 + x^2)(1 + x^3) \cdots (1 + x^m) = \prod_{n=1}^{m} (1 + x^n).$$

In general let $P_1, P_2, P_3, \cdots$ be an infinite sequence of power series each with leading term 1. Then the infinite product $P_1 P_2 P_3 \cdots$ is well-defined if for every positive integer k the power x^k occurs in only a finite number of the power series. For if this condition is satisfied it is clear that the x^m term in the product is determined by a finite product $P_1 P_2 P_3 \cdots P_r$ where r is chosen so that none of the power series $P_{r+1}, P_{r+2}, P_{r+3}, \cdots$ has any term of degree m or lower, except of course the constant term 1 in each series.

The function $(1 - x^n)^{-1}$ has the expansion $\sum_{j=0}^{\infty} x^{jn}$. Taking $n = 1, 2, \cdots, m$ and multiplying we find

$$\begin{aligned}\prod_{n=1}^{m} (1 - x^n)^{-1} &= (1 + x^{1\cdot 1} + x^{2\cdot 1} + \cdots)(1 + x^{1\cdot 2} + x^{2\cdot 2} + \cdots) \\ &\quad \times (1 + x^{1\cdot 3} + x^{2\cdot 3} + \cdots)\cdots(1 + x^{1\cdot m} + x^{2\cdot m} + \cdots) \\ &= \sum_{j_1=0}^{\infty}\sum_{j_2=0}^{\infty}\cdots\sum_{j_m=0}^{\infty} x^{j_1\cdot 1 + j_2\cdot 2 + \cdots + j_m\cdot m} \\ &= \sum_{j=0}^{\infty} c_j x^j\end{aligned}$$

where c_j is the number of solutions of $j_1 \cdot 1 + j_2 \cdot 2 + \cdots + j_m \cdot m = j$ in non-negative integers $j_1, j_2, \cdots, j_m$. That is $c_j = p_m(j)$, and we have

$$\sum_{n=0}^{\infty} p_m(n)x^n = \prod_{n=1}^{m} (1 - x^n)^{-1}.$$

In a similar way we find

$$\sum_{n=0}^{\infty} p(n)x^n = \prod_{n=1}^{\infty} (1 - x^n)^{-1}.$$

The function $\prod_{n=1}^{\infty} (1 - x^n)^{-1}$ is called the generating function for $p(n)$, and application of analytic methods to this function leads to results about $p(n)$. The generating function for $p_m(n)$ is $\prod_{n=1}^{m} (1 - x^n)^{-1}$. Similarly the generating function for $p^o(n)$ is found to be

$$\sum_{n=0}^{\infty} p^o(n)x^n = \prod_{n=1}^{\infty} (1 - x^{2n-1})^{-1},$$

and the generating function for $p^d(n)$ is

$$\sum_{n=0}^{\infty} p^d(n)x^n = \prod_{n=1}^{\infty} (1 + x^n).$$

Theorem 10.2 is equivalent to $\prod_{n=1}^{\infty} (1 + x^n) = \prod_{n=1}^{\infty} (1 - x^{2n-1})^{-1}$. This formula can be proved directly and then used to give another proof of Theorem 10.2. Formally, at least, we have

$$\begin{aligned}(1 - x^{2n-1})(1 + x^{2n-1})(1 + x^{2(2n-1)})(1 + x^{2^2(2n-1)})\cdots \\ &= (1 - x^{2(2n-1)})(1 + x^{2(2n-1)})(1 + x^{2^2(2n-1)})\cdots \\ &= (1 - x^{2^2(2n-1)})(1 + x^{2^2(2n-1)})\cdots \\ &= \cdots \\ &= 1.\end{aligned}$$

Taking $n = 1, 2, 3, \cdots$ and multiplying we find

$$\prod_{n=1}^{\infty} (1 - x^{2n-1}) \prod_{j=1}^{\infty} (1 + x^j) = 1$$

and

$$\prod_{j=1}^{\infty} (1 + x^j) = \prod_{n=1}^{\infty} (1 - x^{2n-1})^{-1}.$$

In a similar way we can multiply out $\prod_{n=1}^{\infty} (1 - x^n)$ formally to get

$$\prod_{n=1}^{\infty} (1 - x^n) = \sum_{n=0}^{\infty} (q^e(n) - q^o(n))x^n.$$

Then Theorem 10.4 implies

$$\prod_{n=1}^{\infty} (1 - x^n) = 1 + \sum_{j=1}^{\infty} (-1)^j (x^{(3j^2+j)/2} + x^{(3j^2-j)/2}).$$

This is known as *Euler's formula.* Whereas here we have proved it only in the formal sense that the coefficients of the power series are identical, an analytic proof is given in Theorem 10.9 with convergence indicated for suitable values of x. Since a variable is never assigned a numerical value in formal power series, questions of convergence never arise.

Theorem 10.5 *Euler's identity. For any positive integer n,*

$$\begin{aligned} p(n) &= p(n-1) + p(n-2) - p(n-5) - p(n-7) \\ &\quad + p(n-12) + p(n-15) - \cdots \\ &= \sum_j (-1)^{j+1} p(n - \tfrac{1}{2}(3j^2 + j)) + \sum_j (-1)^{j+1} p(n - \tfrac{1}{2}(3j^2 - j)), \end{aligned}$$

where each sum extends over all positive integers j for which the arguments of the partition function are non-negative.

Proof. From Euler's formula and the fact that $\Pi\,(1 - x^n)^{-1}$ is the generating function for $p(n)$ we can write

$$\left\{1 + \sum_{j=1}^{\infty} (-1)^j \{x^{(3j^2+j)/2} + x^{(3j^2-j)/2}\}\right\} \sum_{k=0}^{\infty} p(k)x^k = 1,$$

or

$$- x - x^2 + x^5 + x^7 - x^{12} - x^{15} + \cdots\} \sum_{k=0}^{\infty} p(k)x^k = 1.$$

Equating coefficients of x^n on the two sides we get

$$p(n) - p(n-1) - p(n-2) + p(n-5) + p(n-7) - p(n-12) - p(n-15) + \cdots = 0,$$

and thus the theorem is established.

PROBLEMS

1. Show that the infinite product

$$(1 + x_1)(1 + x_1x_2)(1 + x_1x_2x_3) \cdots = 1 + \sum x_1^{a_1}x_2^{a_2} \cdots x_k^{a_k}$$

where $a_i - a_{i+1}$ is 0 or 1, and $a_k = 1$. Count the number of terms in the expansion that are of degree n. Set $x_1 = x_2 = x_3 = \cdots = x$ to show that $(1 + x)(1 + x^2)(1 + x^3) \cdots$ is the generating function for $p'(n)$ of Problem 1, Section 10.2.

2. Compute a short table of the values of $p(n)$, from $n = 1$ to $n = 20$, by use of Theorem 10.5. (Recall that $p(0) = 1$.)

10.4 Euler's Formula

In this section we will prove Euler's formula as an equality between two functions, not just in the formal sense. Formal power series arguments have serious limitations, so it is convenient now to use a few rudimentary facts concerning infinite series and limits. A reader familiar with the theory of analytic functions will recognize that our functions are analytic in $|x| < 1$, and will be able to shorten our proofs.

Theorem 10.6 *Suppose* $0 \leqq x < 1$ *and let* $\phi_m(x) = \prod_{n=1}^{m}(1 - x^n)$. *Then* $\sum_{n=0}^{\infty} p_m(n)x^n$ *converges and*

$$\sum_{n=0}^{\infty} p_m(n)x^n = \frac{1}{\phi_m(x)}.$$

Proof. By Theorem 10.3, $p_m(n)$ is equal to the number of partitions of n into at most m summands. This is the same as the number of partitions into exactly m summands if we allow zero summands. Then each summand is 0 or 1 or 2 or $\cdots$ or n, and we have $p_m(n) \leqq (n + 1)^m$. The series $\sum_{n=0}^{\infty} (n + 1)^m x^n$ converges, by the ratio test, and hence so does $\sum_{n=0}^{\infty} p_m(n)x^n$, by the comparison test.

Now

$$(1 - x^{m!k})^m \phi_m(x)^{-1} = \prod_{n=1}^{m} \frac{1 - x^{m!k}}{1 - x^n} = \prod_{n=1}^{m} \frac{1 - (x^n)^{(m!/n)k}}{1 - x^n}$$

$$= \prod_{n=1}^{m} \sum_{j=0}^{(m!/n)k-1} x^{jn} = \sum_{h} c_h x^h$$

where the last sum is a finite sum and $0 \leqq c_h \leqq p_m(h)$ for all $h = 0, 1, 2, \cdots$, and $c_h = p_m(h)$ if $h < m!k$. Therefore we have

$$\sum_{h=0}^{m!k-1} p_m(h)x^h \leqq (1 - x^{m!k})^m \phi_m(x)^{-1} \leqq \sum_{h=0}^{\infty} p_m(h)x^h.$$

As $k \to \infty$ we have

$$\sum_{h=0}^{m!k-1} p_m(h)x^h \to \sum_{h=0}^{\infty} p_m(h)x^h, \qquad (1 - x^{m!k})^m \to 1,$$

and hence

$$\phi_m(x)^{-1} = \sum_{h=0}^{\infty} p_m(h)x^h.$$

Theorem 10.7 *For $0 \leqq x < 1$, $\lim_{m\to\infty}\phi_m(x)$ exists and is different from zero. We let $\phi(x) = \lim_{m\to\infty} \phi_m(x)$ and define $\prod_{n=1}^{\infty} (1 - x^n)$ to be $\phi(x)$.*

Proof. Since $\phi_m(0) = 1$ the result is obvious for $x = 0$. For $x > 0$ we apply the mean value theorem to the function $\log z$ to obtain a y such that $1 - x^n < y < 1$ and

$$\frac{\log 1 - \log (1 - x^n)}{1 - (1 - x^n)} = \frac{1}{y}.$$

Therefore

$$-\log (1 - x^n) = \frac{x^n}{y}, \qquad -\log (1 - x^n) \leqq \frac{x^n}{1 - x^n} \leqq \frac{x^n}{1 - x}$$

and hence

$$-\log \phi_m(x) = \sum_{n=1}^{m} -\log (1 - x^n) \leqq \sum_{n=1}^{m} \frac{x^n}{1 - x} \leqq \frac{1 - x^{m+1}}{(1 - x)^2} < \frac{1}{(1 - x)^2}.$$

This shows that $-\log \phi_m(x)$, and hence $\phi_m(x)^{-1}$, is bounded for x fixed as $m \to \infty$.

But

$$\phi_m(x)^{-1} = \prod_{n=1}^{m} \frac{1}{1 - x^n}$$

increases monotonically for x fixed as $m \to \infty$. Since $\phi_1(x)^{-1} = 1/(1-x) > 0$ this shows that $\lim_{m\to\infty} \phi_m(x)^{-1}$ exists and is different from zero. Therefore $\lim_{m\to\infty} \phi_m(x)$ exists and is also different from zero.

Theorem 10.8 *For $0 \leqq x < 1$ the series $\sum_{n=0}^{\infty} p(n)x^n$ converges, and*

$$\sum_{n=0}^{\infty} p(n)x^n = \phi(x)^{-1}.$$

Proof. We have, using Theorem 10.6,

$$\sum_{n=0}^{m} p(n)x^n = \sum_{n=0}^{m} p_m(n)x^n \leqq \sum_{n=0}^{\infty} p_m(n)x^n = \phi_m(x)^{-1} \leqq \phi(x)^{-1}.$$

For x fixed, $\sum_{n=0}^{m} p(n)x^n$ increases as $m \to \infty$. Therefore $\sum_{n=0}^{\infty} p(n)x^n = \lim_{m\to\infty} \sum_{n=0}^{m} p(n)x^n$ exists and is $\leqq \phi(x)^{-1}$.

But now

$$\sum_{n=0}^{\infty} p(n)x^n \geqq \sum_{n=0}^{\infty} p_m(n)x^n = \phi_m(x)^{-1}.$$

Letting $m \to \infty$ we have $\sum_{n=0}^{\infty} p(n)x^n \geqq \phi(x)^{-1}$, and hence $\sum_{n=0}^{\infty} p(n)x^n = \phi(x)^{-1}$.

Theorem 10.9 *Euler's formula. For $0 \leqq x < 1$ we have*

$$\phi(x) = 1 + \sum_{j=1}^{\infty} (-1)^j (x^{(3j^2+j)/2} + x^{(3j^2-j)/2}).$$

Proof. The ratio test shows that $\sum_{j=1}^{\infty} x^{(3j^2 \pm j)/2}$ converges; therefore so does the above series. Let $q_m^e(n)$ be the number of partitions of n into an even number of distinct summands no greater than m, and let $q_m^o(n)$ be the number of partitions of n into an odd number of distinct summands no greater than m. As in Definition 10.2 we will take $q_m^e(0) = 1$, $q_m^o(0) = 0$. Then

$$\phi_m(x) = (1-x)(1-x^2)(1-x^3)\cdots(1-x^m) \tag{10.1}$$

$$= \sum_n (q_m^e(n) - q_m^o(n))x^n,$$

a finite sum. But for $n \leqq m$ we have $q_m^e(n) = q(n^e)$, $q_m^o(n) = q^o(n)$, and we also have $q_m^e(n) + q_m^o(n) \leqq p(n)$ for all n. Therefore

$$\left| \phi_m(x) - \sum_{n=0}^{m} (q^e(n) - q^o(n))x^n \right|$$

$$\leqq \sum_{n>m} |q_m^e(n) - q_m^o(n)|\, x^n \leqq \sum_{n=m+1}^{\infty} p(n)x^n.$$

Since $\sum_{n=m+1}^{\infty} p(n)x^n \to 0$ as $m \to \infty$, we get $\sum_{n=0}^{\infty} (q^e(n) - q^o(n))x^n = \phi(x)$ by letting $m \to \infty$. Using Theorem 10.4, we have the present theorem.

We will have occasion to multiply power series. For this we need the following lemma.

Lemma 10.10 *Let* $\sum_{j=0}^{\infty} a_j x^j$ *and* $\sum_{k=0}^{\infty} b_k x^k$ *be absolutely convergent for* $0 \leqq x < 1$. *Then* $\sum_{k=0}^{\infty} (\sum_{j=0}^{h} a_j b_{h-j})x^h$ *converges and has the value* $\sum_{j=0}^{\infty} a_j x_j \sum_{k=0}^{\infty} b_k x^k$ *for* $0 \leqq x < 1$.

Proof. The condition $0 \leqq x < 1$ could be replaced by $|x| < 1$; we take $0 \leqq x < 1$ just to keep the lemma in agreement with our other theorems. The sums $\sum_{j=0}^{m} a_j x^j$ and $\sum_{k=0}^{\infty} b_k x^k$ are polynomials and can be multiplied by the usual rules of algebra. The terms of degree m or less in the product of the polynomials are precisely the terms in

$$\sum_{h=0}^{m}\left(\sum_{j=0}^{h} a_j b_{h-j}\right)x^h.$$

All other terms in the product are of the form $a_j b_k x^{j+k}$ with $j + k > m$. Since $j + k > m$ implies that at least one of j and k exceeds $[m/2]$, we see that

$$\sum_{j=0}^{m} a_j x^j \sum_{k=[m/2]}^{m} b_k x^k + \sum_{j=[m/2]}^{m} a_j x^j \sum_{k=0}^{m} b_k x^k,$$

when multiplied out, will be a sum of terms including all the terms $a_j b_k x^{j+k}$, $j + k > m$, and possibly others. This implies that

$$\left|\sum_{j=0}^{m} a_j x^j \sum_{k=0}^{m} b_k x^k - \sum_{h=0}^{m}\left(\sum_{j=0}^{h} a_j b_{h-j}\right)x^h\right|$$

$$\leqq \sum_{j=0}^{m} |a_j x^j| \sum_{k=[m/2]}^{m} |b_k x^k| + \sum_{j=[m/2]}^{m} |a_j x^j| \sum_{k=0}^{m} |b_k x^k|$$

$$\leqq \sum_{j=0}^{\infty} |a_j x^j| \sum_{k=[m/2]}^{\infty} |b_k x^k| + \sum_{j=[m/2]}^{\infty} |a_j x^j| \sum_{k=0}^{\infty} |b_k x^k|$$

since all four infinite series in this last expression are convergent. Letting $m \to \infty$ we see that $\sum_{k=0}^{\infty} (\sum_{j=0}^{\infty} a_j b_{h-j})x^h$ converges and is equal to $\sum_{j=0}^{\infty} a_j x^j \sum_{k=0}^{\infty} b_k x^k$.

This lemma implies that if $\sum_{j=0}^{\infty} a_j x^j$ and $\sum_{k=0}^{\infty} b_k x^k$ converge absolutely for $0 \leqq x < 1$, then so does $\sum_{h=0}^{\infty} (\sum_{j=0}^{h} a_j b_{h-j})x_k$. Applying it to $\sum_{j=0}^{\infty} |a_j|\, x^j$ and $\sum_{k=0}^{\infty} |b_k|\, x^k$, we find that $\sum_{h=0}^{\infty} (\sum_{j=0}^{\infty} |a_j b_{h-j}|)x^h$ converges for $0 \leqq x < 1$. Since $|\sum_{j=0}^{\infty} a_j b_{h-j}| \leqq \sum_{j=0}^{\infty} |a_j b_{h-j}|$, we see that $\sum_{h=0}^{\infty} (\sum_{j=0}^{h} a_j b_{h-j})x^h$ converges absolutely for $0 \leqq x < 1$. Then the lemma can be extended to the product of any finite number of power series $\sum_{j=0}^{\infty} a_j^{(i)} x^j$ that are absolutely convergent for $0 \leqq x < 1$.

Another fact that we will use is the following:

Lemma 10.11 *If $\sum_{j=0}^{\infty} a_j x^j$ and $\sum_{j=0}^{\infty} b_j x^j$ converge absolutely and $\sum_{j=0}^{\infty} a_j x^j = \sum_{j=0}^{\infty} b_j x^j$ for $0 \leqq x < 1$, then $a_j = b_j$ for all $j = 0, 1, 2, \cdots$.*

Proof. If $c_j = a_j - b_j$, then $|c_j x^j| \leqq |a_j x^j| + |b_j x^j|$, and hence $\sum_{j=0}^{\infty} c_j x^j$ converges absolutely for $0 \leqq x < 1$ and we need only show that $\sum_{j=0}^{\infty} c_j x^j = 0$ implies $c_j = 0$. Setting $x = 0$ we have $c_0 = 0$. Suppose the lemma is false. Then there is some $c_j \neq 0$, and we can let k be the smallest positive integer for which $c_k \neq 0$. Then $\sum_{j=0}^{\infty} c_j x^j = \sum_{j=k}^{\infty} c_j x^j$ and, because this series converges absolutely for $x = \frac{1}{2}$, there is an integer $m > k$ such that $\sum_{j=m+1}^{\infty} |c_j 2^{-j}| < 2^{-k-1} |c_k|$. Now for $0 < x < 1$ we have $\sum_{j=k}^{\infty} c_j x^j = 0$ and, dividing by x^k, we get

$$c_k + \sum_{j=k+1}^{m} c_j x^{j-k} + \sum_{j=m+1}^{\infty} c_j x^{j-k} = 0.$$

Then for $0 < x \leqq \frac{1}{2}$ we can write

$$\begin{aligned} |c_k| &\leqq \left| \sum_{j=k+1}^{m} c_j x^{j-k} \right| + \left| \sum_{j=m+1}^{\infty} c_j x^{j-k} \right| \\ &\leqq \left| \sum_{j=k+1}^{m} c_j x^{j-k} \right| + \sum_{j=m+1}^{\infty} |c_j|\, x^{j-k} \\ &\leqq \left| \sum_{j=k+1}^{m} c_j x^{j-k} \right| + \sum_{j=m+1}^{\infty} |c_j|\, 2^{-j+k} \\ &< |c_{k+1} x + c_{k+2} x^2 + \cdots + c_m x^{m-k}| + 2^k \cdot 2^{-k-1} |c_k|, \end{aligned}$$

and finally

$$\tfrac{1}{2} |c_k| < |c_{k+1} x + c_{k+2} x^2 + \cdots + c_m x^{m-k}|.$$

But $c_{k+1} x + c_{k+2} x^2 + \cdots + c_m x^{m-k}$ is a polynomial, and we can make its value less than the positive number $|c_k|/2$ by choosing x close enough to zero, $0 < x \leqq \frac{1}{2}$. This is a contradiction and therefore the lemma is proved.

The next theorem gives for the sum of divisors function, $\sigma(n)$, an identity similar to that for $p(n)$ in Theorem 10.5.

Theorem 10.12 *For $n \geqq 1$ we have*

$$\sigma(n) - \sigma(n-1) - \sigma(n-2) + \sigma(n-5) + \sigma(n-7) - \sigma(n-12) - \sigma(n-15) + \cdots = \begin{cases} (-1)^{j+1} n \text{ if } n = \dfrac{3j^2 \pm j}{2} \\ 0 \quad \text{otherwise} \end{cases}$$

where the sum extends as far as the arguments are positive.

Proof. Taking the derivative of $\log \phi_m(x) = \log \prod_{n=1}^{m} (1 - x^n)$ we get

$$\frac{\phi'_m(x)}{\phi_m(x)} = \sum_{n=1}^{m} \frac{-nx^{n-1}}{1 - x^n} = \sum_{n=1}^{m} \sum_{j=1}^{\infty} -nx^{jn-1} = \sum_{n=1}^{m} \sum_{k=1}^{\infty} c_{n,k} x^{k-1}$$

for $0 \leqq x < 1$, where

$$c_{n,k} = \begin{cases} -n \text{ if } n \mid k \\ 0 \quad \text{otherwise.} \end{cases}$$

There are m series $\sum_{k=1}^{\infty} c_{n,k} x^{k-1}$ each of which converges absolutely. They can be added term by term to give

$$\frac{\phi'_m(x)}{\phi_m(x)} = \sum_{k=1}^{\infty} \left(\sum_{n=1}^{m} c_{n,k} \right) x^{k-1}. \tag{10.2}$$

Using (10.1) we have $\phi'_m(x) = \sum_n n(q^e_m(n) - q^o_m(n))x^{n-1}$ since $\phi_m(x)$ is a finite sum, a polynomial in x. But we can also write (10.1) in the form of an infinite series,

$$\phi_m(x) = \sum_{n=0}^{\infty} (q^e_m(n) - q^o_m(n))x^n$$

in which all the terms from a certain n on are zero. Then equation (10.2) can be put in the form

$$\sum_n n(q^e_m(n) - q^o_m(n))x^{n-1} = \sum_{n=0}^{\infty} (q^e_m(n) - q^o_m(n))x^n \sum_{j=0}^{\infty} \left(\sum_{i=1}^{m} c_{i,j+1} \right) x^j$$

$$= \sum_{h=0}^{\infty} \left(\sum_{n=0}^{h} (q^e_m(n) - q^o_m(n)) \sum_{i=1}^{m} c_{i,h-n+1} \right) x^h$$

by Lemma 10.10. Then Lemma 10.11 gives us

$$k(q^e_m(k) - q^o_m(k)) = \sum_{n=0}^{k-1} (q^e_m(n) - q^o_m(n)) \sum_{i=1}^{m} c_{i,k-n}.$$

For any given k we can choose $m > k$. Then $q_m^e(k) = q^e(k), q_m^o(k) = q^o(k)$, $q_m^e(n) = q^e(n)$, $q_m^o(n) = q^o(n)$, and $\sum_{i=1}^m c_{i,k-n} = -\sum_{d|k-n} d = -\sigma(k-n)$ for $n \leqq k-1$. This with Theorem 10.4 gives us

$$-\sigma(k) + \sigma(k-1) + \sigma(k-2) - \sigma(k-5) - \sigma(k-7) + \cdots$$

$$= \begin{cases} (-1)^j k \text{ if } k = \dfrac{3j^2 \pm j}{2} \\ 0 \qquad \text{otherwise} \end{cases}$$

and the theorem is proved.

PROBLEM

1. Compute a short table of the values of $\sigma(n)$, from $n = 1$ to $n = 20$, by means of Theorem 10.12. Verify the entries by computing $\sigma(n) = \sum_{d|n} d$ directly.

10.5 Jacobi's Formula

Theorem 10.13 *Jacobi's formula. For* $0 \leqq x < 1$,

$$\phi(x)^3 = \sum_{j=0}^{\infty} (-1)^j (2j+1) x^{(j^2+j)/2}.$$

Proof. The formula is obvious for $x = 0$, so we can suppose $0 < x < 1$. For $0 < q < 1$, $0 < z < 1$, we define

$$f_n(z) = \prod_{k=1}^{n} \{(1 - q^{2k-1}z^2)(1 - q^{2k-1}z^{-2})\} = \sum_{j=-n}^{n} a_j z^{2j} \tag{10.3}$$

where the a_j are polynomials in q. Since $f_n(1/z) = f_n(z)$ we have $a_{-j} = a_j$, and it is easy to see that

$$a_n = (-1)^n q^{1+3+5+\cdots+(2n-1)} = (-1)^n q^{n^2}. \tag{10.4}$$

In order to obtain the other a_j we replace z by qz in (10.3) and find

$$f_n(qz) = \prod_{k=1}^{n} \{(1 - q^{2k+1}z^2)(1 - q^{2k-3}z^{-2})\}$$

$$= \prod_{k=2}^{n+1} (1 - q^{2k-1}z^2) \prod_{j=0}^{n-1} (1 - q^{2j-1}z^{-2})$$

and hence

$$
\begin{aligned}
&qz^2(1 - q^{2n-1}z^{-2})f_n(qz) \\
&\quad = (1 - q^{2n+1}z^2)\left\{\prod_{k=2}^{n}(1 - q^{2k-1}z^2)\right\}qz^2(1 - q^{-1}z^{-2})\prod_{j=1}^{n}(1 - q^{2j-1}z^{-2}) \\
&\quad = -(1 - q^{2n+1}z^2)\prod_{k=1}^{n}(1 - q^{2k-1}z^2)\prod_{j=1}^{n}(1 - q^{2j-1}z^{-2}) \\
&\quad = (q^{2n+1}z^2 - 1)f_n(z).
\end{aligned}
$$

If we write the functions f_n in terms of the a_j, using (10.3), and equate the coefficients of z^{2k}, we find

$$qa_{k-1}q^{2k-2} - q^{2n}a_kq^{2k} = q^{2n+1}a_{k-1} - a_k$$

and then

$$a_{k-1} = \frac{-(1 - q^{2n+2k})}{q^{2k-1}(1 - q^{2n-2k+2})}\,a_k.$$

This, along with (10.4) allows us to find $a_{n-1}, a_{n-2}, \cdots$, in turn. In fact, for $0 < j \leqq n$ we find

$$a_{n-j} = \frac{(-1)^j(1 - q^{4n})(1 - q^{4n-2})\cdots(1 - q^{4n-2j+2})}{(1 - q^2)(1 - q^4)\cdots(1 - q^{2j})}(-1)^nq^{(n-j)^2}$$

and hence

$$(10.5)\quad a_k = \frac{\prod_{h=n+k+1}^{2n}(1 - q^{2h})}{\prod_{h=1}^{n-k}(1 - q^{2h})}(-1)^kq^{k^2} = \frac{\phi_{2n}(q^2)}{\phi_{n+k}(q^2)\phi_{n-k}(q^2)}(-1)^kq^{k^2}.$$

This formula is valid for $0 \leqq k \leqq n$ if we agree to take $\phi_0(q^2) = 1$.

Returning to (10.3), we see that $f_n(z)$ is a product of $2n$ factors, one of which is $(1 - qz^{-2})$, which has the value 0 at $z = q^{1/2}$. Therefore taking the derivative and then setting $z = q^{1/2}$ we have

$$
\begin{aligned}
f_n'(q^{1/2}) &= \prod_{k=1}^{n}(1 - q^{2k-1}q)\left\{\prod_{j=2}^{n}(1 - q^{2j-1}q^{-1})\right\}2qq^{-3/2} \\
&= \frac{2q^{-1/2}}{1 - q^{2n}}\,\phi_n(q^2)^2.
\end{aligned}
$$

On the other hand, we also have, from (10.3),

$$f_n'(q^{1/2}) = \sum_{j=-n}^{n}2ja_jq^{j-1/2} = \sum_{j=1}^{n}2ja_jq^{-1/2}(q^j - q^{-j}).$$

Thus we find

$$\phi_n(q^2)^2 = (1 - q^{2n}) \sum_{j=1}^{n} ja_j(q^j - q^{-j}),$$

and hence, by (10.5),

$$\phi_n(q^2) = (1 - q^{2n}) \prod_{j=1}^{n} (-1)^j j q^{j^2}(q^j - q^{-j}) \frac{\phi_{2n}(q^2)\phi_n(q^2)}{\phi_{n+j}(q^2)\phi_{n-j}(q^2)}.$$

Now

$$0 \leqq \frac{\phi_{2n}(q^2)\phi_n(q^2)}{\phi_{n+j}(q^2)\phi_{n-j}(q^2)} = \prod_{h=n+j+1}^{2n} (1 - q^{2h}) \prod_{k=n-j+1}^{n} (1 - q^{2k}) \leqq 1,$$

and $\sum_{j=1}^{\infty} jq^{j^2} |q^j - q^{-j}|$ converges, so we have for $n > m$

$$\left| \phi_n(q^2)^3 - (1 - q^{2n}) \sum_{j=1}^{m} (-1)^j j q^{j^2}(q^j - q^{-j}) \frac{\phi_{2n}(q^2)\phi_n(q^2)}{\phi_{n+j}(q^2)\phi_{n-j}(q^2)} \right|$$

$$\leqq \sum_{j=m+1}^{n} jq^{j^2} |q^j - q^{-j}| \leqq \sum_{j=m+1}^{\infty} jq^{j^2} |q^j - q^{-j}|.$$

We keep m fixed but arbitrary and let $n \to \infty$. By Theorem 10.7 we have

$$\lim_{n\to\infty} \frac{\phi_{2n}(q^2)\phi_n(q^2)}{\phi_{n+j}(q^2)\phi_{n-j}(q^2)} = \frac{\phi(q^2)^2}{\phi(q^2)^2} = 1$$

and $\lim_{n\to\infty} \phi_n(q^2)^3 = \phi(q^2)^3$ so that we get

$$\left| \phi(q^2)^3 - \sum_{j=1}^{m} (-1)^j j q^{j^2}(q^j - q^{-j}) \right| \leqq \sum_{j=m+1}^{\infty} jq^{j^2} |q^j - q^{-j}|.$$

Now letting $m \to \infty$ we find

$$\phi(q^2)^3 = \sum_{j=1}^{\infty} (-1)^j j q^{j^2}(q^j - q^{-j}) = \sum_{j=1}^{\infty} (-1)^j j q^{j^2+j} + \sum_{j=1}^{\infty} (-1)^{j-1} j q^{j^2-j}$$

where we can make the last step because both series converge. Changing j to $j + 1$, we write the last series as $\sum_{j=0}^{\infty} (-1)^j (j + 1) q^{j^2+j}$ and can then add it to the first series to obtain

$$\phi(q^2)^3 = \sum_{j=0}^{\infty} (-1)^j (2j + 1) q^{j^2+j}.$$

This is our theorem with x replaced by q^2.

PROBLEM

1. Replace z by $q^{1/6}$ in (10.3), multiply by $\phi_n(q^2)$, and use (10.5) to obtain a proof of Euler's formula.

10.6 A Divisibility Property

Theorem 10.14 *If p is a prime and $0 \leqq x < 1$ then*

$$\frac{\phi(x^p)}{\phi(x)^p} = 1 + p \sum_{j=1}^{\infty} a_j x^j$$

where the a_j are integers.

Proof. For $0 \leqq u < 1$ we have the expansion

$$(1-u)^{-p} = 1 + \sum_{j=1}^{\infty} (-1)^j \frac{(-p)(-p-1)\cdots(-p-j+1)}{j!} u^j$$

$$= 1 + \sum_{j=1}^{\infty} \frac{(p+j-1)!}{j!(p-1)!} u^j = \sum_{j=0}^{\infty} b_j u^j,$$

say, and therefore

$$\frac{1-u^p}{(1-u)^p} = (1-u)^{-p} - u^p(1-u)^{-p} = \sum_{j=0}^{\infty} b_j u^j - \sum_{j=0}^{\infty} b_j u^{j+p}$$

$$= \sum_{j=0}^{p-1} b_j u^j + \sum_{j=p}^{\infty} (b_j - b_{j-p}) u^j = \sum_{j=0}^{\infty} c_j u^j,$$

say. But

$$b_j = \frac{(j+1)(j+2)\cdots(j+p-1)}{(p-1)!} \equiv \begin{cases} 1 \pmod{p} \text{ if } j \equiv 0 \pmod{p} \\ 0 \pmod{p} \text{ if } j \not\equiv 0 \pmod{p}, \end{cases}$$

and $b_0 < b_1 < b_2 < \cdots$, so that we have $c_0 = b_0 = 1$, $c_j > 0$, $c_j \equiv 0 \pmod{p}$ for $j > 0$.

Now, for $0 \leqq x < 1$,

$$\frac{\phi_m(x^p)}{\phi_m(x)^p} = \prod_{n=1}^{m} \frac{1-x^{pn}}{(1-x^n)^p} = \sum_{j=0}^{\infty} a_j^{(m)} x^j$$

where $a_i^{(1)} = c_j$ and, by Lemma 10.10,

$$\sum_{h=0}^{\infty} a_h^{(m)} x^h = \sum_{j=0}^{\infty} c_j x^{mj} \sum_{k=0}^{\infty} a_k^{(m-1)} x^k = \sum_{h=0}^{\infty} \sum_{j=0}^{[h/m]} c_j a_{h-mj}^{(m-1)} x^h.$$

By Lemma 10.11 we then have

$$a_h^{(m)} = \sum_{j=0}^{[h/m]} c_j a_{h-mj}^{(m-1)},$$

and hence

$$a_h^{(m)} \equiv a_h^{(m-1)} \equiv a_h^{(1)} \equiv c_h \pmod p,$$

$$a_h^{(m)} \geqq a_h^{(m-1)} \geqq a_h^{(1)} = c_h > 0,$$

$$a_h^{(m)} = a_h^{(m-1)} \text{ if } h \leqq m - 1.$$

Therefore

$$\sum_{h=0}^{m} a_h^{(h)} x^h = \sum_{h=0}^{m} a_h^{(m)} x^h \leqq \sum_{h=0}^{\infty} a_h^{(m)} x^h = \frac{\phi_m(x^p)}{\phi_m(x)^p}.$$

Since the sum on the left increases as $m \to \infty$ we see that $\sum_{h=0}^{\infty} a_h^{(h)} x^h$ converges and

$$\sum_{h=0}^{\infty} a_h^{(h)} x^h \leqq \frac{\phi(x^p)}{\phi(x)^p}.$$

But we also have

$$\sum_{h=0}^{\infty} a_h^{(h)} x^h = \sum_{h=0}^{m} a_h^{(m)} x^h + \sum_{h=m+1}^{\infty} a_h^{(h)} x^h$$

$$\geqq \sum_{h=0}^{m} a_h^{(m)} x^h + \sum_{h=m+1}^{\infty} a_h^{(m)} x^h = \frac{\phi_m(x^p)}{\phi_m(x)^p},$$

$$\sum_{h=0}^{\infty} a_h^{(h)} x^h \geqq \frac{\phi(x^p)}{\phi(x)^p},$$

and finally

$$\sum_{h=0}^{\infty} a_h^{(h)} x^h = \frac{\phi(x^p)}{\phi(x)^p}.$$

Since $a_0^{(0)} = c_0 = 1$ and $a_h^{(h)} \equiv c_h \equiv 0 \pmod p$ for $h \geqq 1$, the theorem is proved.

Theorem 10.15 *For* $0 \leqq x < 1$ *we have* $x\phi(x)^4 = \sum_{m=1}^{\infty} b_m x^m$ *where the* b_m *are integers and* $b_m \equiv 0 \pmod 5$ *if* $m \equiv 0 \pmod 5$.

Proof. We can write Theorem 10.9 in the form

$$\phi(x) = \sum_{k=0}^{\infty} c_k x^k, \qquad c_k = \begin{cases} (-1)^j \text{ if } k = (3j^2 \pm j)/2 \\ 0 \qquad \text{otherwise} \end{cases}$$

and Theorem 10.13 as

$$\phi(x)^3 = \sum_{n=0}^{\infty} d_n x^n, \qquad d_n = \begin{cases} (-1)^j(2j+1) \text{ if } n = (j^2+j)/2 \\ 0 \qquad \text{otherwise,} \end{cases}$$

and then apply Lemma 10.9 to obtain

$$\begin{aligned} x\phi(x)^4 &= x\phi(x)\phi(x)^3 \\ &= x \sum_{h=0}^{\infty} \left(\sum_{k=0}^{h} c_k d_{h-k} \right) x^h = \sum_{m=1}^{\infty} b_m x^m. \end{aligned}$$

Then $b_m = \sum_{k=0}^{m-1} c_k d_{m-1-k}$ can be written as $\sum c_k d_n$ summed over all $k \geqq 0$, $n \geqq 0$, such that $k + n = m - 1$. But d_n is 0 unless $n = (j^2 + j)/2$, $j = 0, 1, 2, \cdots$, in which case it is $(-1)^j(2j+1)$. Furthermore we can describe c_k by saying that it is 0 unless $k = (3i^2 + i)/2$, $i = 0, \pm 1, \pm 2, \cdots$, in which case it is $(-1)^i$. Then we can write

$$(10.6) \qquad b_m = \sum (-1)^i(-1)^j(2j+1) = \sum (-1)^{i+j}(2j+1)$$

summed over all i and j such that $j \geqq 0$ and $(3i^2 + i)/2 + (j^2 + j)/2 = m - 1$. But

$$2(i+1)^2 + (2j+1)^2 = 8\left(1 + \frac{3i^2 + i}{2} + \frac{j^2 + j}{2}\right) - 10i^2 - 5$$

so that if $m \equiv 0 \pmod 5$, the terms in (10.6) will have to be such that $2(i+1)^2 + (2j+1)^2 \equiv 0 \pmod 5$. That is $(2j+1)^2 \equiv -2(i+1)^2 \pmod 5$. However, -2 is a quadratic nonresidue modulo 5, so this condition implies $2j + 1 \equiv 0 \pmod 5$, and hence $b_m \equiv 0 \pmod 5$ if $m \equiv 0 \pmod 5$.

Theorem 10.16 *We have* $p(5m + 4) \equiv 0 \pmod 5$.

Proof. By Theorems 10.15, 10.14, and 10.8 we have

$$\begin{aligned} \sum_{n=0}^{\infty} p(n)x^{n+1} &= \frac{x}{\phi(x)} = x\phi(x)^4 \frac{\phi(x^5)}{\phi(x)^5} \frac{1}{\phi(x^5)} \\ &= \sum_{m=1}^{\infty} b_m x^m \left(1 + 5\sum_{j=1}^{\infty} a_j x^j\right) \sum_{k=0}^{\infty} p(k)x^{5k} \end{aligned}$$

where the a_j and b_m are integers and $b_m \equiv 0 \pmod 5$ for $m \equiv 0 \pmod 5$. Using Lemmas 10.10 and 10.11 we find that

$$p(n-1) \equiv \sum_{k=0}^{[n/5]} p(k) b_{n-5k} \pmod 5$$

and hence $p(5m+4) \equiv 0 \pmod 5$ since $b_{5m+5-5k} \equiv 0 \pmod 5$.

This theorem is only one of several divisibility properties of the partition function. The methods of this section can be used to prove that $p(7n+5) \equiv 0 \pmod 7$. With the aid of more extensive analysis, it can be shown that $p(5^k n + r) \equiv 0 \pmod{5^k}$ if $24r \equiv 1 \pmod{5^k}$, $k = 2, 3, 4, \cdots$, and there are still other congruences related to powers of 5. There are somewhat similar congruences related to powers of 7, but it is an interesting fact that $p(7^k n + r) \equiv 0 \pmod{7^k}$ if $24r \equiv 1 \pmod{7^k}$ is valid for $k = 1, 2$ but is false for $k = 3$. There are also divisibility properties related to the number 11. An identity typical of several connected with the divisibility properties is

$$\sum_{n=0}^{\infty} p(5n+4)x^n = 5\frac{\phi(x^5)^5}{\phi(x)^6}, \qquad |x| < 1.$$

PROBLEMS

1. Write Euler's formula as

$$\phi(x) = \sum_{j=-\infty}^{\infty} (-1)^j x^{(3j^2+j)/2}.$$

Use Jacobi's formula as in Theorem 10.13, multiply $x\phi(x)\phi(x)^3$ out formally and verify (10.6).

2. Obtain a congruence similar to that in Theorem 10.16 but for the modulus 35, using Theorem 10.16 and $p(7n+5) \equiv 0 \pmod 7$.

NOTES ON CHAPTER 10

For a more extensive discussion of the methods of Section 10.3, including a proof of Theorem 10.12 avoids all questions of convergence, see I. Niven, Formal Power Series, *Amer. Math. Monthly*, **76,** 871–889 (1969). This paper also treats recurrence functions in a quite different way from that in Section 4.5.

For some of the original basic work on congruence properties of partitions, see S. Ramanujan, *Collected Papers*, Cambridge Press, 1927.

11

The Density of Sequences of Integers

In order even to define what is meant by the density of a sequence of integers it is necessary to use certain concepts from analysis. In this chapter it is assumed that the reader is familiar with the ideas of the limit inferior of a sequence of real numbers and the greatest lower bound, or infimum, of a set of real numbers. Also, in Section 11.2 we make use of the fact that $\sum_{n=1}^{\infty} 1/n^2 = \pi^2/6$. These are discussed in many texts, for example in *Mathematical Analysis*, by Apostol.*

Two common types of density are considered in this chapter, asymptotic density and Schnirelman density. The first is discussed in Sections 11.1 to 11.3, and the second in Section 11.4. Density will be defined for a set A of distinct positive integers. We will think of the elements of A as being arranged in a sequence according to size,

$$a_1 < a_2 < a_3 < \cdots, \tag{11.1}$$

and we will also denote A by $\{a_i\}$. Furthermore we will use both the terms set and sequence to describe A. The set A may be infinite or finite. That is, it may contain infinitely many elements or only a finite number of elements. It may even be empty, in which case it will be denoted by 0. If an integer m is an element of A we write $m \in A$, if not we write $m \notin A$. The set A is contained in B, $A \subset B$ or $B \supset A$, if every element of A is an element of B. We

* Tom M. Apostol, *Mathematical Analysis*, Addison-Wesley, 1957.

write $A = B$ if $A \subset B$ and $B \subset A$, that is if A and B have precisely the same elements. The *union* $A \cup B$ of two sets A and B is the set of all elements m such that $m \in A$ or $m \in B$. The *intersection* $A \cap B$ of A and B is the set of all m such that $m \in A$ and $m \in B$. Thus, for example, $A \cup A = A \cap A = A$, $A \cup 0 = A, A \cap 0 = 0$. If A and B have no element in common, $A \cap B = 0$, A and B are said to be *disjoint*. By the *complement* $\bar{A}$ of A we mean the set of all positive integers that are not elements of A. Thus $A \cap \bar{A} = 0$ and $\bar{0}$ is the set of all positive integers.

11.1 Asymptotic Density

The number of positive integers in a set A that are less than or equal to x is denoted by $A(x)$. For example, if A consists of the even integers $2, 4, 6, \cdots$, then $A(1) = 0$, $A(2) = 1$, $A(6) = 3$, $A(7) = 3$, $A(15/2) = 3$; in fact $A(x) = [x/2]$ if $x \geqq 0$. On the other hand, for any set $A = \{a_i\}$ we have $A(a_j) = j$.

Definition 11.1 *The asymptotic density of a set A is*

$$\delta_1(A) = \liminf_{n\to\infty} \frac{A(n)}{n}. \tag{11.2}$$

In case the sequence $A(n)/n$ has a limit, we say that A has a natural density, $\delta(A)$. Thus

$$\delta(A) = \delta_1(A) = \lim_{n\to\infty} \frac{A(n)}{n} \tag{11.3}$$

if A has a natural density. If A is a finite sequence, it is clear that $\delta(A) = 0$.

Theorem 11.1 *If A is an infinite sequence, then*

$$\delta_1(A) = \liminf_{n\to\infty} \frac{n}{a_n}.$$

If $\delta(A)$ exists, then $\delta(A) = \lim_{n\to\infty} n/a_n$.

Proof. The sequence k/a_k is a subsequence of $A(n)/n$ and hence

$$\liminf_{n\to\infty} \frac{A(n)}{n} \leqq \liminf_{k\to\infty} \frac{k}{a_k}.$$

If n is any integer $\geqq a_1$ and a_k is the smallest integer in A that exceeds n, then $a_{k-1} \leqq n < a_k$ and

$$\frac{k}{a_k} - \frac{A(n)}{n} = \frac{k}{a_k} - \frac{k-1}{n} < \frac{k}{n} - \frac{k-1}{n} = \frac{1}{n}.$$

It follows that

$$\frac{k}{a_k} < \frac{A(n)}{n} + \frac{1}{n}, \qquad \liminf_{k\to\infty} \frac{k}{a_k} \leqq \liminf_{n\to\infty} \frac{A(n)}{n}$$

and so the theorem is proved.

PROBLEMS

1. Prove that each of the following sets has a natural density, and find its value:
(*a*) the set of even positive integers;
(*b*) the set of odd positive integers;
(*c*) the positive multiples of 3;
(*d*) the positive integers of the form $4k + 2$;
(*e*) all positive integers a satisfying $a \equiv b \pmod{m}$, where b and $m > 1$ are fixed;
(*f*) the set of primes;
(*g*) the set $\{ar^n\}$ with $n = 1, 2, 3, \cdots$ and fixed $a \geqq 1$, fixed $r > 1$;
(*h*) the set of all perfect squares;
(*i*) the set of all positive cubes;
(*j*) the set of all positive powers, that is, all numbers of the form a^n with $a \geqq 1, n \geqq 2$.
2. If the natural density $\delta(A)$ exists, prove that $\delta(\bar{A})$ also exists and that $\delta(A) + \delta(\bar{A}) = 1$.
3. Prove that $\delta(A)$ exists if and only if $\delta_1(A) + \delta_1(\bar{A}) = 1$.
4. For any set A, prove that $\delta_1(A) + \delta_1(\bar{A}) \leqq 1$.
5. Define A_n as the set of all a such that $(2n)! \leqq a < (2n + 1)!$ and let A be the union of all sets A_n, $n = 1, 2, 3, \cdots$. Prove that $\delta_1(A) + \delta_1(\bar{A}) = 0$.
6. Let A^* be the set remaining after a finite number of integers are deleted from a set A. Prove that $\delta_1(A) = \delta_1(A^*)$, and that $\delta(A)$ exists if and only if $\delta(A^*)$ exists.
7. If two sets A and B are identical beyond a fixed integer n, prove that $\delta_1(A) = \delta_1(B)$.
8. Given any set $A = \{a_j\}$ and any integer $b \geqq 0$, define $B = \{b + a_j\}$. Prove that $\delta_1(A) = \delta_1(B)$.
9. Let A be the set of all even positive integers, B_1 the set of all even positive integers with an even number of digits to base ten, and B_2 the set of all odd positive integers with an odd number of digits. Define $B = B_1 \cup B_2$, and prove that $\delta(A)$ and $\delta(B)$ exist, but that $\delta(A \cup B)$ and $\delta(A \cap B)$ do not exist.
10. If $A \cap B = 0$, prove that $\delta_1(A \cup B) \geqq \delta_1(A) + \delta_1(B)$.
11. Let S denote any finite set of positive integers $a_1, a_2, \cdots, a_m$. Prove that the set A of all positive integers not divisible by any member of S has natural density

$$1 - \sum_{i=1}^{m} \frac{1}{a_i} + \sum_{i<j} \frac{1}{[a_i, a_j]} - \sum_{i<j<k} \frac{1}{[a_i, a_j, a_k]} + \cdots + \frac{(-1)^m}{[a_1, a_2, \cdots, a_m]}.$$

12. Let A be a set of positive integers such that for every integer m, the equation $x + y = m$ has at most one solution not counting order with x and y in A. Prove that A has density zero. Even more, prove that $A(n) \leqq 2\sqrt{n}$.

13. Define $A = \{a_j\}$ as follows. With $a_1 = 1$, define a_{k+1} as the least positive integer that is different from all the numbers $a_h + a_i - a_j$, with $1 \leqq h \leqq k$, $1 \leqq i \leqq k$, $1 \leqq j \leqq k$. Prove that A satisfies the inequality of the preceding problem, and that $A(n) \geqq \sqrt[3]{n} - 1$.

14. Let P be the set of integers $\{m^k\}$ with $m = 1, 2, 3, \cdots$ and $k = 2, 3, 4, \cdots$. Let P_1 be the subset with $k = 3, 4, \cdots$. Prove that

$$\lim_{n\to\infty} \frac{P_1(n)}{P(n)} = 0.$$

15. Find the asymptotic density of the set positive of integers having an odd number of digits in the base ten representation.

11.2 Square-Free Integers

An integer is square-free if it is divisible by no perfect square $a^2 > 1$. We will prove that the set of square-free integers has natural density $6/\pi^2$.

Lemma 11.2 *The function $\tau(n)$, representing the number of positive divisors of n, satisfies the inequality $\tau(n) \leqq 2\sqrt{n}$ for $n \geqq 1$.*

Proof. Consider the positive divisors, d, of n. Corresponding to each $d \geqq \sqrt{n}$ is the distinct divisor $d' = n/d$, and $1 \leqq d' \leqq \sqrt{n}$. Therefore $\tau(n)$ cannot exceed twice the number of divisors d such that $1 \leqq d \leqq \sqrt{n}$. Clearly the number of these d cannot exceed $\sqrt{n}$, and we have $\tau(n) \leqq 2\sqrt{n}$.

Theorem 11.3 *We have*

$$\sum_{n=1}^{\infty} \frac{\mu(n)}{n^2} \sum_{n=1}^{\infty} \frac{1}{n^2} = 1.$$

Proof. Writing

$$P_m = \sum_{n=1}^{m} \frac{\mu(n)}{n^2} \sum_{n=1}^{m} \frac{1}{n^2} = R_m S_m, \tag{11.4}$$

we consider first a fixed integer $j \leqq m$. If d is any divisor of j, say $j = dq$, then $\mu(d)/d^2$ is a term of R_m and $1/q^2$ is a term of S_m and $\mu(d)/(dq)^2 = \mu(d)/j^2$ occurs in P_m. Then $1/j^2$ occurs in P_m with coefficient

$$\sum_{d \mid j} \mu(d) = \begin{cases} 1 \text{ if } j = 1 \\ 0 \text{ if } j > 1 \end{cases}$$

by Theorem 4.6. In case $j > m$ the product $\mu(d)/(dq)^2$ may appear in P_m for some divisors d of j. Therefore we can write

$$P_m = \sum_{j=1}^{m}\left(\sum_{d\mid j}\mu(d)\right)\frac{1}{j^2} + \sum_{j=m+1}^{m^2}\left(\sum_{d\mid j}{}' \mu(d)\right)\frac{1}{j^2}$$

where $\sum'$ denotes a sum over the appropriate divisors d of j. Thus we have

$$P_m - 1 = \sum_{j=m+1}^{m^2}\sum_{d\mid j}{}' \frac{\mu(d)}{j^2} = \sum_{j=m+1}^{m^2}\frac{c_j}{j^2}, \qquad c_j = \sum_{d\mid j}{}' \mu(d),$$

and, using Lemma 11.2, we observe that

$$|c_j| \leqq \sum_{d\mid j}{}' |\mu(d)| \leqq \sum_{d\mid j}|\mu(d)| \leqq \sum_{d\mid j} 1 = \tau(j) \leqq 2\sqrt{j}.$$

Now we have

$$|P_m - 1| \leqq \sum_{j=m+1}^{m}\frac{|c_j|}{j^2} \leqq \sum_{j=m+1}^{m^2}\frac{2\sqrt{j}}{j^2} = \sum_{j=m+1}^{m^2}\frac{2}{j^{3/2}}.$$

Applying Cauchy's condition to the convergent series $\sum 2/j^{3/2}$ we see that $|P_m - 1|$ tends to zero as m tends to infinity. In view of (11.4) this establishes the theorem.

Corollary 11.4 *We have*

$$\sum_{d=1}^{\infty}\frac{\mu(d)}{d^2} = \frac{6}{\pi^2}.$$

Proof. It is well known from elementary results in the theory of Fourier series that $\sum_{n=1}^{\infty} 1/n^2 = \pi^2/6$. For instance, it follows by setting $x = 0$ in the result

$$\frac{x^2}{2} = \pi x - \frac{\pi^2}{3} + 2\sum_{n=1}^{\infty}\frac{\cos nx}{n^2},$$

which is valid for x in the range $0 \leqq x \leqq 2\pi$. This result is to be found in Apostol's *Mathematical Analysis.** This with Theorem 11.3 proves the corollary.

Theorem 11.5 *The set of square-free integers has natural density* $6/\pi^2$.

Proof. Let S denote the sequence $1, 2, 3, 5, 6, 7, 10, \cdots$ of square-free integers. For any positive integer n let $p_1, p_2, \cdots, p_r$ denote all the primes

* Tom M. Apostol, *Mathematical Analysis*, Addison-Wesley, 1957, p. 501.

such that $p_j^2 \leqq n$. We first wish to prove

$$S(n) = \sum (-1)^{\alpha_1+\alpha_2+\cdots+\alpha_r}\left[\frac{n}{(p_1^{\alpha_1}p_2^{\alpha_2}\cdots p_r^{\alpha_r})^2}\right] \tag{11.5}$$

where the sum ranges over the 2^r terms obtained by setting each $\alpha_j = 0$ or 1. Now $[n/t^2]$ is the number of integers $\leqq n$ that are divisible by t^2, and we can interpret each term on the right side of (11.5) as a count of those integers $m \leqq n$ that are divisible by $(p_1^{\alpha_1}p_2^{\alpha_2}\cdots p_r^{\alpha_r})^2$.

If m is square-free, $1 \leqq m \leqq n$, then m is counted by the term $[n]$ and by no other terms. If $1 \leqq m \leqq n$ and m is divisible by p_1^2 but by no other p_j^2, then m is counted by the terms $[n]$ and $-[n/p_1^2]$, once positively and once negatively, a net count of $1 - 1 = 0$. To take the general case, consider an integer m, $1 \leqq m \leqq n$, that is divisible by $p_{j_1}^2, p_{j_2}^2, \cdots, p_{j_s}^2$, $s \geqq 1$, but not by any of the other p_j^2. Then m is counted by the terms

$$(-1)^{\alpha_{j1}+\alpha_{j2}+\cdots+\alpha_{js}}\left[\frac{n}{(p_{j_1}^{\alpha_{j1}}p_{j_2}^{\alpha_{j2}}\cdots p_{j_s}^{\alpha_{js}})^2}\right].$$

The net count for this m is thus

$$\sum (-1)^{\alpha_{j_1}+\alpha_{j_2}+\cdots+\alpha_{j_s}} = \sum (-1)^{\alpha_{j_1}} \sum (-1)^{\alpha_{j_2}} \cdots \sum (-1)^{\alpha_{j_s}} = 0$$

since

$$\sum (-1)^{\alpha_j} = 1 + (-1) = 0.$$

This establishes (11.5).

Next we note that (11.5) can be written as

$$S(n) = \sum_{d \mid p_1p_2\cdots p_r} \mu(d)\left[\frac{n}{d^2}\right]. \tag{11.6}$$

In this sum any term for which $d^2 > n$ has the factor $[n/d^2] = 0$, and we can restrict d in (11.6) to be such that $d^2 \leqq n$. In fact, we have

$$(n) = \sum_{d^2 \leqq n} \mu(d)\left[\frac{n}{d^2}\right], \tag{11.7}$$

where d ranges over all positive integers such that $d^2 \leqq n$, since any term in (11.7) that is not in (11.6) will belong to a value of d that is not square-free. In this case the term has the factor $\mu(d) = 0$.

Using Corollary 11.4 and (11.7) we find

$$S(n) - \frac{6}{\pi^2} = \sum_{d^2 \leqq n} \mu(d)\left(\left[\frac{n}{d^2}\right] - \frac{n}{d^2}\right) - \sum_{d^2 > n} \mu(d)\frac{n}{d^2},$$

and hence

$$\left|\frac{S(n)}{n}-\frac{6}{\pi^2}\right| \leqq \frac{1}{n}\sum_{d^2\leqq n}\left|\left[\frac{n}{d^2}\right]-\frac{n}{d^2}\right|+\sum_{d^2>n}\frac{1}{d^2}. \tag{11.8}$$

But

$$\frac{1}{n}\sum_{d^2\leqq n}\left|\left[\frac{n}{d^2}\right]-\frac{n}{d^2}\right| \leqq \frac{1}{n}\sum_{d^2\leqq n}1 \leqq \frac{1}{\sqrt{n}}\to 0 \qquad \text{as } n\to\infty$$

and

$$\sum_{d^2>n}\frac{1}{d^2}\to 0 \qquad \text{as } n\to\infty$$

since $\sum 1/d^2$ converges. Therefore the right side of (11.8) tends to zero, and we have $S(n)/n \to 6/\pi^2$ as $n\to\infty$.

PROBLEMS

1. Find the density of the set of integers divisible by no square >4.
2. Find the density of the set of integers divisible by no square >100.
3. (*a*) In Corollary 11.4 separate the infinite series into two parts S_1 and S_2, where S_1 is the sum over odd values of d, and S_2 over even values. Prove that $S_1 = 8/\pi^2$. (*b*) Find the density of the set of positive integers divisible by no odd square >1.

11.3 Sets of Density Zero

We shall need the following well-known result from the theory of infinite products. For convenience we prove it here.

Lemma 11.6 *Let $\sum c_j$ be a divergent series with $0 < c_j < 1$ for $j = 1, 2, \cdots$. Then, given any real number $\varepsilon > 0$, there is an integer N such that $\prod_{j=1}^{n}(1-c_j) < \varepsilon$ for every integer $n \geqq N$.*

Proof. We note that

$$e^{-c_j} = (1-c_j) + \left(\frac{c_j^2}{2!}-\frac{c_j^3}{3!}\right) + \left(\frac{c_j^4}{4!}-\frac{c_j^5}{5!}\right)\cdots > 1-c_j$$

and hence that

$$\prod_{j=1}^{n}(1-c_j) < \prod_{j=1}^{n} e^{-c_j} = e^{-\sum_{j=1}^{n} c_j}$$

Since $\sum c_j$ diverges we can choose N so that

$$e^{-\Sigma_{j=1}^{N} c_j} < \varepsilon,$$

and the lemma follows.

In this section we shall use one item of special notation. For any set of integers A and any prime p, A_p will denote the set of those elements a of A such that $p \mid a$ but $p^2 \nmid a$.

Theorem 11.7 *If there is a set of primes $\{p_i\}$ such that $\sum p_i^{-1}$ diverges and A_{p_i} has natural density zero for $i = 1, 2, 3, \cdots$, then A has natural density zero.*

Proof. Let I denote the set of all positive integers, let $C^{(r)} = I_{p_1} \cup I_{p_2} \cup \cdots \cup I_{p_r}$, and let $B^{(r)} = \overline{C^{(r)}}$. Then $A \cap I_{p_i} = A_{p_i}$, $A \cap C^{(r)} = A_{p_1} \cup A_{p_2} \cup \cdots \cup A_{p_i}$, and hence

$$A \subset B^{(r)} \cup A_{p_1} \cup A_{p_2} \cup \cdots \cup A_{p_r}. \tag{11.9}$$

Now $B^{(r)}$ consists of all positive integers except those such that $p_j \mid n$, $p_j^2 \nmid n$ for at least one $j = 1, 2, \cdots, r$, and we will prove that

$$B^{(r)}(n) = \sum (-1)^{\alpha_1+\alpha_2+\cdots+\alpha_r} \left[\frac{n}{p_1^{\alpha_1} p_2^{\alpha_2} \cdots p_r^{\alpha_r}}\right] \tag{11.10}$$

where the sum extends over the 3^r terms obtained by taking each $\alpha_i = 0$, 1, or 2. The proof is similar to that of (11.5). Any positive integer $m \leqq n$ can be written as

$$m = p_1^{\beta_1} p_2^{\beta_2} \cdots p_r^{\beta_r} k, \qquad (k, p_1 p_2 \cdots p_r) = 1, \qquad \beta_i \geqq 0.$$

Let $\gamma_i = \beta_i$ if $\beta_i = 0$ or 1, and $\gamma_i = 2$ if $\beta_i \geqq 2$. Then m is counted by the terms on the right side of (11.10) for which $0 \leqq \alpha_i \leqq \gamma_i$, $i = 1, 2, \cdots, r$, and it is counted with the sign $(-1)^{\alpha_1+\alpha_2+\cdots+\alpha_r}$. The net count for m is then

$$\sum_{\alpha_1=0}^{\gamma_1} \sum_{\alpha_2=0}^{\gamma_2} \cdots \sum_{\alpha_r=0}^{\gamma_r} (-1)^{\alpha_1+\alpha_2+\cdots+\alpha_r} = \prod_{i=1}^{r} \left(\sum_{\alpha_i=0}^{\gamma_i} (-1)^{\alpha_i} \right).$$

But

$$\sum_{\alpha_i=0}^{\gamma_i} (-1)^{\alpha_i} = \begin{cases} 1 & = 1 \text{ if } \gamma_i = 0 \\ 1 - 1 & = 0 \text{ if } \gamma_i = 1 \\ 1 - 1 + 1 & = 1 \text{ if } \gamma_i = 2 \end{cases}$$

and we see that m has a count of 0 if any $\gamma_i = 1$ and otherwise has a count of 1. Since $\gamma_i = 1$ if and only if $\beta_i = 1$, the right side of (11.10) counts the $m \leqq n$ that are in $B^{(r)}$, and (11.10) is established.

Removing the greatest integer symbol in (11.10) we get the inequality

$$B^{(r)}(n) \leqq \sum (-1)^{\alpha_1+\alpha_2+\cdots+\alpha_r} \frac{n}{p_1^{\alpha_1} p_2^{\alpha_2} \cdots p_r^{\alpha_r}} + 3^r,$$

hence

$$\text{(11.11)} \qquad \frac{B^{(r)}(n)}{n} \leqq \prod_{i=1}^{r} \left(1 - \frac{1}{p_i} + \frac{1}{p_i^2}\right) + \frac{3^r}{n}.$$

To prove the theorem we must show that for any real $\varepsilon > 0$ there is an N such that $A(n)/n < \varepsilon$ for $n \geqq N$. First we choose r so that

$$\text{(11.12)} \qquad \prod_{i=1}^{r} \left(1 - \frac{1}{p_i} + \frac{1}{p_i^2}\right) < \frac{\varepsilon}{4},$$

which can be done by Lemma 11.6 since $\sum p_i^{-1}$, and hence also $\sum (p_i^{-1} - p_i^{-2})$, diverge. The sets A_{p_i} have natural density zero so we can find an integer N_1 such that

$$\text{(11.13)} \qquad \frac{A_p(n)}{n} < \frac{\varepsilon}{2r}, \qquad i = 1, 2, \cdots, r$$

if $n \geqq N_1$. Taking $N \geqq N_1$, $N \geqq 3^r \cdot 4/\varepsilon$, and using (11.9), (11.11), (11.12), and (11.13) we see that

$$\frac{A(n)}{n} \leqq \frac{B^{(r)}(n)}{n} + \sum_{i=1}^{r} \frac{A_{p_i}(n)}{n} < \frac{\varepsilon}{4} + \frac{\varepsilon}{4} + r\left(\frac{\varepsilon}{2r}\right) = \varepsilon$$

if $n \geqq N$.

Theorem 11.8 *Let k be a fixed positive integer. If each integer in a set A is divisible by k or fewer distinct prime factors, then $\delta(A) = 0$.*

Proof. We let $D^{(k)}$ denote the set of all positive integers having k or fewer distinct prime factors. Then $A \subset D^{(k)}$, $A(n) \leqq D^{(k)}(n)$, and we need only prove the theorem for $D^{(k)}$. The proof is by induction on k. For $k = 1$ the set $D^{(1)}$ consists of all prime powers, $D^{(1)} = \{p^s\}$. We apply Theorem 11.7, taking the p_i to be all primes. The series $\sum p_i^{-1}$ diverges by Theorem 8.2, and $\delta(D_{p_i}^{(1)}) = 0$ since $D_{p_i}^{(1)}$ consists of the single element p_i. Thus $\delta(D^{(1)}) = 0$.

Turning to general k, we assume that the theorem holds in the case $k - 1$. The elements of $D_p^{(k)}$ are the positive integers that are divisible by p, but not by p^2, and that have k or fewer distinct prime factors. If $a \in D_p^{(k)}$, then $a/p \in D^{(k-1)}$. Therefore $D_p^{(k)}(n) \leqq D^{(k-1)}(n/p)$, and hence $\delta(D^{(k-1)}) = 0$ implies $\delta(D_p^{(k)}) = 0$. Now we can apply Theorem 11.7 as before, and it follows that $\delta(D^{(k)}) = 0$.

As another application of Theorem 11.7 we prove the following:

Theorem 11.9 *The set of integers* $\{\phi(m)\}$, $m = 1, 2, 3, \cdots$, *has natural density zero.*

Proof. Denote the set under consideration by A. Given $\varepsilon > 0$, we choose k so that $2^{-k} < \varepsilon/2$ and separate A into two disjoint sets B and C where B consists of those members of A that are divisible by 2^k. Hence $B(n) \leqq 2^{-k}n$ for all n.

Now C consists of the numbers $\phi(m)$ of A that are not divisible by 2^k. We let C^* denote the set of m for which $\phi(m) \in C$. If $q_1, q_2, \cdots, q_r$ are the distinct prime factors of m, then

$$\phi(m) = \frac{m}{q_1 q_2 \cdots q_r}(q_1 - 1)(q_2 - 1)\cdots(q_r - 1),$$

which shows that $2^{r-1} \mid \phi(m)$ since all but one of the q_i must be odd. Therefore, if $m \in C^*$ then $r \leqq k$ and

$$\phi(m) = m \prod_{i=1}^{r}\left(1 - \frac{1}{q_i}\right) \geqq m\left(1 - \frac{1}{2}\right)\left(1 - \frac{1}{3}\right)\cdots\left(1 - \frac{1}{p_k}\right) = mc_k,$$

say. Hence if $\phi(m) \in C$ and $\phi(m) \leqq n$, then $m \leqq n/c_k$ and hence $C(n) \leqq C^*(n/c_k)$. But now the elements of the set C^* have k or fewer distinct prime factors, and so by Theorem 11.8 we see that there is an integer N such that $C^*(m)/m < \varepsilon c_k/2$ for $m \geqq N$. Therefore $C(n) \leqq C^*(n/c_k) < \varepsilon n/2$ if $n \geqq c_k N$, and $B(n) \leqq 2^{-k}n < \varepsilon n/2$. Finally we have $A(n) = B(n) + C(n) < n\varepsilon$ if $n \geqq c_k N$, and this implies $\delta(A) = 0$.

PROBLEMS

1. Let k be a fixed integer. Prove that $\delta(A) = 0$ if every integer in A has the form

$$p_1 p_2 \cdots p_r p_{r+1}^{\alpha_1} p_{r+2}^{\alpha_2} \cdots p_{r+s}^{\alpha_s}$$

where the p_i are any primes, $0 \leqq r \leqq k$, s is arbitrary, and $\alpha_i \geqq 2$ for $i = 1, 2, \cdots, s$.

2. If a sequence of integers $A = \{a_i\}$ has the property that $\sum a_i^{-1}$ converges, prove that $\delta(A) = 0$. Prove that the converse is false.

3. Assuming the proposition that the set of primes $\{q_i\}$ of the form $4n + 3$ has the property that $\sum q_i^{-1}$ diverges, prove that the set of integers each of which is representable as a sum of two squares has density zero.

11.4 Schnirelmann Density and the αβ Theorem

Definition 11.2 *The Schnirelmann density $d(A)$ of a set A of non-negative integers is*

$$d(A) = \inf_{n \geqq 1} \frac{A(n)}{n},$$

where $A(n)$ is the number of positive integers $\leqq n$ in the set A.

Comparing this with Definition 11.1 we immediately see that $0 \leqq d(A) \leqq \delta_1(A) \leqq 1$. Schnirelmann density differs from asymptotic density in that it is sensitive to the first terms in the sequence. Indeed if $1 \notin A$ then $d(A) = 0$, if $2 \notin A$ then $d(A) \leqq \frac{1}{2}$, whereas it is easy to see that $\delta_1(A)$ is unchanged if the numbers 1 or 2 are removed from or adjoined to A. Also, $d(A) = 1$ if and only if A contains all the positive integers.

Until now we have been considering sets A consisting only of positive integers. However, Definition 11.2 is worded in such a way that A can contain 0, but it should be noted that the number 0 is not counted by $A(n)$.

Definition 11.3 *Assume that $0 \in A$ and $0 \in B$. The sum $A + B$ of the sets A and B is the collection of all integers of the form $a + b$ where $a \in A$ and $b \in B$.*

Note that $A \subset A + B$, $B \subset A + B$. As an example let us take S to be the set of squares $0, 1, 4, 9, \cdots$ and I the set of all non-negative integers. Then by Theorem 5.6 we see that $S + S + S + S = I$.

The sum $A + B$ has not been defined unless $0 \in A$ and $0 \in B$. We will assume that 0 is in both A and B in the rest of this chapter. However, the sum could be defined for all A and B as the sum of the sets obtained from A and B by adjoining the number 0 to each. This is equivalent to defining the sum as the collection $\{a, b, a + b\}$ with $a \in A$, $b \in B$.

The result that is proved in the remainder of this section is the $\alpha\beta$ theorem of H. B. Mann, which was conjectured about 1931, with proofs attempted subsequently by many mathematicians. The theorem states that if A and B are sets of non-negative integers, each containing 0, and if α, β, γ are the Schnirelmann densities of A, B, $A + B$, then $\gamma \geqq \min(1, \alpha + \beta)$. In other words $\gamma \geqq \alpha + \beta$ unless $\alpha + \beta > 1$, in which case $\gamma = 1$.

Actually we will prove a somewhat stronger result, Theorem 11.15, from which we will deduce the $\alpha\beta$ theorem. We start by considering any positive integer g and two sets A_1 and B_1 of non-negative integers not exceeding g. We assume throughout that A_1 and B_1 are such sets and that 0 belongs to both A_1 and B_1. Denoting $A_1 + B_1$ by C_1, we observe that C_1 may have elements $>g$ even though A_1 and B_1 do not. We also assume that for some θ, $0 < \theta \leqq 1$,

$$A_1(m) + B_1(m) \geqq \theta m, \qquad m = 1, 2, \cdots, g. \tag{11.14}$$

Our idea is to first replace A_1 and B_1 by two new sets, A_2 and B_2, in such a way that (11.14) holds for A_2 and B_2, that $C_2 = A_2 + B_2 \subset C_1$, and that $B_2(g) < B_1(g)$.

Lemma 11.10 *Let A_1 and B_1 satisfy* (11.14). *If $B_1 \not\subset A_1$, then there exist sets A_2 and B_2 with $C_2 = A_2 + B_2$ such that $C_2 \subset C_1$, $B_2(g) < B_1(g)$ and $A_2(m) + B_2(m) \geqq \theta m$ for $m = 1, 2, \cdots, g$.*

Proof. We merely shift to A_1 all elements of B_1 that are not already in A_1. Define $B' = B_1 \cap \bar{A}_1$, $A_2 = A_1 \cup B'$, $B_2 = B_1 \cap \bar{B}'$, where by $\bar{A}_1$ we mean the complement of A_1, now the set of all non-negative integers not in A_1. Thus 0 belongs to both A_2 and B_2. Then $A_2(m) = A_1(m) + B'(m)$ and $B_2(m) = B_1(m) - B'(m)$ so we have $A_2(m) + B_2(m) = A_1(m) + B_1(m) \geqq \theta m$ for $m = 1, 2, \cdots, g$. Now consider any $h \in C_2$. Then $h = a + b$ with $a \in A_2$ and $b \in B_2$. Noting that B_2 is contained in both A_1 and B_1 and that $A_2 = A_1 \cup B'$, we have either $a \in A_1$ or $a \in B' \subset B_1$. In the first case we can write $h = a + b, a \in A_1, b \in B_1$; in the second case $h = b + a, b \in A_1, a \in B_1$; hence in both cases we have $h \in C_1$. Thus we have $C_2 \subset C_1$. Since it is obvious that $B_2(g) < B_1(g)$, the lemma is proved.

We will get a similar result for the case $B_1 \subset A_1$, but it is a little more complicated. We assume $B_1(g) > 0$, which implies that there is some integer $b > 0$ in B_1. Then if a is the largest integer in A_1, the sum $a + b$ is certainly not in A_1. There may be other pairs $a \in A_1$, $b \in B_1$ such that $a + b \notin A_1$. We let a_0 denote the smallest $a \in A_1$ such that there is a $b \in B_1$ for which $a + b \notin A_1$. Since $B_1 \subset A_1$ we see that $a_0 \neq 0$. Before defining A_2 and B_2 we will obtain two preliminary results.

Lemma 11.11 *Let A_1 and B_1 satisfy $B_1 \subset A_1$ and $B_1(g) > 0$. Let a_0 be defined as above. Suppose that there are integers b and z such that $b \in B_1$ and $z - a_0 < b \leqq z \leqq g$. Then for each $a \in A_1$ such that $1 \leqq a \leqq z - b$, we have $a + b \in A_1$, and*

$$A_1(z) \geqq A_1(b) + A_1(z - b). \tag{11.15}$$

Proof. We have $a \leqq z - b < a_0$ and $a + b \leqq z \leqq g$, hence $a + b \in A_1$ because a_0 is minimal. Now there are $A_1(z - b)$ positive integers a belonging to A_1 with $a \leqq z - b$, and to each such a the corresponding $a + b$ also belongs to A_1. Furthermore each such $a + b$ satisfies $b < a + b \leqq z$, and hence $A_1(z) - A_1(b) \geqq A_1(z - b)$, and we have (11.15).

Lemma 11.12 *Let A_1 and B_1 satisfy* (11.14), $B_1 \subset A_1$, *and* $B_1(g) > 0$. *Define a_0 as before. If there is an integer $y \leqq g$ such that $A_1(y) < \theta y$, then $y > a_0$.*

Proof. Let z be the least integer such that $A_1(z) < \theta z$. Then $y \geqq z \geqq 1$. Since $A_1(z) + B_1(z) \geqq \theta z$ we have $B_1(z) > 0$, and hence there is a $b \in B_1$ such that $0 < b \leqq z \leqq g$. If $z \leqq a_0$, we would have $z - a_0 < b \leqq z \leqq g$, and we could apply Lemma 11.11 to get $A_1(z) \geqq A_1(b) + A_1(z - b)$. Now $b \in B_1 \subset A_1$, so we have $A_1(b) = A_1(b - 1) + 1 \geqq \theta(b - 1) + 1$ since $b - 1 < z$. Also, $A_1(z - b) \geqq \theta(z - b)$, and we are led to the contradiction $A_1(z) \geqq \theta(b - 1) + 1 + \theta(z - b) = \theta(z - 1) + 1 \geqq \theta z$. Therefore we have $z > a_0$, and hence $y > a_0$.

Lemma 11.13 *Let A_1 and B_1 satisfy $B_1 \subset A_1$ and $B_1(g) > 0$. Let B' denote the set of all $b \in B_1$ such that $a_0 + b \notin A_1$, and let A' denote the set of all integers $a_0 + b$ such that $b \in B'$ and $a_0 + b \leqq g$. Finally let $A_2 = A_1 \cup A'$ and $B_2 = B_1 \cap \overline{B}'$. Then $C_2 \subset C_1$ and $B_2(g) < B_1(g)$.*

Proof. Note that $0 \in A_2$ and $0 \in B_2$, so that the sum C_2 is well defined. If $h \in C_2$, then $h = a + b$, $a \in A_1 \cup A'$, $b \in B_1 \cap \overline{B}'$. If $a \in A_1$, then $h = a + b \in C_1$ since $a \in A_1, b \in B_1$. If $a \in A'$, then $a = a_0 + b_1$ for some $b_1 \in B'$, and we have $h = a_0 + b + b_1$. Here $a_0 + b \in A_1$ since otherwise we would have $b \in B'$. Since $b_1 \in B_1$, we again have $h \in C_1$. Finally $B_2(g) < B_1(g)$, since the definition of a_0 ensures that $B'(g) > 0$.

Lemma 11.14 *For A_1, B_1, A_2, B_2 as in Lemma* 11.13, *if A_1, B_1 satisfy* (11.14) *then*

$$A_2(m) + B_2(m) \geqq \theta m \quad \textit{for} \quad m = 1, 2, \cdots, g. \tag{11.16}$$

Proof. From the way A', B', A_2, B_2 were chosen we have

$$\begin{aligned} A_2(m) &= A_1(m) + A'(m), \\ B_2(m) &= B_1(m) - B'(m), \\ A'(m) &= B'(m - a_0), \\ A_2(m) + B_2(m) &= A_1(m) + B_1(m) - (B'(m) - B'(m - a_0)), \end{aligned}$$

for $m = 1, 2, \cdots, g$. Therefore (11.16) holds for all m for which $B'(m) = B'(m - a_0)$. Consider any $m \leqq g$ for which $B'(m) > B'(m - a_0)$. Then $B_1(m) - B_1(m - a_0) \geqq B'(m) - B'(m - a_0) > 0$, and we let b_0 denote the smallest element of B_1 such that $m - a_0 < b_0 \leqq m$. Therefore

$$\begin{aligned} A_2(m) + B_2(m) &\geqq A_1(m) + B_1(m) - (B_1(m) - B_1(m - a_0)) \\ &= A_1(m) + B_1(m - a_0) \\ &= A_1(m) + B_1(b_0 - 1). \end{aligned} \tag{11.17}$$

Now $m - a_0 < b_0 \leqq m \leqq g$, so we can apply Lemma 11.11 with $b = b_0$ and $z = m$ to get

$$A_1(m) \geqq A_1(b_0) + A_1(m - b_0).$$

We also have $m - b_0 < a_0$ so Lemma 11.12 shows that

$$A_1(m - b_0) \geqq \theta(m - b_0).$$

Thus we can reduce (11.17) to

$$A_2(m) + B_2(m) \geqq A_1(b_0) + \theta(m - b_0) + B_1(b_0 - 1).$$

But $b_0 \in B_1 \subset A_1$ so we have $A_1(b_0) = A_1(b_0 - 1) + 1$. Using this and (11.14) we have,

$$\begin{aligned} A_2(m) + B_2(m) &\geqq A_1(b_0 - 1) + B_1(b_0 - 1) + 1 + \theta(m - b_0) \\ &\geqq \theta(b_0 - 1) + 1 + \theta(m - b_0) \\ &\geqq \theta m. \end{aligned}$$

Theorem 11.15 *For any positive integer g let A_1 and B_1 denote fixed sets of non-negative integers $\leqq g$. Let 0 belong to both sets A_1 and B_1, and write C_1 for $A_1 + B_1$. If for some θ such that $0 < \theta \leqq 1$,*

$$A_1(m) + B_1(m) \geqq \theta m, \qquad m = 1, 2, \cdots, g,$$

then $C_1(g) \geqq \theta g$.

Proof. If $B_1(g) = 0$, then B_1 consists of the single integer 0, $C_1 = A_1$, and $C_1(g) = A_1(g) = A_1(g) + B_1(g) \geqq \theta g$. We prove the theorem for general sets by mathematical induction. Suppose $k \geqq 1$ and that the theorem is true for all A_1, B_1 with $B_1(g) < k$. If $A_1(m) + B_1(m) \geqq \theta m$ for $m = 1, 2, \cdots, g$, and if $B_1(g) = k$, then Lemma 11.10 or Lemmas 11.13 and 11.14 supply us with sets A_2, B_2 such that $B_2(g) < k$, $C_2 \subset C_1$, and $A_2(m) + B_2(m) \geqq \theta m$ for $m = 1, 2, \cdots, g$. Therefore, by our induction hypothesis, we have $C_2(g) \geqq \theta g$, which implies $C_1(g) \geqq \theta g$.

Theorem 11.16 *The $\alpha\beta$ theorem. Let A and B be any sets of non-negative integers, each containing 0, and let α, β, γ denote the Schnirelmann densities of A, B, $A + B$ respectively. Then $\gamma \geqq \min(1, \alpha + \beta)$.*

Proof. Let A_1 and B_1 consist of the elements of A and B, respectively, that do not exceed g, an arbitrary positive integer. Then $A_1(m) \geqq \alpha m$ and $B_1(m) \geqq \beta m$ for $m = 1, 2, \cdots, g$. If we take $\theta = \min(1, \alpha + \beta)$, the conditions of Theorem 11.15 are satisfied and we conclude that $C_1(g) \geqq \theta g$. Since $C_1(g) \geqq \theta g$ for every positive integer g, we have $\gamma \geqq \theta = \min(1, \alpha + \beta)$.

PROBLEMS

1. What is the Schnirelmann density of the set of positive odd integers? The set of positive even integers? The set of positive integers $\equiv 1 \pmod 3$? The set of positive integers $\equiv 1 \pmod m$?
2. Prove that the analogue of Theorem 11.1 for Schnirelmann density, namely, $d(A) = \inf n/a_n$, is false.
3. Prove that the analogue of Theorem 11.16 for asymptotic density is false. *Suggestion:* take A as the set of all positive even integers, and consider $A + A$.
4. Prove that if $d(A) = \alpha$, then $A(n) \geqq \alpha n$ for every positive integer n. Prove that the analogue of this for asymptotic density is false.
5. Establish that Theorem 11.16 does not imply Theorem 11.15 by considering the sets $A = \{0, 1, 2, 4, 6, 8, 10, \cdots\}$, $B = \{0, 2, 4, 6, 8, 10, \cdots\}$. Theorem 11.16 asserts that the density of $A + B$ is $\geqq \frac{1}{2}$, whereas Theorem 11.15 says much more.
6. Exhibit two sets A and B such that $d(A) = d(B) = 0$, $d(A + B) = 1$.
7. For any two sets A and B of non-negative integers, write $\alpha = d(A)$, $\beta = d(B)$, $\gamma = d(A + B)$. Prove that $\gamma \geqq \alpha + \beta - \alpha\beta$.
8. Consider a set A with positive Schnirelmann density. Prove that for some positive integer n

$$nA = (n + 1)A = (n + 2)A = \cdots = I,$$

where I is the set of all non-negative integers, and $nA = A + A + \cdots + A$ with n summands.

NOTES ON CHAPTER 11

For further reading on the subject of this chapter, see the book by Halberstam and Roth listed in the General References on page 269 and the following:

E. Artin and P. Scherk, "On the sums of two sets of integers," *Ann. Math.* (2) **44**, 138–142 (1943).

F. J. Dyson, "A theorem on the densities of sets of integers," *J. London Math. Soc.* **20**, 8–14 (1945).

H. B. Mann, "A proof of the fundamental theorem on the density of sums of sets of positive integers," *Ann. Math.* (2) **43**, 523–527 (1942).

Miscellaneous Problems

Several of these problems are not easy, and so can be regarded by the reader as "research" problems rather than routine exercises.

1. Let L_n denote the least common multiple $[1, 2, \cdots, n]$ of the first n positive integers. Prove that $\sum L_n^{-1}$ converges, where the sum is taken over all positive integers n.

2. Given any integers a, b, c, and any prime p not a divisor of ab, prove that $ax^2 + by^2 \equiv c \pmod{p}$ is solvable.

3. If g is a divisor of each of ab, cd, and $ac + bd$, prove that it also divides ac and bd, where a, b, c, d are integers.

4. Let p and q be *twin primes*, i.e., primes satisfying $q = p + 2$. Prove that there is an integer a such that $p \mid (a^2 - q)$ if and only if there is an integer b such that $q \mid (b^2 - p)$. (There is a famous unsolved problem to prove that the number of pairs of twin primes is infinite. What *is* known is that the sum of the reciprocals of all twin primes is, if not a finite sum, certainly a convergent series; this result can be contrasted with Theorem 8.3. A proof of this result can be found in Chapter 15 of the book by Hans Rademacher listed in the General References on page 269.)

5. Prove that $x^2 \equiv 1 \pmod{p^k}$ has exactly 2 solutions if p is an odd prime and k is a positive integer, but exactly 4 solutions if $p = 2$ and $k \geqq 3$. Hence prove that for any integer $m > 1$ the solutions of $x^2 \equiv 1 \pmod{m}$ form a multiplicative group, and determine its order.

6. For any positive integers a, b, and n, prove that if n is a divisor of $a^n - b^n$ then n is a divisor of $(a^n - b^n)/(a - b)$. *Suggestion:* Write g for g.c.d. $(a - b, n)$ and prove that $g^2 \mid (a^n - b^n)$.

7. For any positive integers n and k prove that $\sum_{j=n}^{n+k} j^{-1}$ is not an integer. (This is a generalization of Problem 31 of Section 1.3.)

8. Find all positive integers n such that $\phi(n) \mid n$.

9. For any positive integer n prove that $\phi(n) + \sigma(n) \geqq 2n$, with equality if and only if $n = 1$ or n is a prime.

10. Let $\rho(n)$ denote the number of positive integers $a \leqq n$ such that $a^x \equiv a \pmod{n}$ for some integer $x > 1$. Prove that $\rho(n)$ is a multiplicative function. (This function is akin to the Euler ϕ-function which can be defined in an analogous way but with the congruence replaced by $a^x \equiv 1 \pmod{n}$.)

11. Prove that an odd integer $n \geqq 1$ is a prime if and only if it is not expressible as a sum of three or more consecutive positive integers.

12. Let $q = 4^n + 1$ where n is a positive integer. Prove that q is a prime if and only if $3^{(q-1)/2} \equiv -1 \pmod{q}$.

13. Prove that n is a prime if and only if all the coefficients of $(x + y)^n$ except the first and last are divisible by n.

14. Let d be the greatest common divisor of the coefficients of $(x + y)^n$ except the first and last, where n is any positive integer >1. Prove that $d = p$ if n is a power of a prime p, and that $d = 1$ otherwise.

15. Prove that n is a divisor of the binomial coefficient $\binom{n}{k}$ if $(n, k) = 1$.

16. If a positive integer n is written to base b prove that the digit in the jth position, counting from the right hand end of the digital formulation, is $[n/b^{j-1}] - b[n/b^j]$.

17. Given any prime p and positive integers m and r with $m > r$, say that the triple p, m, r has property Q if the following condition holds: if r and $m - r$ are written to base p there is at least one digit a in the expression for r such that if b is the corresponding digit in $m - r$ then $a + b \geqq p$. Prove that p is a divisor of the binomial coefficient $\binom{m}{r}$ if and only if Q holds. (The property Q amounts to saying that if r and $m - r$, written to base p, are added as in elementary arithmetic, there is at least one "carry" in the addition.)

18. Evaluate the integral $\int_0^1\int_0^1\int_0^1 [x + y + z]\, dx\, dy\, dz$ where the square brackets denote the greatest integer function. Generalize to n-dimensions, with an n-fold integral.

19. Prove that of the two equations

$$[\sqrt{n} + \sqrt{n+1}] = [\sqrt{n} + \sqrt{n+2}],$$

$$[\sqrt[3]{n} + \sqrt[3]{n+1}] = [\sqrt[3]{n} + \sqrt[3]{n+2}],$$

the first holds for every positive integer n, but the second does not, where $[x]$ is the greatest integer notation.

20. Let j and k be positive integers. Prove that

$$[(j+k)\alpha] + [(j+k)\beta] \geqq [j\alpha] + [j\beta] + [k\alpha + k\beta]$$

for all real numbers α and β if and only if $j = k$. *Suggestion:* It suffices to take $0 \leqq \alpha < 1$, $0 \leqq \beta < 1$, and g.c.d. $(j, k) = 1$. (This generalizes Problem 16 of §4.1.)

21. Prove that among any ten consecutive positive integers at least one is relatively prime to the product of the other nine. (*Remark:* if the "ten" in this statement is replaced by "n," the result is true for every positive integer $n \leqq 16$, but false for $n > 16$. This is not easy to prove; cf. Ronald J. Evans, *Amer. Math. Monthly*, **76**, 48–49 (1969).)

22. Prove that the sum of the first n natural numbers is a perfect square for infinitely many values of n.

23. Prove that $2n^2 + 2n + 1$ is a perfect square for infinitely many integer values n.

24. If any $[n/2] + 1$ integers are selected from $1, 2, 3, \cdots, n$, prove that two of them are relatively prime. Establish that if only $[n/2]$ integers are selected, it does not follow that two of them must be relatively prime.

25. Given any set of n positive integers, prove that there is a non-empty subset whose sum is a multiple of n.

26. Let n and k be positive integers with $k < n$ and $(k, n) = 1$. Prove that if k distinct integers are selected at random from $1, 2, \cdots, n$, the probability that their sum is divisible by n is $1/n$.

27. Let $f(x)$ and $g(x)$ be polynomials in $Z[x]$, i.e., polynomials with integral coefficients. Suppose that $g(m) \mid f(m)$ for infinitely many positive integers m. Prove that $g(x) \mid f(x)$ in $Q[x]$, that is, there exists a quotient polynomial $q(x)$ with rational coefficients such that $f(x) = g(x) \cdot q(x)$. *Suggestion:* After applying Theorem 9.1 to get polynomials $q(x)$ and $r(x)$ in $Q[x]$, multiply by a suitable positive integer k so that $kq(x)$ and $kr(x)$ have integer coefficients, and use the fact that $g(m) > kr(m)$ for sufficiently large integers m. (*Remark:* The example $g(x) = 2x + 2$, $f(x) = x^2 - 1$ with m odd shows that $q(x)$ need not have integral coefficients.)

28. Let $f(x)$ and $g(x)$ be primitive nonconstant polynomials in $Z[x]$ such that the greatest common divisor $(f(m), g(m)) > 1$ for infinitely many positive integers m. Construct an example to show that such polynomials exist with g.c.d. $(f(x), g(x)) = 1$ in the polynomial sense.

29. For any odd prime p except 3 define $f(x) = (x - 1)(x - 2) \cdots (x - p + 1)$. Prove that $p^2 \mid f'(0)$. An alternative formulation of this problem is that if $1 + \frac{1}{2} + \frac{1}{3} + \cdots + \dfrac{1}{p - 1}$ is added to give the rational fraction a/b, then $p^2 \mid a$.

30. Given any nonconstant polynomial $f(x)$ with integer coefficients, prove that there are infinitely many primes p such that $f(x) \equiv 0 \pmod{p}$ is solvable. *Suggestion:* If there were only finitely many such primes p, let P be their product, define $x_0 = P^n f(0)$, and examine $f(x_0)$ with n large.

31. If $A = \{a_1, a_2, a_3, \cdots\}$ is an increasing sequence of positive integers with positive natural density, prove that $\lim (a_n - a_{n-1})/a_n = 0$ as n tends to infinity.

32. Let $B_1, B_2, B_3, \cdots$ be an infinite sequence of disjoint sets of positive integers, with asymptotic densities $\beta_1, \beta_2, \beta_3, \cdots$ respectively. Prove that $\sum \beta_j$ converges, and that $\delta_1(B) \geqq \sum \beta_j$ where B is the set $B_1 \cup B_2 \cup B_3 \cup \cdots$. (This is a generalization of Problem 10 in Section 11.1.)

33. Prove that $\tau^2 * \mu = \mu^2 * \tau$, where μ is the Moebius function, τ is the number of divisors function, and the equation contains both kinds of multiplication of arithmetic functions given in Section 4.4, so that $\tau^2(n)$ means $\tau(n) \cdot \tau(n)$. *Suggestion:* The product of multiplicative functions is multiplicative for both ordinary products and Dirichlet products. Cf. Problem E 2235, *Amer. Math. Monthly*, **78,** 406, 407 (1971) for the source.

34. Let n be a given positive integer. What is the number of elements in a minimal set S of distinct integers having the property that every residue class modulo n occurs at least once among the sums of the non-empty subsets of S? (For example, if $n = 6$ the answer is 3 because $\{1, 3, 5\}$ is a minimal set, with every residue class modulo 6 occurring among $1, 3, 5, 1 + 3, 1 + 5, 3 + 5, 1 + 3 + 5$.)

35. Prove that no n points with rational coordinates (x, y) can be chosen in the Euclidean plane to form the vertices of a regular polygon with n sides, except in the case $n = 4$. *Suggestions:* If $n = 3$, the area of such a triangle can be shown to be rational by the use of one standard elementary formula, but irrational by another. For values of n other than 3, 4, or 6, a similar contradiction can be obtained by applying the law of cosines to a triangle formed by two adjacent vertices and the center of the polygon.

Special Topics

With the miscellaneous topics below, only outlines of proofs are given but in sufficient detail so that the reader should find no great difficulty filling in the gaps. Thus these sketches may be regarded as elaborate problems arising from extensions of the theory, with detailed suggestions for their solution.

Periodic Decimals

We now sketch the proofs of two basic results about periodic or repeating decimals to base ten, namely decimals of the sort $0.06272727\cdots$, which is written $0.06\overline{27}$ where the bar indicates that the digital pair 27 is repeated indefinitely. The least number of digits in the repeating block is called the period of the repeating decimal, where the word "least" must be included so as to avoid writing $0.06\overline{27}$ in the form $0.06\overline{2727}$, and so interpreting the period here as 4 when of course it is 2. A terminating decimal is one with a finite number of nonzero digits following the decimal point, such as 0.25. Any terminating decimal can be written as an infinite periodic decimal with an endless succession of nines, as for example $0.25 = 0.24\overline{9}$.

The first basic result is that *any rational number is expressible as a terminating decimal or an infinite periodic decimal, and conversely*. The proof is straightforward. First, given any rational number a/b, the decimal expansion is obtained by dividing b into a, with a continuing division to create the decimal expansion. If the remainder is zero at some stage, the decimal expansion terminates; if not, the remainder must repeat, since the possible remainders are $1, 2, \cdots, b-1$, and so the periodic decimal is formed. Conversely, any terminating decimal is obviously expressible as a rational number. Any infinite periodic decimal can be written, if we omit the integer part, as

$$.\,a_1a_2a_3\cdots a_r\overline{b_1b_2\cdots b_s}. \tag{1}$$

Denoting this number by β we note that $10^{r+s}\beta - 10^r\beta$ is an integer, and the result follows.

It will be convenient to write every decimal expansion (1) with r as the minimal length of the nonrepeating part and s the minimal length of the repeating part. Thus for $0.26\overline{3}$ the values of r and s are 2 and 1, not 3 and 1 as suggested by $0.263\overline{3}$, and not 3 and 2 as suggested by $0.263\overline{33}$.

The second result relates the length of the period to the rational number. We confine attention to rational numbers between 0 and 1, because the extension to all other cases is apparent; for example the decimal expansion of 1/6 and 13/6 are the same to the right of the decimal point. Furthermore we set aside all rational numbers c/d with denominators d divisible by no primes other than 2 and 5; such rational numbers have terminating decimal expansions, and are the only rational numbers (in lowest terms) whose decimal expansions terminate. The result is as follows. *Let c/d be a rational number with $0 < c < d$, $(c, d) = 1$, and such that some prime $p \mid d$ with $p \neq 2$, $p \neq 5$. If c/d has the decimal expansion* (1), *where r and s are the minimal lengths of the nonrepeating and the repeating parts, then r and s are the least integers such that*

$$d \mid 10^r(10^s - 1), \qquad r \geqq 0, \qquad s \geqq 1. \tag{2}$$

Another way of stating this is to write $d = 2^\alpha 5^\beta k$ where $(k, 10) = 1$, with the conclusion that $r = \max(\alpha, \beta)$, and s is the exponent to which 10 belongs modulo k. To prove this result we note that $(10^{r+s} - 10^r)c/d$ is an integer because of (1), and so d is a divisor of $10^r(10^s - 1)$. To prove that r and s are the least values satisfying (2), suppose that $d \mid 10^j(10^t - 1)$ with $0 \leqq j \leqq r$ and $1 \leqq t \leqq s$. Then it must be established that $j = r$ and $t = s$. Define q as $10^j(10^t - 1)/d$ so that

$$10^j(10^t - 1)c/d = cq$$

is an integer. Thus $10^{j+t}(c/d)$ and $10^j(c/d)$ have the same decimal part Substituting the decimal expansion (1) for c/d here, we conclude that $j = r$ and $t = s$ because r was the minimal length of the nonrepeating part and s is the minimal length of the repeating part of the decimal expansion (1).

Unit Fractions

A *unit fraction*, or *Egyptian fraction*, is one of the form $1/n$, where n is a positive integer. *Every positive rational number is expressible as a sum of a finite number of distinct unit fractions.* We now sketch a proof of this result, beginning with rationals <1. Consider the rational number a/b with $0 < a < b$ and $(a, b) = 1$. To prove that a/b is expressible as a sum of distinct unit fractions we use induction on a. If $a = 1$ the result is obvious. Next suppose the proposition holds for all fractions with numerator $a - 1$ or less. Now dividing $b + a$ by a to get a quotient q and a remainder r, we note that $r \neq 0$ because $(a, b) = 1$. Hence $b + a = aq + r$ with $0 < r < a$, and $a/b = 1/q + (a - r)/bq$. The last fraction has positive numerator $<a$, and so can be expressed as a sum of distinct unit fractions, each with denominator larger than q because $(a - r)/bq < 1/q$.

Next if a/b is any rational >1, define the integer n by

$$1 + \frac{1}{2} + \frac{1}{3} + \cdots + \frac{1}{n} \leqq \frac{a}{b} < 1 + \frac{1}{2} + \frac{1}{3} + \cdots + \frac{1}{n} + \frac{1}{n+1},$$

so that

$$0 \leqq \frac{c}{d} < \frac{1}{n+1} \quad \text{where} \quad \frac{c}{d} = \frac{a}{b} - 1 - \frac{1}{2} - \frac{1}{3} - \cdots - \frac{1}{n}.$$

By the first part of the proof c/d, if not 0, is expressible as a sum of distinct unit fractions each with denominator $>n + 1$.

The Equation $\mathbf{x^{-n} + y^{-n} = z^{-n}}$

For any positive integer n the equation $x^{-n} + y^{-n} = z^{-n}$ has a solution in positive integers if and only if $x^n + y^n = z^n$ does. This is easy to establish because the latter equation is equivalent to $(yz)^{-n} + (xz)^{-n} = (xy)^{-n}$, and the former is equivalent to $(yz)^n + (xz)^n = (xy)^n$.

For $n = 1$ and 2 it is not difficult to characterize all solutions in positive integers. If a solution has $(x, y, z) > 1$ we can remove the greatest common divisor to get a primitive solution, i.e., one with $(x, y, z) = 1$. This condition is now assumed, and we begin with $x^{-1} + y^{-1} = z^{-1}$ by noting that it is equivalent to $yz + xz = xy$. From this it follows that $z \mid xy$, $y \mid xz$, and $x \mid yz$. Write $a = (x, z)$, $b = (y, z)$, and $c = (x, y)$ and it follows that $(a, b) = (b, c) = (c, a) = 1$.

Also, if p is a prime that divides x and z, then $p \mid a$ but $p \nmid y$. Then the divisibility conditions $x \mid yz$ and $z \mid xy$ imply that the same exact power of p that divides x also divides z. This holds for the other pairs x, y and z, y, and it follows that $x = ac$, $y = bc$, $z = ab$. Substituting these in $x^{-1} + y^{-1} = z^{-1}$ we arrive at $a + b = c$ and so $x = a(a + b)$, $y = b(a + b)$, $z = ab$ is the solution of the equation.

Every solution in positive integers of $x^{-1} + y^{-1} = z^{-1}$ *with* $(x, y, z) = 1$ *is given by taking arbitrary positive integers* a, b *with* $(a, b) = 1$ *and writing* $x = a^2 + ab$, $y = b^2 + ab$, $z = ab$.

As an immediate consequence of this we note that for all solutions of $x^{-1} + y^{-1} = z^{-1}$ in positive integers, the result

$$(x, z) + (y, z) = (x, y) > 1$$

holds.

Turning to the case $x^{-2} + y^{-2} = z^{-2}$, it follows from the results above that any primitive solution of this equation in positive integers has $z^2 = ab$, $x^2 = a(a + b)$, $y^2 = b(a + b)$ for some positive integers a and b with $(a, b) = 1$. Hence a, b, and $a + b$ are perfect squares, say $a = c^2$, $b = d^2$ with $(c, d) = 1$. Also $c^2 + d^2$ is a perfect square, so by Theorem 5.1 there are integers r and s such that

$$(*) \quad c = r^2 - s^2, \qquad d = 2rs, \qquad r > s > 0,$$

$$(r, s) = 1, \qquad r + s \equiv 1 \pmod 2.$$

By assigning d the value $2rs$ we are taking, in effect, y to be the even value and x the odd, it being clear that x and y must have opposite parity.

Every solution in positive integers of $x^{-2} + y^{-2} = z^{-2}$ *with* $(x, y, z) = 1$ *and* y *even is given by*

$$x = r^4 - s^4, \qquad y = 2rs(r^2 + s^2), \qquad z = 2rs(r^2 - s^2)$$

where r *and* s *are arbitrary integers subject only to the conditions imposed in* $(*)$ *above.*

Gauss' Generalization of Fermat's Theorem

(i) First establish that if p is any prime and a is any integer not divisible by p, then $a^{mp} - a^m$ is divisible by p^k if m is divisible by p^{k-1}. (ii) Hence prove the result of Gauss that if a and n are any positive integers, then n is a divisor of

$$\sum_{d \mid n} a^d \mu(n/d).$$

If n is a prime, this is the theorem of Fermat. *Suggestions:* Prove the property separately for each prime factor p^k of n. If p does not divide a use the property (i), noting that the sum in (ii) can be separated into a collection of pairs of terms as in (i).

A Primitive Root mod p by Group Theory

The existence of a primitive root modulo p, a prime, can be approached by group theory, to give an alternative proof of the first assertion in Theorem 2.25. By showing the existence of one primitive root g, we can conclude without difficulty that g^r is a primitive root for integral values of r from 1 to $p - 1$ if and only if $(r, p - 1) = 1$, and hence it follows that there are $\phi(p - 1)$ primitive roots modulo p.

The group theoretic language for the existence of a primitive root is that there is an element g of order $p - 1$ in the multiplicative group modulo p. This can also be stated in the form that the multiplicative group modulo p is cyclic, and this can be proved as follows.

(i) Denote by ord a the order of any element a of a finite abelian group. If $a^r = e$ for some integer r, where e is the identity element of the group then r is divisible by ord a. (ii) If ord $a = h$ and $d \mid h$ then ord $(a^d) = h/d$. (iii) If ord $a = h$, ord $b = k$ and $(h, k) = 1$ then ord $(ab) = hk$. (iv) Let h be the maximum order of any element of a finite abelian group, say ord $b = h$. Then if a is any element, $a^h = e$ because h is divisible by ord a by the following argument. Write k for ord a. If $k \nmid h$ then there is a prime p such that $p^\alpha \parallel k$, $p^\beta \parallel h$, and $\alpha > \beta \geqq 0$. Write r for k/p^α. Then ord $a^r = p^\alpha$ and ord $b^{p^\beta} = h/p^\beta$. These orders are relatively prime, so by (iii) we see that ord $(a^r b^{p^\beta}) = hp^{\alpha-\beta} > h$, thus contradicting the maximal property of h.

(v) Consider now the multiplicative group $\{1, 2, \cdots, p - 1\}$ modulo p. If h denotes the maximum order of any element then $x^h \equiv 1 \pmod{p}$ for every element in the group, by (iv). But by Theorem 2.20 this congruence, having $p - 1$ solutions, must have degree at least $p - 1$, so $h \geqq p - 1$. Hence $h = p - 1$ because $x^{p-1} \equiv 1 \pmod{p}$ for every element x in the group by Fermat's theorem.

The Group of Rational Points on the Unit Circle

Let x and y be rational numbers satisfying $x^2 + y^2 = 1$, so that (x, y) is a rational point on the unit circle. (Thus (x, y) here is a pair of coordinates, not the greatest common divisor symbol.) (i) Prove that these points form a group G under the multiplication

$$(x_1, y_1) \cdot (x_2, y_2) = (x_1x_2 - y_1y_2, x_1y_2 + x_2y_1),$$

where this multiplication has been suggested by the product of the complex numbers $x_1 + iy_1$ and $x_2 + iy_2$. (ii) Prove that if $0 \leqq y_1 < y_2 \leqq 1$, then a rational number $r = u/v$ can be chosen so that

$$y_1 < \frac{u^2 - v^2}{u^2 + v^2} = \frac{r^2 - 1}{r^2 + 1} < y_2.$$

(iii) Prove that the elements (x, y) in the group G are dense on the unit circle (meaning that between any two points of the group there lies another) by taking $x = 2uv/(u^2 + v^2)$, $y = (u^2 - v^2)/(u^2 + v^2)$ in part (ii) above.

The Day of the Week from the Date

Any date, such as December 25, 1984, can be separated into integer parts N, M, C, Y as follows. Let N be the number of the day in the month, so $N = 25$. Let M be the number of the month *beginning with March*, so that $M = 10$ for December, $M = 1$ for March, $M = 11$ for January, $M = 12$ for February. (The peculiarity of starting with March is needed because the extra leap year day is added at the end of February.) Let C be the hundreds in the year, and Y the rest, so that $C = 19$ and $Y = 84$ for 1984. If d denotes the day of the week, where $d = 0$ for Sunday, $d = 1$ for Monday, $\cdots$, $d = 6$ for Saturday, then

$$d \equiv N + [2.6M - 0.2] + Y + [Y/4] + [C/4] - 2C - (1 + L)[M/11] \pmod 7,$$

where $L = 1$ for a leap year and $L = 0$ for a nonleap year. For example, in the case of December 25, 1984 we have $L = 1$ and so

$$d \equiv 25 + [25.8] + 84 + [21] + [19/4] - 38 - 2[10/11] \equiv 2 \pmod 7,$$

so Christmas day in 1984 falls on a Tuesday.

This formula holds for any date after 1582, following the adoption of the Gregorian calendar at that time. The leap years are those divisible by 4, except the years divisible by 100 which are leap years only if divisible by 400. For example, 1984, 2000, 2004, 2400 are leap years, but 1900, 1901, 2100, 2401 are not.

Verify the correctness of the formula by establishing (i) that if it is correct for any date then it is also correct for the date of the next succeeding day, and (ii) that it *is* correct for one particular day selected from the current calendar.

Some Number Theoretic Determinants

The purpose here is to establish some results about determinants of order n such that the element in the i, j position (intersection of ith row and jth column) is some expression related to the g.c.d. (i, j). For example, in Theorem 2 below we evaluate such a determinant with (i, j) itself in the i, j position. For brevity of notation we shall write the functional notation $f(i, j)$ in place of the technically correct $f((i, j))$ to denote the value of the function $f(x)$ for $x =$ g.c.d. (i, j). We state a succession of results with sketches of proofs.

Theorem 1. *For any arithmetic function $f(m)$, let $g(m)$ be defined for all positive integers m by*

$$g(m) = \sum_{d|m} \mu(d) f(m/d), \qquad f(m) = \sum_{d|m} g(d), \tag{1}$$

these two equations being equivalent by Theorems 4.7 *and* 4.8. *Let M be the square matrix of order n with the element $f(i, j)$ in the ith row, jth column position. Then the determinant of M is given by*

$$\det M = \prod_{j=1}^{n} g(j). \tag{2}$$

Proof. Define $a_{ij} = 1$ if $j \mid i$, and $a_{ij} = 0$ otherwise. Let A be the square matrix of order n with a_{ij} in the i, j position, and B the matrix of order n with $g(j)a_{ij}$ in the i-j position. Denoting the transpose of A by A^t, we see that BA^t has the element in the i, j position

$$\sum_{k=1}^{n} g(k) a_{ik} a_{jk} = \sum_{k|(i,j)} g(k) = f(i, j),$$

the last two formulations being readily calculated. Also by evaluating the determinants we get

$$\det A = 1, \qquad \det B = \prod_{j=1}^{n} g(j),$$

and the theorem follows because $M = BA^t$ implies $\det M = \det B \cdot \det A$.

Corollary. *If f is totally multiplicative and $f(p) \neq 0$ for every prime p, then*

$$\det M = \prod_{j=1}^{n} f(j) \prod_{p|j} \left\{1 - \frac{1}{f(p)}\right\}.$$

Proof. We note that

$$g(j) = f(j) \sum_{d|j} \frac{\mu(d)}{f(d)} = f(j) \prod_{p|j} \left\{1 - \frac{1}{f(p)}\right\}.$$

Theorem 2. *Let M_1 be the matrix of order n with the element* g.c.d. (i, j) *in the i,j position. Then*

$$\det M_1 = \phi(1)\phi(2)\phi(3)\cdots\phi(n).$$

This follows from the Corollary to Theorem 1 by taking $f(m) = m$.

Theorem 3. *Let M_2 be the matrix of order n with the element* l.c.m. $[i, j]$ *in the i,j position. Then*

$$\det M_2 = \prod_{k=1}^{n} \phi(k) \prod_{p|k} (-p).$$

Proof. Define $f(m) = 1/m$, and note that

$$g(m) = \frac{1}{m}\prod_{p|m}(1 - p) = \frac{\phi(m)}{m^2}\prod_{p|m}(-p).$$

Then use the Corollary to Theorem 1, and $1/(i, j) = [i, j]/ij$, and the fact that a common multiplier of the elements of a row or column of a determinant can be factored out.

Theorem 4. *Let M_3 be the matrix of order n with $\tau(i, j)$ in the i,j position, where $\tau(m)$ is the number of positive divisors of m. Then* $\det M_3 = 1$.

This follows from Theorem 1 with $g(m) = 1$.

Theorem 5. *Let M_4 be the matrix of order n with $\sigma(i, j)$ in the i,j position, where σ is the sum of divisors function. Then* $\det M_4 = n!$.

This follows from Theorem 1 with $g(m) = m$.

Theorem 6. *Let M_5 be the matrix of order n with $\mu(i, j)$ in the i,j position. Then for* $n = 1, 2, 3, \cdots, 7$ *we have*

$$\det M_5 = 1, -2, 4, 4, -8, -32, 64$$

respectively, and $\det M_5 = 0$ *if* $n \geqq 8$.

To prove this we use Theorem 1 with $f(m) = \mu(m)$, so that $g(m)$ is multiplicative by Theorem 4.12, and

$$g(p) = -2, \qquad g(p^2) = 1, \qquad g(p^r) = 0 \quad \text{if} \quad r \geqq 3.$$

Next we generalize Theorem 2.

Theorem 7. *Let M_6 be the matrix of order n with the element $(i, j)^\alpha$ in the i, j position, where α is any real number. Then*

$$\det M_6 = (n!)^\alpha \prod_{j=1}^{n} \prod_{p|j} \left(1 - \frac{1}{p^\alpha}\right).$$

This can be proved by taking $f(m) = m^\alpha$ in the Corollary to Theorem 1.

From Theorem 7 an analogous result can be found at once for the case with $[i, j]^\beta$ in the i, j position, where β is any real number. This can be done by replacing α by $-\beta$ in Theorem 7, next substituting $[i, j]^\beta / i^\beta j^\beta$ for $(i, j)^{-\beta}$ in the determinant M_6, and then factoring out the i^β and j^β multipliers from the denominators of the elements in the rows and columns. We omit the details.

Finally, we use Theorem 1 in the case where $f(m)$ is defined as the largest square-free divisor of m, so that

$$(3) \qquad f(m) = \prod_{p \mid m} p, \qquad g(p) = p - 1, \qquad g(p^r) = 0 \quad \text{if} \quad r > 1.$$

Theorem 8. *Let M_7 be the matrix of order n with $f(i, j)$ in the i,j position, with f as defined in* (3) *above. Then* det $M_7 = 1, 1, 2$ *if* $n = 1, 2, 3$ *respectively, and* det $M_7 = 0$ *if* $n \geqq 4$.

Gaussian Integers as Sums of Squares

Let a and b be rational integers. Prove that the Gaussian integer $a + 2bi$ is expressible as a sum of two squares of Gaussian integers if and only if not both $a/2$ and b are odd integers, by the following steps. First if $a/2$ and b are odd integers we look at the possibility that $a + 2bi$ is a sum of two squares,

$$a + 2bi = (r + si)^2 + (u + vi)^2, \quad a = r^2 + u^2 - s^2 - v^2, \qquad b = rs + uv.$$

The impossibility in integers of the last two equations here can be shown by a simple examination of the evenness and oddness of r, s, u, v.

Conversely, assume that $a/2$ and b are not both odd integers. Then we seek Gaussian integers ξ, η such that $\xi^2 + \eta^2 = a + 2bi$ by factoring $\xi^2 + \eta^2$ into $(\xi + \eta i)(\xi - \eta i)$ and writing assorted possible corresponding factorings of $a + 2bi$:

$$\text{(i)} \quad \xi + \eta i = a + 2bi, \qquad \xi - \eta i = 1;$$

$$\text{(ii)} \quad \xi + \eta i = \tfrac{1}{2}a + bi, \qquad \xi - \eta i = 2;$$

$$\text{(iii)} \quad \xi + \eta i = (\tfrac{1}{2}a + bi)(1 + i), \qquad \xi - \eta i = 1 - i.$$

If a is odd, (i) can serve because the equations in (i) can be solved for Gaussian integers ξ and η, for example $\xi = (a + 1)/2 + bi$. If a is even, we separate the cases where $a/2$ and b are both even on the one hand, and where $a/2$ and b are of opposite parity on the other. If $a/2$ and b are even integers, (ii) can be used because these equations can be solved for Gaussian integers ξ and η. Finally, if $a/2$ and b are integers of opposite parity, then (iii) will do, again yielding Gaussian integers ξ and η.

The above proof was given by L. J. Mordell, *Math. Magazine*, **40,** 209 (1967).

Unique Factorization in Gaussian Integers

An alternative proof of the case $m = -1$ of Theorem 9.27 can be constructed by analogy with the second proof given of Theorem 1.16, the unique factorization theorem, as follows. First, given any two nonzero Gaussian integers α and β, establish that some associate β' of β, perhaps β itself, can be chosen so that the triangle formed by the three points α, β', 0 in the complex plane has an angle $\leqq \pi/4$ at the origin 0. Hence prove that if $N(\beta) \leqq N(\alpha)$ then $N(\alpha - \beta') < N(\alpha)$. With the use of this result, unique factorization of integers in $Q(i)$ can be proved by presuming that if unique factorization fails, it fails for some integer with a least norm. This integer then plays the role of n in the second proof of Theorem 1.16.

The Eisenstein Irreducibility Criterion

Let p be a prime, and let

$$f(x) = a_0 + a_1x + a_2x^2 + \cdots + a_nx^n$$

be a polynomial with integer coefficients such that p is a divisor of all coefficients except a_n, so $p \nmid a_n$. If also $p^2 \nmid a_0$ then $f(x)$ is irreducible in that it cannot be factored into two polynomials with integer coefficients, excluding of course the trivial case where one factor is a constant. By Theorem 9.7 it follows that $f(x)$ cannot be factored into two polynomials with rational coefficients, excluding again trivial cases.

To prove this result, suppose on the contrary that $f(x) = g(x)h(x)$ where

$$g(x) = \sum_{j=0}^{r} b_jx^j, \qquad h(x) = \sum_{j=0}^{k} c_jx^j, \qquad k + r = n, \qquad k \geqq 1, \qquad r \geqq 1,$$

where the coefficients are integers. Now $a_0 = b_0c_0$ and a_0 is divisible by p but not by p^2, so we may presume that $p \mid b_0$ but $p \nmid c_0$. Next $a_1 = b_0c_1 + b_1c_0$ and $p \mid a_1$ so $p \mid b_1$. Similarly $p \mid a_2$ implies $p \mid b_2$, and by induction we establish that p is divisor of every coefficient of $g(x)$, the last step being

$$p \mid a_r, \qquad p \mid (b_0c_r + b_1c_{r-1} + \cdots + b_rc_0), \quad \text{so} \quad p \mid b_r.$$

It follows that p is also a divisor of every coefficient of $f(x)$.

General References

J. W. S. Cassels, *An Introduction to Diophantine Approximation*, Cambridge Tract 45, 1957.

K. Chandrasekharan, *Introduction to Analytic Number Theory*, New York, Springer-Verlag, 1968.

L. E. Dickson, *History of the Theory of Numbers*, Washington, Carnegie Institution of Washington, 1919; reprinted, New York, Chelsea, 1950.

L. E. Dickson, *Introduction to the Theory of Numbers*, Chicago, University of Chicago Press, 1929.

L. E. Dickson, *Modern Elementary Theory of Numbers*, Chicago, University of Chicago Press, 1939.

Emil Grosswald, *Topics from the Theory of Numbers*, New York, Macmillan, 1966.

H. Halberstam and K. F. Roth, *Sequences*, vol. 1, Oxford, Clarendon Press, 1966.

G. H. Hardy and E. M. Wright, *An Introduction to the Theory of Numbers*, 4th ed., Oxford, Clarendon Press, 1960.

B. W. Jones, *The Arithmetic Theory of Quadratic Forms*, Carus Monograph 10, New York, John Wiley and Sons, 1950.

D. H. Lehmer, *Guide to Tables in the Theory of Numbers*, Washington, Bulletin, National Research Council, No. 105, 1941.

W. J. LeVeque, *Topics in Number Theory*, vols. I and II, Reading, Mass., Addison-Wesley, 1956.

L. J. Mordell, *Diophantine Equations*, New York, Academic Press, 1969.

Ivan Niven, *Irrational Numbers*, Carus Monograph 11, New York, John Wiley and Sons, 1956.

O. Ore, *Number Theory and its History*, New York, McGraw-Hill, 1949.

Hans Rademacher, *Lectures on Elementary Number Theory*, New York, Blaisdell Publishing Company, 1964.

Daniel Shanks, *Solved and Unsolved Problems in Number Theory*, Washington, D.C., Spartan, 1962.

W. Sierpinski, *Elementary Theory of Numbers*, New York, Hafner, 1964.

Harold M. Stark, *An Introduction to Number Theory*, Chicago, Markham, 1970.

J. V. Uspensky and M. H. Heaslet, *Elementary Number Theory*, New York, McGraw-Hill, 1939.

I. M. Vinogradov, *Elements of Number Theory*, translation of 5th Russian edition, New York, Dover, 1954.

Answers

Section 1.2, p. 9

1. (*a*) 77, (*b*) 1, (*c*) 7, (*d*) 1.
2. $g = 17$, $x = 71$, $y = -36$.
3. (*a*) $x = 9, y = -11$, (*b*) $x = 31, y = 44$, (*c*) $x = 3, y = -2$. (*d*) $x = 7, y = 8$, (*e*) $x = 1, y = 1, z = -1$.
4. (*a*) 3374 (*b*) 3660.
5. 128.
7. 6, 10, 15.
17. 1; $n(n + 1)$.
18. a; b.
25. $x = 100n + 5$, $y = 95 - 100n$, $n = 1, 2, \cdots$, will do.
27. $a = 10, b = 100$ is a solution in positive integers. All solutions are given by $a = \pm 10, b = \pm 100$; $a = \pm 20, b = \pm 50$; $a = \pm 100, b = \pm 10$; $a = \pm 50, b = \pm 20$; with all arrangements of signs. There are 16 solutions in all.
28. $a = 10, b = 100, c = 10, 20, 50$, or 100; $a = 20, b = 50, c = 10, 20, 50$, or 100; and all permutations of these, 36 answers in all.

Section 1.3, p. 15

1. $\alpha_j \leqq \beta_j$ for every $j = 1, 2, \cdots, r$; $\alpha_j = 0$ or $\beta_j = 0$ for every j.
2. 3; 7.
16. p, p^2; p, p^2, p^3; p^2, p^3.
17. p^3, p.
18. $2 \mid \alpha_j$; $3 \mid \alpha_j$; $\alpha_j \leqq \beta_j$; $\alpha_j \leqq \beta_j$; for all j, $1 \leqq j \leqq r$ in each part.

24. Counterexamples for false statements are
    (1) $a = 2, b = 6, c = 10$.
    (8) $a = 8, c = 4$.
    (10) $p = 5, a = 2, b = 1, c = 3$.
    (13) $a = 2, b = 5$.
25. All n not of the form $p - 1$ where p is an odd prime.
39. $a, a, \cdots, a$ or $a, a, \cdots, a, 2a$ or $a, 2a, 3a$.

Section 2.1, p. 25

1. 7, 24, 41, 58, 75, 92.
2. 0, 18, 36, 3, 21, 39, 6, 24, 42, 9, 27, 45, 12, 30, 48, 15, 33.
3. 1, 5, 7, 11 (mod 12); 1, 7, 11, 13, 17, 19, 23, 29 (mod 30).
4. $y \equiv 1 \pmod 2$; $z \equiv 1 \pmod 6$.
5. $x \equiv 5 \pmod{12}$.
10. $m = 1, 2, 3, 4, 5, 6, 7, 8, 9, 10, 11, 12.$
    $\phi(m) = 1, 1, 2, 2, 4, 2, 6, 4, 6, 4, 10, 4.$
11. $x = 5$.
13. 1, 9, 3, 81, 243, 27.
14. $x = 9 + 11j$.
16. $a = 2, 3, 4, 5, 6, 7, 8, 9, 10.$
    $x = 6, 4, 3, 9, 2, 8, 7, 5, 10.$
25. 1.
26. 6.
27. 0, 1.
35. Primitive solutions with $a \leqq b \leqq c$ are $a = b = 1$, c any positive integer.
36. Solutions such that $(a, b, c) = 1$, $c \geqq |b| \geqq |a|$ are
    $a = -b = \pm 1$, $c = 1$ or 2;
    $a = -1, b = 2, c = 3$;
    $a = b = \pm 1$ with any $c > 0$;
    $a = 1, b = 1 - c$ with any $c > 2$;
    $a = 2, b = -2n + 1, c = 2n + 1$ with any $n > 1$.

Section 2.2, p. 29

4. $x(x + 1)(x + 2) \cdots (x + m - 1) \equiv 0 \pmod m$.

Section 2.3, p. 31

1. (*a*) no solution
   (*b*) no solution
   (*c*) $x \equiv -82 \pmod{400}$.

2. (*a*) 5, (*b*) 0, (*c*) 5.
3. $x = 106$.
4. $23 + 30j$.
5. $x \equiv 33 \pmod{84}$.
6. $60j - 2$.
7. $\frac{73}{105}, \frac{4}{7}$.

12. $x \equiv 42 \pmod{125}$.

Section 2.4, p. 36

1. 1, 2.
2. 960.
3. 2640.
4. 1920.
5. 6720.

10. n odd.
11. n even.
12. $n = 5^k$, $k = 1, 2, \cdots$ will do.
13. 35, 39, 45, 52, 56, 70, 72, 78, 84, 90.

15. 3, 1, 2, 4.

Section 2.5, p. 38

1. $x \equiv 1, 2, 6 \pmod 9$
   $x \equiv 1, 3 \pmod 5$
   $x \equiv 1, 6, 11, 28, 33, 38 \pmod{45}$.
2. No solution.
3. $x \equiv 1, 3, 5 \pmod{503}$.
4. $x \equiv 1, 3, 5, 14, 16, 27, 122, 133, 135 \pmod{143}$.

Section 2.6, p. 42

6. No solution.
7. $x \equiv 4 \pmod{5^3}$.
8. $x \equiv 7, 15, 16, 24 \pmod{36}$.
9. $x \equiv 15 \pmod{3^3}$.
10. No solution.
11. $x \equiv 23 \pmod{7^3}$.

Section 2.7, p. 45

1. (*a*) $x^5 + x^2 + 5 \equiv 0 \pmod 7$,
   (*b*) $x^2 + 3x - 2 \equiv 0 \pmod 7$,
   (*c*) $x^4 - x^3 - 4x + 3 \equiv 0 \pmod 7$.

Section 2.8, p. 46

1. (*a*) $(4x + 1)^2 \equiv 2 \pmod 5$
   (*b*) $(x + 1)^2 \equiv 4 \pmod 7$
   (*c*) $(4x + 7)^2 \equiv 8 \pmod{11}$
   (*d*) $(2x + 1)^2 \equiv 5 \pmod{13}$.

Section 2.9, p. 50

1. 2, 2, 3, 2, 2.
2. 5.
3. 4.
4. 1, 3, 6, 3, 6, 2.
   1, 10, 5, 5, 5, 10.
7. $p - 1$. 0.
8. (*a*) 4, (*b*) 0, (*c*) 4, (*d*) 1.
9. $x^2 \equiv 1$, $x^2 \equiv 2$, $x^2 \equiv 4$, $x^2 \equiv 8$, $x^2 \equiv 9$, $x^2 \equiv 13$, $x^2 \equiv 15$, $x^2 \equiv 16 \pmod{17}$.

Section 2.10, p. 54

1. (*a*), (*e*), (*f*), (*h*), (*i*).

3.

	7	−2	17	30	8	3
7	8	17	30	7	3	−2
−2	17	8	3	−2	30	7
17	30	3	−2	17	7	8
30	7	−2	17	30	8	3
8	3	30	7	8	−2	17
3	−2	7	8	3	17	30

	1	4	5	0	2	3
1	2	5	0	1	3	4
4	5	2	3	4	0	1
5	0	3	4	5	1	2
0	1	4	5	0	2	3
2	3	0	1	2	4	5
3	4	1	2	3	5	0

Section 2.11, p. 60

6. 8.

13.

$\oplus$	0	1	2	3	4	5	6
0	0	1	2	3	4	5	6
1	1	2	3	4	5	6	0
2	2	3	4	5	6	0	1
3	3	4	5	6	0	1	2
4	4	5	6	0	1	2	3
5	5	6	0	1	2	3	4
6	6	0	1	2	3	4	5

$\odot$	0	1	2	3	4	5	6
0	0	0	0	0	0	0	0
1	0	1	2	3	4	5	6
2	0	2	4	6	1	3	5
3	0	3	6	2	5	1	4
4	0	4	1	5	2	6	3
5	0	5	3	1	6	4	2
6	0	6	5	4	3	2	1

23. 0, 1, 6, 10, 15, 16, 21, 25.
    0, 2, 3, 4, 5, 6, 8, 9, 10, 12, 14, 15, 16, 18, 20, 21, 22, 24, 25, 26, 27, 28.
24. (*a*) and (*c*) are integral domains; (*b*) is an integral domain if and only if m is a prime.

Section 3.1, p. 66

1. 1, −2, 3, −7, 0.
4. $\left(\frac{-1}{11}\right) = -1,\ \left(\frac{-1}{13}\right) = +1,\ \left(\frac{-1}{17}\right) = +1,$

   $\left(\frac{2}{11}\right) = -1,\ \left(\frac{2}{13}\right) = -1,\ \left(\frac{2}{17}\right) = +1,$

   $\left(\frac{-2}{11}\right) = +1,\ \left(\frac{-2}{13}\right) = -1,\ \left(\frac{-2}{17}\right) = +1,$

   $\left(\frac{3}{11}\right) = +1,\ \left(\frac{3}{13}\right) = +1,\ \left(\frac{3}{17}\right) = -1.$
5. $x \equiv \pm 1 \pmod{11}$, $x \equiv \pm 5 \pmod{11}$, $x \equiv \pm 2 \pmod{11}$, $x \equiv \pm 4 \pmod{11}$, $x \equiv \pm 3 \pmod{11}$.
   $x \equiv \pm 1 \pmod{11^2}$, $x \equiv \pm 27 \pmod{11^2}$, $x \equiv \pm 2 \pmod{11^2}$, $x \equiv \pm 48 \pmod{11^2}$, $x \equiv \pm 3 \pmod{11^2}$.
6. $1, 2, 4 \pmod 7$, $\pm 1, \pm 3, \pm 4 \pmod{13}$, $\pm 1, \pm 2, \pm 4, \pm 8 \pmod{17}$, $\pm 1, \pm 4, \pm 5, \pm 6, \pm 7, \pm 9, \pm 13 \pmod{29}$, $\pm 1, \pm 3, \pm 4, \pm 7, \pm 9, \pm 10, \pm 11, \pm 12, \pm 16 \pmod{37}$.
7. (*d*) 2, (*h*) 2.
8. (*a*) 2, (*b*) 0, (*c*) 4, (*d*) 0, (*e*) 2, (*f*) 0.

Section 3.2, p. 70

4. (*b*), (*c*), (*d*), (*e*,) (*f*).

5. $\left(\frac{7}{227}\right) = +1, \left(\frac{7}{229}\right) = -1, \left(\frac{7}{1009}\right) = +1,$

$$\left(\frac{11}{227}\right) = +1, \left(\frac{11}{229}\right) = +1, \left(\frac{11}{1009}\right) = -1,$$

$$\left(\frac{13}{227}\right) = -1, \left(\frac{13}{229}\right) = -1, \left(\frac{13}{1009}\right) = -1.$$

6. Yes.
7. $p = 2, p = 13$, and $p \equiv 1, 3, 4, 9, 10, 12 \pmod{13}$.
8. $p \equiv \pm 1, \pm 3, \pm 9, \pm 13 \pmod{40}$.
9. Odd primes $p \equiv 2, 3 \pmod 5$.
10. $p \equiv 1, 3 \pmod 8$.

Section 3.3, p. 74

1. $\left(\frac{-23}{83}\right) = -1, \left(\frac{51}{71}\right) = -1, \left(\frac{71}{73}\right) = +1, \left(\frac{-35}{97}\right) = +1.$

2. (*b*).
3. (*c*).
10. $p = 2$ and $p \equiv 1 \pmod 4$.
11. 2 and p^a for $p \equiv 1 \pmod 4$ and $a = 1, 2, 3, \cdots$.
12. $n = 2^{a_1} p_2^{a_2} \cdots p_k^{a_k}$, $a_1 = 0$ or 1, $p_j \equiv 1 \pmod 4$, $a_j = 1, 2, 3, \cdots$.

Section 4.1, p. 82

1. 529, 263, 263, 263, 87.
2. 24.
3. (*a*) all x such that $x - [x] < \frac{1}{2}$,
   (*b*) all x,
   (*c*) all integers,
   (*d*) all x such that $x - [x] \geqq \frac{1}{2}$,
   (*e*) all x such that $1 \leqq x < 10/9$.

5. (*a*)

$$e = \begin{cases} \sum_{i=1}^{\infty} \left[\frac{n}{p^i}\right] & \text{if } p \text{ is odd,} \\ n + \sum_{i=1}^{\infty} \left[\frac{n}{2^i}\right] & \text{if } p = 2, \end{cases}$$

(*b*)

$$e = \begin{cases} \sum_{i=1}^{\infty} \left(\left[\frac{2n}{p^i}\right] - \left[\frac{n}{p^i}\right]\right) & \text{if } p \text{ is odd,} \\ 0 \text{ if } p = 2. \end{cases}$$

12. $a - m(a - 1)/m$.

Section 4.2, p. 87

1. 7.
2. 12.
3. 2, 1, 12, 24.
4. 6.
8. $\sigma_k(p_1^{e_1} \cdots p_r^{e_r}) = \prod_{i=1}^{r} \frac{p_i^{k(e_i+1)} - 1}{p_i^k - 1}$.
10. $f(n) = n$ will do.
13. $x = p^{n-1}$ will do, where p is any prime.
16. 6, 28, 496.

Section 4.3, p. 90

1. $n = 33$ will do.
3. 1.
7. $\sum_{d|n} \mu(d)\sigma(d) = (-1)^k p_1 p_2 \cdots p_k$.

Section 4.4, p. 96

2. No; no.

Section 4.5, p. 98

1. $x_n = a^n x_0$.

$$x_n = \begin{cases} b^{n/2}x_0 \text{ if } n \text{ is even,} \\ b^{(n-1)/2}x_1 \text{ if } n \text{ is odd.} \end{cases}$$

2. $x_n = n$. $x_n = 1$. $x_n = (3^n - (-1)^n)/4$. $x_n = (3^n + (-1)^n)/2$.
3. 0, 1, 1, 2, 3, 5, 8, 13, 21, 34.
11. (*b*) $-[-n/3]$ if n is odd.
    (*d*) $S_3 = \{1\}$, $S_9 = \{2, 3, 5, 7\}$.
12.

$n =$ 1	2	3	4	5	6	7	8	9	10
$f(n) =$ 0	1	2	4	6	9	12	16	20	25

$f(5 + 3) - f(5 - 3) = f(8) - f(2) = 16 - 1 = 15 = 5 \cdot 3$, for example.

13. $$x_n = 1 + 2^{n-2} - (-2)^{n-2} = \begin{cases} 1 \text{ if } n \text{ is even,} \\ 1 + 2^{n-1} \text{ if } n \text{ is odd.} \end{cases}$$

Section 5.2, p. 101

2. $x = 1 + 7t$, $y = -1 + 10t$.
9. $(b_1 - b_2, c_1 - c_2) \mid (d_1 - d_2)$.

Section 5.3, p. 103

1. (*a*) $x = 8, y = 4$; $x = 5, y = 9$; $x = 2, y = 14$.
   (*b*) $x = 6, y = 3$.
   (*c*) $x = 2, y = 7$.
   (*d*) $x = 2, y = 5$.
   (*e*) no solution.
   (*f*) no solution.
   (*g*) no solution.
10. ab.

Section 5.4, p. 105

1. (*a*) $x = 1 + u$, $y = -2u + 3v$, $z = u - 2v$,
   (*b*) $x = 10 + u$, $y = -2u + 3v$, $z = u - 2v$,
   (*c*) $x = 1 + 2t$, $y = 6 + 15t - 2v$, $z = -2 - 5t + v$,

(*d*) $x = 2 + 2t$, $y = 15t - 2v$, $z = -5t + v$,
(*e*) $x = t$, $y = -11 + 3t + 5v$, $z = -11 + 3t + 6v$.
(*f*) no solution.

Section 5.5, p. 107

1. $x = 3,\ y = 4,\ z = 5,\quad x = 4,\ y = 3,\ z = 5,$
$x = 15, y = 8,\ z = 17,\quad x = 8,\ y = 15, z = 17,$
$x = 5,\ y = 12, z = 13,\quad x = 12, y = 5,\ z = 13,$
$x = 21, y = 20, z = 29,\quad x = 20, y = 21, z = 29,$
$x = 7,\ y = 24, z = 25.\quad x = 24, y = 7,\ z = 25.$
3. (*a*) $x = 3k$, $y = 4k$, $z = 5k$,
$x = 4k$, $y = 3k$, $z = 5k$.
(*b*) none.
5. $n \equiv 0, 1, 3 \pmod 4$.

Section 5.7, p. 111

4. 1, 2, 3, 4, 5, 7, 8, 10, 11, 13, 16, 19.

Section 5.10, p. 117

1. $N(100) = 12$, $P(100) = 0$, $Q(100) = 0$,
$N(101) = 8$, $P(101) = 2$, $Q(101) = 8$,
$N(102) = 0$, $P(102) = 0$, $Q(102) = 0$.

Section 5.13, p. 126

1. (*a*) all $n = 2^e p_1^{e_1} \cdots p_k^{e_k}$, $e \geqq 1$, e_i even if $p_i \equiv 3 \pmod 4$.
(*b*) all $n \equiv 0, 2, 6 \pmod 8$.
(*c*) all integers.
(*d*) all non-negative even integers.
2. All perfect squares.
9. $x = 2, y = -3,\quad x = -2, y = 3,$
$x = 4, y = -1,\quad x = -4, y = 1,$
$x = 6, y = -5,\quad x = -6, y = 5,$
$x = 8, y = -5,\quad x = -8, y = 5.$
(*b*) no solution.
10. 1.

Section 5.14, p. 132

2. (*a*) $x^2 + xy + 3y^2$, (*c*) $2x^2 + xy + 6y^2$,
   (*b*) $x^2 + xy + 2y^2$, (*d*) $x^2 + xy + 3y^2$.
4. $x^2 + xy + 5y^2$.

Section 6.1, p. 137

6. $a = b = d = 1$, $c = 0$ will do.

Section 7.1, p. 152

1. $17/3 = \langle 5, 1, 2\rangle$, $3/17 = \langle 0, 5, 1, 2\rangle$, $8/1 = \langle 8\rangle$.
3. $\langle 2, 1, 4\rangle = 14/5$, $\langle -3, 2, 12\rangle = -63/25$, $\langle 0, 1, 1, 100\rangle = 101/201$.

Section 7.2, p. 153

1. The following conditions are necessary and sufficient. In case $a_j = b_j$ for $0 \leqq j \leqq n$, then n must be even. Otherwise define r as the least value of j such that $a_j \neq b_j$. In case $r \leqq n - 1$, then for r even we require $a_r < b_r$, but for r odd, $a_r > b_r$. In case $r = n$, then for n even we require $a_n < b_n$, but for n odd we require $a_n > 1 + b_n$, or $a_n = 1 + b_n$ with $b_{n+1} > 1$.

Section 7.3, p. 157

1. $(1 + \sqrt{5})/2$.
2. $(3 + \sqrt{5})/2$, $(25 - \sqrt{5})/10$.
3. (*a*) $1 + \sqrt{2}$, (*b*) $(1 + \sqrt{3})/2$, (*c*) $1 + \sqrt{3}$, (*d*) $3 - \sqrt{3}$.
4. $$h_n/h_{n-1} = \begin{cases} \langle a_n, a_{n-1}, \cdots, a_2, a_1, a_0\rangle \text{ if } a_0 \neq 0, \\ \langle a_n, a_{n-1}, \cdots, a_4, a_3, a_2\rangle \text{ if } a_0 = 0. \end{cases}$$

Section 7.4, p. 159

1. $\sqrt{2} = \langle 1, 2, 2, 2, \cdots\rangle$, $\sqrt{2} - 1 = \langle 0, 2, 2, 2, \cdots\rangle$,
   $\sqrt{2}/2 = \langle 0, 1, 2, 2, 2 \cdots\rangle$
   $\sqrt{3} = \langle 1, 1, 2, 1, 2, 1, 2, \cdots\rangle$, $\dfrac{1}{\sqrt{3}} = \langle 0, 1, 1, 2, 1, 2, 1, 2, \cdots\rangle$.

Section 7.6, p. 166

1. 1/1, 3/2 will do.
2. 3/1, 22/7 will do.

Section 7.7, p. 172

1. $c = 1, 2, \cdots, 2[\sqrt{d}]$.

Section 8.2, p. 184

1. log 9 (base 10).

Section 8.3, p. 187

5. 1, 2, 3, 4, 6, 8, 12, 18, 24, 30.

Section 9.2, p. 195

1. $x - 7$, $x^3 - 7$, $x^3 - 3x^2/2 + 3x/4 - 1$, $x^4 - 4x^3 - 4x^2 + 16x - 8$. 7, $\sqrt[3]{7}$, $1 + \sqrt{2} + \sqrt{3}$ are algebraic integers.

Section 9.4, p. 201

3. Yes; no, for example $\alpha = \frac{1}{2}(1 + i\sqrt{3})$.

Section 9.5, p. 203

6. $\alpha = (1 + 7i)/5$ will do.
7. The suggestion also works in case $m = -2$. The other special cases can be handled by such numbers as: $\dfrac{1 + 4\sqrt{-3}}{7}$, $\dfrac{9 + 4\sqrt{2}}{7}$, $\dfrac{27 + \sqrt{3}}{11}$, $\dfrac{4 + 10\sqrt{5}}{11}$.

Section 10.3, p. 227

2. $n = 1, 2, 3, 4, 5, 6, 7, 8, 9, 10, 11, 12.$
$p(n) = 1, 2, 3, 5, 7, 11, 15, 22, 30, 42, 56, 77.$
$n = 13, 14, 15, 16, 17, 18, 19, 20.$
$p(n) = 101, 135, 176, 231, 297, 385, 490, 627.$

Section 10.4, p. 233

1. $n = 1, 2, 3, 4, 5, 6, 7, 8, 9, 10, 11, 12.$
$\sigma(n) = 1, 3, 4, 7, 6, 12, 8, 15, 13, 18, 12, 28.$
$n = 13, 14, 15, 16, 17, 18, 19, 20.$
$\sigma(n) = 14, 24, 24, 31, 18, 39, 20, 42.$

Section 10.6, p. 239

2. $p(35m + 19) \equiv 0 \pmod{35}$.

Section 11.1, p. 242

1. (*a*) $\frac{1}{2}$, (*b*) $\frac{1}{2}$, (*c*) $\frac{1}{3}$, (*d*) $\frac{1}{4}$, (*e*) $\dfrac{1}{m}$, (*f*) 0, (*g*) 0, (*h*) 0, (*i*) 0, (*j*) 0.

15. $\frac{1}{11}$.

Section 11.2, p. 246

1. $15/(2\pi^2)$.

2. $\dfrac{6}{\pi^2}\sum_{j=1}^{10}\dfrac{1}{j^2}$.

3. (*b*) $8/\pi^2$.

Section 11.4, p. 254

1. $1/2, 0, 1/3, 1/m$.

Miscellaneous Problems, p. 255

5. If m has exactly b distinct odd prime factors, the order is 2^b, 2^{b+1}, or 2^{b+2} according as $m/4$ is not an integer, is an odd integer, or is an even integer.
8. $n = 1, 2^j$ or 2^j3^k with $j \geqq 1, k \geqq 1$.
18. $1; (n - 1)/2$.
34. $1 + [\log_2 n]$.

Index